RISK NEUTRAL PRICING AND
FINANCIAL MATHEMATICS

RISK NEUTRAL PRICING AND FINANCIAL MATHEMATICS

A Primer

Peter M. Knopf

John L. Teall

AMSTERDAM • BOSTON • HEIDELBERG • LONDON
NEW YORK • OXFORD • PARIS • SAN DIEGO
SAN FRANCISCO • SINGAPORE • SYDNEY • TOKYO

Academic Press is an imprint of Elsevier

Academic Press is an imprint of Elsevier
525 B Street, Suite 1800, San Diego, CA 92101-4495, USA
225 Wyman Street, Waltham, MA 02451, USA
The Boulevard, Langford Lane, Kidlington, Oxford OX5 1GB, UK

Notices

Knowledge and best practice in this field are constantly changing. As new research and experience broaden our understanding, changes in research methods, professional practices, or medical treatment may become necessary.

Practitioners and researchers may always rely on their own experience and knowledge in evaluating and using any information, methods, compounds, or experiments described herein. In using such information or methods they should be mindful of their own safety and the safety of others, including parties for whom they have a professional responsibility.

To the fullest extent of the law, neither the Publisher nor the authors, contributors, or editors, assume any liability for any injury and/or damage to persons or property as a matter of products liability, negligence or otherwise, or from any use or operation of any methods, products, instructions, or ideas contained in the material herein.

ISBN 978-0-12-801534-6

Library of Congress Cataloging-in-Publication Data
A catalog record for this book is available from the Library of Congress

British Library Cataloguing-in-Publication Data
A catalogue record for this book is available from the British Library

For information on all Academic Press publications
visit our website at http://store.elsevier.com/

Working together
to grow libraries in
developing countries

www.elsevier.com • www.bookaid.org

Publisher: Nikki Levy
Acquisition Editor: J. Scott Bentley
Editorial Project Manager: Susan Ikeda
Production Project Manager: Melissa Read
Designer: Mark Rogers

Printed and bound in the United States of America

Dedication

to the lovely and supportive ladies in our lives,
Anne and Arli,
and the many others who shall remain nameless here.

Contents

About the Authors

Peter Knopf obtained his Ph.D. from Cornell University and subsequently taught at Texas A&M University and Rutgers University. He is currently Professor of Mathematics at Pace University. He has numerous research publications in both pure and applied mathematics. His recent research interests have been in the areas of difference equations and stochastic delay equation models for pricing securities.

John L. Teall is a visiting professor of finance at the LUISS Business School, LUISS Guido Carli University. He was Jackson Tai '72 Professor of Practice at Rensselaer Polytechnic Institute, and also served on the faculties of New York University, Cornell University, Pace University, Dublin City University, University of Melbourne as well as other institutions in the United States, Europe and Asia. His primary areas of research and publication have been related to corporate finance and financial institutions. He is the author/co-author of 5 books. Dr. Teall obtained his Ph.D. from the Stern School of Business at New York University and is a former member of the American Stock Exchange. Dr. Teall has consulted with numerous financial institutions including Deutsche Bank, Goldman Sachs, National Westminster Bank and Citicorp.

Preface

It's no easy task to write a book that makes quantitative finance seem easy. Writing such a book with two bull-headed authors can be a real battle, at least at times. But, now that we're finished, or at least ready to begin pondering the book's eventual second edition, we both are surprised and quite pleased with the result. Our various pre-publication readers claim to like the chapters, and seem to believe that the book well suits our intended audience (we'll get to this shortly). Competition, skepticism, and stubbornness do seem to have a place in a joint writing endeavor.

Somewhere in the middle of an early draft of Chapter 7, a passage appeared that sparked the following exchange between the coauthors of this book, altered somewhat to soften the language:

> This paragraph doesn't mean anything!
> That's because you're dense.
> (exasperated) "You're such a pinhead. Why did I ever agree to write this book?"

Actually, we knew from the start exactly why we needed to write this book. We've seen a nice assortment of books in financial mathematics, written by mathematicians, that are nicely suited to readers with solid technical backgrounds in engineering, the physical sciences and math. There are a smaller number of financial math books geared towards undergraduates and MBA students who really aren't interested in the stochastic processes underlying models for pricing derivatives and fixed income instruments. Your coauthors sought to fill a rapidly growing gap between these two groups of readers: ambitious financial mathematics students with relatively modest mathematics backgrounds, say with one or two calculus courses along with a course or two in linear mathematics and another in statistics. While a stronger math background would only help the reader, we have provided a brief review (Chapter 1 and much of Chapter 2) for readers who are a bit rusty or have gaps in their undergraduate preparation. We believe that readers with technical backgrounds such as in mathematics and engineering as well as readers with nontechnical backgrounds such as studies in undergraduate finance and economics will be able to follow and benefit from the presentations of this book. After all, the book is still focused on finance, and not even most mathematicians will have seen at least some of the mathematics used here and in the financial industry more generally. We also believe that the large number of solved end-of-chapter exercises and online materials will boost reader comfort with the material and deepen the learning experience.

In order to keep the price of the textbook down, certain materials such as additional readings, proofs and verifications have been placed on the authors' website at http://www.jteall.com/books.html (click the "Resources" link), and are also available online at the Elsevier textbook site (see overleaf for address). These readings, proofs and verifications offer a wealth of insight into many of the processes used to model financial instruments and to the mathematics underlying valuation and hedging. For example, in the additional readings for Chapter 6, we derive the solution to the Black-Scholes differential equation. Of similar importance to Chapter 8, the text website contains a derivation of the Vasicek single factor model for pricing bonds. These website readings are presented to improve the balance between rigor and pedagogy in order for the student to gain as much intuition and understanding of the most important derivations in quantitative finance without losing the ability to apply the important results.

Introduction to Risk-Neutral Pricing and Financial Mathematics: A Primer seeks to introduce financial mathematics to students in quantitative finance, financial engineering, actuarial science and computational finance. The central theme of the book is risk-neutral (martingale) pricing, though it does venture into a number of other areas. The text endeavors to provide a foundation in financial mathematics for use in an introductory financial engineering, financial modeling, or financial mathematics course, primarily to students whose math backgrounds do not extend much beyond two semesters of calculus, linear math and statistics. Students with stronger mathematics preparations may find this book somewhat easier to follow, but most will not find significant amounts of material repeating what they have seen elsewhere. The book will present and apply topics such as stochastic processes, equivalent martingales, Radon—Nikodym derivatives, and stochastic calculus, but it will start from the perspective that reader mathematical training does not extend beyond basic calculus, linear math and statistics.

Writing a textbook is never a solo effort, or even an effort undertaken by only two coauthors. As have most authors, we benefitted immensely from the help from many people, including friends, family, students, colleagues and practitioners. For example, we owe special thanks to some of our favorite students, including Xiao (Sean) Tang, T.J. Wu, Alban Leung, Haoyu Li, Victor Shen, and Matthew Spector. A number of our colleagues were also most helpful, including Professors Steven Kalikow and Maury Bramson for valuable input concerning stochastic processes and Professor Matthew Hyatt for programming and use of software tools. We are particularly grateful to Hong Tu Yan for his careful reading of the text and competent coauthoring of supplemental materials. We thank Ryan Cummings for her delightful cartoon appearing on the text companion website, and Ying Sue Huang and Imon Palit for their thoughtful comments. And then there's Chris Stone, always there when we need help, and always going above and beyond the call of duty, such as for her 11th hour sacrifice of vacation time to ensure that we met our manuscript submission deadline. Of course, we are also grateful to Scott Bentley, Melissa Read, Susan Ikeda, and Mckenna Bailey at Academic Press/Elsevier. And finally, we enjoyed the unfailing support and encouragement from Arli Epton and the unique culinary skills of Anne Teall. Emily Teall's unbounded interest in everything related to finance boosted our morale as well. Unfortunately, a number of errors will surely persist even after the book goes into production. We apologize sincerely for these errors, and we blame our parents for them.

Additional readings, proofs and verifications can be found on the companion website.
Go online to access it at:
http://booksite.elsevier.com/9780128015346

1

Preliminaries and Review

1.1 FINANCIAL MODELS

A model can be characterized as an artificial structure describing the relationships among variables or factors. Practically all of the methodology in this book is geared toward the development and implementation of financial models to solve financial problems. For example, valuation models provide a foundation for investment decision-making and models describing stochastic processes provide an important tool to account for risk in decision-making.

The use of models is important in finance because "real world" conditions that underlie financial decisions are frequently extraordinarily complicated. Financial decision-makers frequently use existing models or construct new ones that relate to the types of decisions they wish to make. Models proposing decisions that ought to be made are called normative models.[1]

The purpose of models is to simulate or behave like real financial situations. When constructing financial models, analysts exclude the "real world" conditions that seem to have little or no effect on the outcomes of their decisions, concentrating on those factors that are most relevant to their situations. In some instances, analysts may have to make unrealistic assumptions in order to simplify their models and make them easier to analyze. After simple models have been constructed with what may be unrealistic assumptions, they can be modified to match more closely "real world" situations. A good financial model is one that accounts for the major factors that will affect the financial decision (a good model is complete and accurate), is simple enough for its use to be practical (inexpensive to construct and easy to understand), and can be used to predict actual outcomes. A model is not likely to be useful if it is not able to project an outcome with an acceptable degree of accuracy. Completeness and simplicity may directly conflict with one another. The financial analyst must determine the appropriate trade-off between completeness and simplicity in the model he wishes to construct.

In finance, mathematical models are usually the easiest to develop, manipulate, and modify. These models are usually adaptable to computers and electronic spreadsheets. Mathematical models are obviously most useful for those comfortable with math; the primary purpose of this book is to provide a foundation for improving the quantitative preparation of the less mathematically oriented analyst. Other models used in finance include those based on graphs and those involving simulations. However, these models are often based on or closely related to mathematical models.

P.M. Knopf & J.L. Teall: Risk Neutral Pricing and Financial Mathematics: A Primer.
DOI: http://dx.doi.org/10.1016/B978-0-12-801534-6.00001-1

The concepts of *market efficiency* and *arbitrage* are essential to the development of many financial models. Market efficiency is the condition in which security prices fully reflect all available information. Such efficiency is more likely to exist when wealth-maximizing market participants can instantaneously and costlessly execute transactions as information is revealed. Transactions costs, irrationality, and poor execution systems reduce efficiency. Arbitrage, in its simplest scenario, is the simultaneous purchase and sale of the same asset, or more generally, the nearly simultaneous purchase and sale of assets generating nearly identical cash flow structures. In either case, the arbitrageur seeks to produce a profit by purchasing at a price that is less than the selling price. Proceeds of the sales are used to finance purchases such that the portfolio of transactions is self-financing, and that over time, no additional capital is devoted to or lost from the portfolio. Thus, the portfolio is assured a non-negative profit at each time period. The arbitrage process is riskless if purchase and sale prices are known at the times they are initiated. Arbitrageurs frequently seek to profit from market inefficiencies. The existence of arbitrage profits is inconsistent with market efficiency.

1.2 FINANCIAL SECURITIES AND INSTRUMENTS

A security is a tradable claim on assets. Real assets contribute to the productive capacity of the economy; securities are financial assets that represent claims on real assets or other securities. Most securities are marketable to the general public, meaning that they can be sold or assigned to other investors in the open marketplace. Some of the more common types of securities and tradable instruments are briefly introduced in the following:

1. *Debt securities*: Denote creditorship of an individual, firm or other institution. They typically involve payments of a fixed series of interest (often known as *coupon payments*) or amounts towards principal along with principal repayment (often known as *face value*). Examples include:
   * *Bonds*: Long-term debt securities issued by corporations, governments, or other institutions. Bonds are normally of the coupon variety (they make periodic interest payments on the principal) or *pure discount* (they are *zero coupon instruments* that are sold at a discount from *face value*, the bond's final maturity value).
   * *Treasury securities:* Debt securities issued by the Treasury of the United States federal government. They are often considered to be practically free of default risk.
2. *Equity securities (stock)*: Denote ownership in a business or corporation. They typically permit for dividend payments if the firm's debt obligations have been satisfied.
3. *Derivative securities*: Have payoff functions derived from the values of other securities, rates, or indices. Some of the more common derivative securities are:
   * *Options*: Securities that grant their owners rights to buy (call) or sell (put) an underlying asset or security at a specific price (exercise price) on or before its expiration date.
   * *Forward* and *futures contracts*: Instruments that oblige their participants to either purchase or sell a given asset or security at a specified price (settlement price) on the future settlement date of that contract. A *long* position obligates the investor to purchase the given asset on the settlement date of the contract and a *short* position obligates the investor to sell the given asset on the settlement date of the contract.
   * *Swaps*: Provide for the exchange of cash flows associated with one asset, rate, or index for the cash flows associated with another asset, rate, or index.

4. *Commodities*: Contracts, including futures and options on physical commodities such as oil, metals, corn, etc. Commodities are traded in *spot markets*, where the exchange of assets and money occurs at the time of the transaction or in forward and futures markets.
5. *Currencies*: Exchange rates denote the number of units of one currency that must be given up for one unit of a second currency. Exchange transactions can occur in either spot or forward markets. As with commodities, in the spot market, the exchange of one currency for another occurs when the agreement is made. In a forward market transaction, the actual exchange of one currency for another actually occurs at a date later than that of the agreement. Spot and forward contract participants take one position in each of two currencies:
   * *Long*: An investor has a "long" position in that currency that he will accept at the later date.
   * *Short*: An investor has a "short" position in that currency that he must deliver in the transaction.
6. *Indices*: Contracts pegged to measures of market performance such as the Dow Jones Industrials Average or the S&P 500 Index. These are frequently futures contracts on portfolios structured to perform exactly as the indices for which they are named. Index traders also trade options on these futures contracts.

This list of security types is far from comprehensive; it only reflects some of those instruments that will be emphasized in this book. In addition, most of the instrument types will have many different variations.

1.3 REVIEW OF MATRICES AND MATRIX ARITHMETIC

A *matrix* is simply an ordered rectangular array of numbers. A matrix is an entity that enables one to represent a series of numbers as a single object, thereby providing for convenient systematic methods for completing large numbers of repetitive computations. Such objects are essential for the management of large data structures. Rules of matrix arithmetic and other matrix operations are often similar to rules of ordinary arithmetic and other operations, but they are not always identical. In this text, matrices will usually be denoted with bold uppercase letters. When the matrix has only one row or one column, bold lowercase letters will be used for identification. The following are examples of matrices:

$$\mathbf{A} = \begin{bmatrix} 4 & 2 & 6 \\ 3 & 7 & 4 \\ 8 & -5 & 9 \end{bmatrix} \quad \mathbf{B} = \begin{bmatrix} 2 & -3 \\ 3/4 & -1/2 \end{bmatrix} \quad \mathbf{c} = \begin{bmatrix} 1 \\ 5 \\ 7 \end{bmatrix} \quad \mathbf{d} = [4]$$

The dimensions of a matrix are given by the ordered pair $m \times n$, where m is the number of rows and n is the number of columns in the matrix. The matrix is said to be of *order* $m \times n$ where, by convention, the number of rows is listed first. Thus, $\mathbf{A}$ is 3×3, $\mathbf{B}$ is 2×2, $\mathbf{c}$ is 3×1, and $\mathbf{d}$ is 1×1. Each number in a matrix is referred to as an element. The symbol $a_{i,j}$ denotes the element in Row i and Column j of Matrix $\mathbf{A}$, $b_{i,j}$ denotes the element in Row i and Column j of Matrix $\mathbf{B}$, and so on. Thus, $a_{3,2}$ is -5 and $c_{2,1} = 5$.

There are specific terms denoting various types of matrices. Each of these particular types of matrices has useful applications and unique properties for working with. For example, a *vector* is a matrix with either only one row or one column. Thus, the dimensions of a vector are $1 \times n$ or $m \times 1$. Matrix $\mathbf{c}$ above is a column vector; it is of order 3×1. A $1 \times n$ matrix is a row vector with n elements. The column vector has one column and the row vector has one row. A *scalar* is a 1×1 matrix with exactly one entry, which means that a scalar is simply a number. Matrix $\mathbf{d}$ is a scalar, which we normally write as simply the number 4. A *square matrix* has the same number of rows and columns ($m = n$). Matrix $\mathbf{A}$ is square and of order 3. The set of elements extending from the upper leftmost corner to the lower rightmost corner in a square matrix is said to be on the *principal diagonal*. For a square matrix $\mathbf{A}$, each of these elements are those of the form $a_{i,j}$, $i = j$. Principal diagonal elements of Square Matrix $\mathbf{A}$ in the example above are $a_{1,1} = 4$, $a_{2,2} = 7$, and $a_{3,3} = 9$. Matrices $\mathbf{B}$ and $\mathbf{d}$ are also square matrices.

A *symmetric matrix* is a square matrix where $c_{i,j}$ equals $c_{j,i}$ for all i and j. This is equivalent to the condition kth row equals the kth column for every k. Scalar $\mathbf{d}$ and matrices $\mathbf{H}$, $\mathbf{I}$, and $\mathbf{J}$ below are all symmetric matrices. A *diagonal matrix* is a symmetric matrix whose elements off the principal diagonal are zero, where the *principal diagonal* contains the series of elements where $i = j$. Scalar $\mathbf{d}$ and Matrices $\mathbf{H}$ and $\mathbf{I}$ below are all diagonal matrices. An *identity* or *unit* matrix is a diagonal matrix consisting of ones along the principal diagonal. Matrix $\mathbf{I}$ below is the 3×3 identity matrix:

$$
\mathbf{H} = \begin{bmatrix} 13 & 0 & 0 \\ 0 & 11 & 0 \\ 0 & 0 & 10 \end{bmatrix} \quad \mathbf{I} = \begin{bmatrix} 1 & 0 & 0 \\ 0 & 1 & 0 \\ 0 & 0 & 1 \end{bmatrix} \quad \mathbf{J} = \begin{bmatrix} 1 & 7 & 2 \\ 7 & 5 & 0 \\ 2 & 0 & 4 \end{bmatrix}
$$

1.3.1 Matrix Arithmetic

Matrix arithmetic provides for standard rules of operation just as conventional arithmetic. Matrices can be added or subtracted if their dimensions are identical. Matrices $\mathbf{A}$ and $\mathbf{B}$ add to $\mathbf{C}$ if $a_{i,j} + b_{i,j} = c_{i,j}$ for all i and j:

$$
\underset{\mathbf{A}}{\begin{bmatrix} a_{1,1} & a_{1,2} & \cdots & a_{1,n} \\ a_{2,1} & a_{2,2} & \cdots & a_{2,n} \\ \vdots & \vdots & \vdots & \vdots \\ a_{m,1} & a_{m,2} & \cdots & a_{m,n} \end{bmatrix}} + \underset{\mathbf{B}}{\begin{bmatrix} b_{1,1} & b_{1,2} & \cdots & b_{1,n} \\ b_{2,1} & b_{2,2} & \cdots & b_{2,n} \\ \vdots & \vdots & \vdots & \vdots \\ b_{m,1} & b_{m,2} & \cdots & b_{m,n} \end{bmatrix}} = \underset{\mathbf{C}}{\begin{bmatrix} c_{1,1} & c_{1,2} & \cdots & c_{1,n} \\ c_{2,1} & c_{2,2} & \cdots & c_{2,n} \\ \vdots & \vdots & \vdots & \vdots \\ c_{m,1} & c_{m,2} & \cdots & c_{m,n} \end{bmatrix}}
$$

Note that each of the three matrices is of dimension 3×3 and that each of the elements in Matrix $\mathbf{C}$ is the sum of corresponding elements in Matrices $\mathbf{A}$ and $\mathbf{B}$. The process of subtracting matrices is similar, where $d_{i,j} - e_{i,j} = f_{i,j}$ for $\mathbf{D} - \mathbf{E} = \mathbf{F}$:

$$
\underset{\mathbf{D}}{\begin{bmatrix} d_{1,1} & d_{1,2} & \cdots & d_{1,n} \\ d_{2,1} & d_{2,2} & \cdots & d_{2,n} \\ \vdots & \vdots & \vdots & \vdots \\ d_{m,1} & d_{m,2} & \cdots & d_{m,n} \end{bmatrix}} - \underset{\mathbf{E}}{\begin{bmatrix} e_{1,1} & e_{1,2} & \cdots & e_{1,n} \\ e_{2,1} & e_{2,2} & \cdots & e_{2,n} \\ \vdots & \vdots & \vdots & \vdots \\ e_{m,1} & e_{m,2} & \cdots & e_{m,n} \end{bmatrix}} = \underset{\mathbf{F}}{\begin{bmatrix} f_{1,1} & f_{1,2} & \cdots & f_{1,n} \\ f_{2,1} & f_{2,2} & \cdots & f_{2,n} \\ \vdots & \vdots & \vdots & \vdots \\ f_{m,1} & f_{m,2} & \cdots & f_{m,n} \end{bmatrix}}
$$

Now consider a third matrix operation. The *transpose* $\mathbf{A}^{\mathrm{T}}$ of Matrix $\mathbf{A}$ is obtained by interchanging the rows and columns of Matrix $\mathbf{A}$. Each $a_{i,j}$ becomes $a_{j,i}$. The following represent Matrix $\mathbf{A}$ and its transpose $\mathbf{A}^{\mathrm{T}}$:

$$
\begin{bmatrix}
a_{1,1} & a_{1,2} & \cdots & a_{1,n} \\
a_{2,1} & a_{2,2} & \cdots & a_{2,n} \\
\vdots & \vdots & \vdots & \vdots \\
a_{m,1} & a_{m,2} & \cdots & a_{m,n}
\end{bmatrix}
,
\begin{bmatrix}
a_{1,1} & a_{2,1} & \cdots & a_{m,1} \\
a_{1,2} & a_{2,2} & \cdots & a_{m,2} \\
\vdots & \vdots & \vdots & \vdots \\
a_{1,n} & a_{2,n} & \cdots & a_{m,n}
\end{bmatrix}
$$

$$\mathbf{A} \qquad\qquad\qquad \mathbf{A}^{\mathrm{T}}$$

Two matrices $\mathbf{A}$ and $\mathbf{B}$ can be multiplied to obtain the product $\mathbf{AB} = \mathbf{C}$ if the number of columns in the first Matrix $\mathbf{A}$ equals the number of rows $\mathbf{B}$ in the second.[2] If Matrix $\mathbf{A}$ is of dimension $m \times n$ and Matrix $\mathbf{B}$ is of dimension $n \times q$, the dimensions of the product Matrix $\mathbf{C}$ will be $m \times q$. Each element $c_{i,k}$ of Matrix $\mathbf{C}$ is determined by the following sum:

$$c_{i,k} = \sum_{j=1}^{n} a_{i,j} b_{j,k}$$

$$
\begin{bmatrix}
a_{1,1} & a_{1,2} & \cdots & a_{1,n} \\
a_{2,1} & a_{2,2} & \cdots & a_{2,n} \\
\vdots & \vdots & \vdots & \vdots \\
a_{m,1} & a_{m,2} & \cdots & a_{m,n}
\end{bmatrix}
\times
\begin{bmatrix}
b_{1,1} & b_{1,2} & \cdots & b_{1,q} \\
b_{2,1} & b_{2,2} & \cdots & b_{2,q} \\
\vdots & \vdots & \vdots & \vdots \\
b_{n,1} & b_{n,2} & \cdots & b_{n,q}
\end{bmatrix}
$$

$$\mathbf{A} \qquad \times \qquad \mathbf{B}$$

$$
=
\begin{bmatrix}
\sum_{j=1}^{n} a_{1,j} b_{j,1} & \sum_{j=1}^{n} a_{1,j} b_{j,2} & \cdots & \sum_{j=1}^{n} a_{1,j} b_{j,q} \\
\sum_{j=1}^{n} a_{2,j} b_{j,1} & \sum_{j=1}^{n} a_{2,j} b_{j,2} & \cdots & \sum_{j=1}^{n} a_{2,j} b_{j,q} \\
\vdots & \vdots & \vdots & \vdots \\
\sum_{j=1}^{n} a_{m,j} b_{j,1} & \sum_{j=1}^{n} a_{m,j} b_{j,m} & \cdots & \sum_{j=1}^{n} a_{m,j} b_{j,q}
\end{bmatrix}
=
\begin{bmatrix}
c_{1,1} & c_{1,2} & \cdots & c_{1,q} \\
c_{2,1} & c_{2,2} & \cdots & c_{2,q} \\
\vdots & \vdots & \vdots & \vdots \\
c_{m,1} & c_{m,2} & \cdots & c_{m,q}
\end{bmatrix}
$$

$$\mathbf{A} \times \mathbf{B} \qquad\qquad = \qquad\qquad \mathbf{C}$$

Notice that the number of columns (n) in Matrix $\mathbf{A}$ equals the number of rows in Matrix $\mathbf{B}$. Also note that the number of rows in Matrix $\mathbf{C}$ equals the number of rows in Matrix $\mathbf{A}$; the number of columns in $\mathbf{C}$ equals the number of columns in Matrix $\mathbf{B}$. One additional detail on matrix multiplication is that scalar multiplication is the product of a real number c with a matrix $\mathbf{A}$:

$$
c\mathbf{A} =
\begin{bmatrix}
ca_{1,1} & ca_{1,2} & \cdots & ca_{1,n} \\
ca_{2,1} & ca_{2,2} & \cdots & ca_{2,n} \\
\vdots & \vdots & \vdots & \vdots \\
ca_{m,1} & ca_{m,2} & \cdots & ca_{m,n}
\end{bmatrix}
$$

Matrix Arithmetic Illustration:

Consider the following matrices **A** and **B** below:

$$\mathbf{A} = \begin{bmatrix} 3 & 0 \\ -2 & -1 \end{bmatrix}, \quad \mathbf{B} = \begin{bmatrix} 5 & 2 \\ -6 & 4 \end{bmatrix}$$

We find $\mathbf{A}^T, 4\mathbf{A}, \mathbf{A} + \mathbf{B}, \mathbf{AB}$, and $\mathbf{BA}$ as follows:

$$\mathbf{A}^T = \begin{bmatrix} 3 & -2 \\ 0 & -1 \end{bmatrix}, \quad 4\mathbf{A} = \begin{bmatrix} 4(3) & 4(0) \\ 4(-2) & 4(-1) \end{bmatrix} = \begin{bmatrix} 12 & 0 \\ -8 & -4 \end{bmatrix}$$

$$\mathbf{A} + \mathbf{B} = \begin{bmatrix} 3 + 5 & 0 + 2 \\ -2 - 6 & -1 + 4 \end{bmatrix} = \begin{bmatrix} 8 & 2 \\ -8 & 3 \end{bmatrix}$$

$$\mathbf{AB} = \begin{bmatrix} 3(5) + 0(-6) & 3(2) + 0(4) \\ (-2)(5) + (-1)(-6) & -2(2) + (-1)(4) \end{bmatrix} = \begin{bmatrix} 15 & 6 \\ -4 & -8 \end{bmatrix}$$

$$\mathbf{BA} = \begin{bmatrix} 5(3) + 2(-2) & 5(0) + 2(-1) \\ (-6)(3) + 4(-2) & -6(0) + 4(-1) \end{bmatrix} = \begin{bmatrix} 11 & -2 \\ -26 & -4 \end{bmatrix}$$

1.3.1.1 Matrix Arithmetic Properties

It is useful to note that matrices have certain algebraic properties that are similar to the algebraic properties of real numbers. Here are a few of their properties:

1. $\mathbf{A} + \mathbf{B} = \mathbf{B} + \mathbf{A}$ (commutative property of addition)
2. $\mathbf{A}(\mathbf{B} + \mathbf{C}) = \mathbf{AB} + \mathbf{AC}$ (distributive property)
3. $\mathbf{AI} = \mathbf{IA} = \mathbf{A}$ where $\mathbf{I}$ is the identity matrix
4. $(\mathbf{AB})^T = \mathbf{B}^T\mathbf{A}^T$

However, it is important to observe that, unlike real numbers, the commutative property of multiplication does not hold for matrices; that is, in general, $\mathbf{AB} \neq \mathbf{BA}$.

1.3.1.2 The Inverse Matrix

An *inverse* Matrix $\mathbf{A}^{-1}$ exists for the square Matrix $\mathbf{A}$ if the products $\mathbf{AA}^{-1}$ or $\mathbf{A}^{-1}\mathbf{A}$ equal the identity Matrix $\mathbf{I}$:

$$\mathbf{AA}^{-1} = \mathbf{I}$$
$$\mathbf{A}^{-1}\mathbf{A} = \mathbf{I}$$

One means for finding the inverse Matrix $\mathbf{A}^{-1}$ for Matrix $\mathbf{A}$ is through the use of a process called the *Gauss–Jordan method*.

ILLUSTRATION: THE GAUSS–JORDAN METHOD

An *inverse* Matrix $\mathbf{A}^{-1}$ exists for the square Matrix $\mathbf{A}$ if the product $\mathbf{A}^{-1}\mathbf{A}$ or $\mathbf{A}\mathbf{A}^{-1}$ equals the identity Matrix $\mathbf{I}$. Consider the following product:

$$\text{A.}\quad \begin{bmatrix} 2 & 4 \\ 8 & 1 \end{bmatrix} \begin{bmatrix} \dfrac{-1}{30} & \dfrac{2}{15} \\ \dfrac{4}{15} & \dfrac{-1}{15} \end{bmatrix} = \begin{bmatrix} 1 & 0 \\ 0 & 1 \end{bmatrix}$$

$$\mathbf{A} \qquad\qquad \mathbf{A}^{-1} \qquad = \qquad \mathbf{I}$$

To construct $\mathbf{A}^{-1}$ given a square matrix $\mathbf{A}$, we will use the Gauss–Jordan method. We illustrate the method with the example above. First, augment $\mathbf{A}$ with the 2×2 identity matrix as follows:

$$\text{B.}\quad \left[\begin{array}{cc:cc} 2 & 4 & 1 & 0 \\ 8 & 1 & 0 & 1 \end{array} \right]$$

For the sake of convenience, call the above augmented Matrix $\mathbf{B}$. Now, a series of *elementary row operations* (involves addition, subtraction, and multiplication of rows, as described below) will be performed such that the identity matrix replaces the original Matrix $\mathbf{A}$ (on the left side). The right-side elements will comprise the inverse Matrix $\mathbf{A}^{-1}$. Thus, in our final augmented matrix, we will have ones along the principal diagonal on the left side and zeros elsewhere; the right side of the matrix will comprise the inverse of $\mathbf{A}$. Allowable elementary row operations include the following:

1. Multiply a given row by any constant. Each element in the row must be multiplied by the same constant.
2. Add a given row to any other row in the matrix. Each element in a row is added to the corresponding element in the same column of another row.
3. Subtract a given row from any other row in the matrix. Each element in a row is subtracted from the corresponding element in the same column of another row.
4. Any combination of the above. For example, a row may be multiplied by a constant before it is subtracted from another row.

Our first row operation will serve to replace the upper left corner value with a one. We multiply Row 1 in $\mathbf{B}$ by .5:

$$\mathbf{B} = \left[\begin{array}{cc:cc} \mathbf{2} & 4 & 1 & 0 \\ 8 & 1 & 0 & 1 \end{array} \right] \quad \overset{(\text{row1}) \,\times\, .5}{\longrightarrow} \quad \left[\begin{array}{cc:cc} 1 & 2 & .5 & 0 \\ 8 & 1 & 0 & 1 \end{array} \right] = \mathbf{C}$$

Now we obtain a zero in the lower left corner by multiplying Row 2 in $\mathbf{C}$ by 1/8 and subtracting the result from Row 1 of $\mathbf{C}$ as follows:

$$\mathbf{C} = \begin{bmatrix} 1 & 2 & \vdots & .5 & 0 \\ 8 & 1 & \vdots & 0 & 1 \end{bmatrix} \quad \begin{matrix} \text{row1} - 1/8(\text{row2}) \\ \longrightarrow \end{matrix} \quad \begin{bmatrix} 1 & 2 & \vdots & .5 & 0 \\ 0 & \dfrac{15}{8} & \vdots & .5 & \dfrac{-1}{8} \end{bmatrix} = \mathbf{D}$$

Next, we obtain a 1 in the lower right corner of the left side of the matrix by multiplying Row 2 of matrix $\mathbf{D}$ by 8/15:

$$\mathbf{D} = \begin{bmatrix} 1 & 2 & \vdots & .5 & 0 \\ 0 & \dfrac{15}{8} & \vdots & .5 & \dfrac{-1}{8} \end{bmatrix} \quad \begin{matrix} (\text{row2}) \times \dfrac{8}{15} \\ \longrightarrow \end{matrix} \quad \begin{bmatrix} 1 & 2 & \vdots & .5 & 0 \\ 0 & 1 & \vdots & \dfrac{4}{15} & \dfrac{-1}{15} \end{bmatrix} = \mathbf{E}$$

We obtain a zero in the upper right corner of the left-side matrix by multiplying Row 2 of matrix $\mathbf{E}$ above by 2 and subtracting from Row 1 in $\mathbf{E}$:

$$\mathbf{E} = \begin{bmatrix} 1 & 2 & \vdots & .5 & 0 \\ 0 & 1 & \vdots & \dfrac{4}{15} & \dfrac{-1}{15} \end{bmatrix} \quad \begin{matrix} \text{row1} - (\text{row2}) \times 2 \\ \longrightarrow \end{matrix} \quad \begin{bmatrix} 1 & 0 & \vdots & \dfrac{-1}{30} & \dfrac{2}{15} \\ 0 & 1 & \vdots & \dfrac{4}{15} & \dfrac{-1}{15} \end{bmatrix} = \mathbf{F}$$

The left side of augmented Matrix $\mathbf{F}$ is the identity matrix; the right side of $\mathbf{F}$ is $\mathbf{A}^{-1}$.

ILLUSTRATION: SOLVING SYSTEMS OF EQUATIONS

Matrices can be very useful in arranging systems of equations. Consider, for example, the following system of equations:

$$.05x_1 + .12x_2 = .05$$

$$.10x_1 + .30x_2 = .08$$

This system of equations can be represented as follows:

$$\begin{bmatrix} .05 & .12 \\ .10 & .30 \end{bmatrix} \qquad \begin{bmatrix} x_1 \\ x_2 \end{bmatrix} = \begin{bmatrix} .05 \\ .08 \end{bmatrix}$$
$$\qquad \mathbf{C} \qquad\quad \times \quad \mathbf{x} \quad = \quad \mathbf{s}$$

Thus, we can express this system of equations as the matrix equation $\mathbf{Cx} = \mathbf{s}$, where in general $\mathbf{C}$ is a given $n \times n$ matrix, $\mathbf{s}$ is a given $n \times 1$ column vector, and $\mathbf{x}$ is the unknown $n \times 1$ column vector for which we wish to solve. In ordinary algebra, if we had the real-valued equation $Cx = s$, we would solve for s by dividing both sides of the equation by C, which is equivalent to multiplying both sides of the equation by the inverse of C. Here we show the algebra, so that we see that this process with real numbers is essentially equivalent for the process with matrices:

$$Cx = s, \ C^{-1}Cx = C^{-1}s, \ 1(x) = C^{-1}s, \ x = C^{-1}s$$

With matrices, the process is:

$$\mathbf{Cx} = \mathbf{s}, \ \mathbf{C}^{-1}\mathbf{Cx} = \mathbf{C}^{-1}\mathbf{s}, \ \mathbf{Ix} = \mathbf{C}^{-1}\mathbf{s}, \ \mathbf{x} = \mathbf{C}^{-1}\mathbf{s}$$

Of course, in ordinary algebra, it is trivial to find the inverse of a number C, which is simply its reciprocal $1/C$. To find the inverse of a matrix $\mathbf{C}$, we use the Gauss–Jordan method described above. We begin by augmenting the matrix $\mathbf{C}$ by placing its corresponding identity matrix $\mathbf{I}$ immediately to its right:

A. $\begin{bmatrix} .05 & .12 & \vdots & 1 & 0 \\ .10 & .30 & \vdots & 0 & 1 \end{bmatrix}$

We will reduce this matrix using the allowable elementary row operations described earlier to the form with the identity matrix $\mathbf{I}$ on the left replacing $\mathbf{C}$, and to the right will be the inverse of $\mathbf{C}$:

B. $\begin{bmatrix} 1 & 2.4 & \vdots & 20 & 0 \\ 0 & .6 & \vdots & -20 & 10 \end{bmatrix}$ Row $B1 = A1 \cdot 20$
Row $B2 = (10 \cdot A2) - B1$

C. $\begin{bmatrix} 1 & 0 & \vdots & 100 & -40 \\ 0 & 1 & \vdots & \dfrac{-100}{3} & \dfrac{50}{3} \end{bmatrix}$ Row $C1 = B1 - (2.4 \cdot C2)$
Row $C2 = B2 \cdot 5/3$

$$\mathbf{I} \qquad \mathbf{C}^{-1}$$

Thus, we obtain Vector $\mathbf{x}$ with the following product:

D. $\begin{bmatrix} x_1 \\ x_2 \end{bmatrix} = \begin{bmatrix} 100 & -40 \\ \dfrac{-100}{3} & \dfrac{50}{3} \end{bmatrix} \begin{bmatrix} .05 \\ .08 \end{bmatrix} = \begin{bmatrix} 1.8 \\ \dfrac{-1}{3} \end{bmatrix}$

$$\mathbf{x} \qquad = \qquad \mathbf{C}^{-1} \qquad \times \quad \mathbf{s}$$

Thus, we find that $x_1 = 1.8$ and $x_2 = -1/3$.

1.3.2 Vector Spaces, Spanning, and Linear Dependence

$\mathbb{R}^n$ is defined as the set of all vectors (may be represented as a column or row vectors) with n real-valued entries or coordinates. The row vector $\mathbf{x}^T = (x_1, x_2, \ldots, x_n)$ or column vector $\mathbf{x} = (x_1, x_2, \ldots, x_n)^T$ can be regarded as a point in the n-dimensional space $\mathbb{R}^n$ and x_i is the ith coordinate of the point (vector) $\mathbf{x}$.

The set $\mathbb{R}^n$ with the operations of vector addition and scalar multiplication (discussed earlier) makes $\mathbb{R}^n$ an n-dimensional vector space. A *linear combination* of vectors is accomplished through either or both of the following:

- Multiplication of any vector by a scalar (real number)
- Addition of any combination of vectors either before or after multiplication by scalars

1.3.2.1 *Linear Dependence and Linear Independence*

If a vector in $\mathbb{R}^n$ can be expressed as a linear combination of a set of other vectors in $\mathbb{R}^n$, then that set of vectors including the first is said to be *linearly dependent*. Suppose we are given a set of m vectors: $\{x_1, x_2, \ldots, x_m\}$ with each vector x_i in $\mathbb{R}^n$. An equivalent definition of linear dependence of the set of vectors $\{x_1, x_2, \ldots, x_m\}$ is that there exists m scalars: $\alpha_1, \alpha_2, \ldots, \alpha_m$ so that:

$$\alpha_1 x_1 + \alpha_2 x_2 + \alpha_3 x_3 + \cdots + \alpha_m x_m = 0$$

where at least one of the scalars α_i is non-zero and $0 = (0,0,\ldots,0)$ or $0 = (0,0,\ldots,0)^T$ depending upon whether the vectors x_i are expressed as row or column vectors. We note that 0 is called the zero vector. The set of vectors $\{x_1, x_2, \ldots, x_m\}$ is said to be linearly independent when the only set of scalars $\{\alpha_1, \alpha_2, \ldots, \alpha_m\}$ that satisfy the equation above is when $\alpha_i = 0$ for all $i = 1, 2, \ldots, m$. When the set of vectors $\{x_1, x_2, \ldots, x_m\}$ is linearly independent, then no vector in this set can be expressed as a linear combination of the other vectors in the set. If we denote the $n \times m$ matrix $X = [x_1, x_2, \ldots, x_m]$ and the $m \times 1$ column vector of scalars $\alpha = (\alpha_1, \alpha_2, \ldots, \alpha_m)^T$, then we can express the above equation as the matrix equation $X\alpha = 0$, where $0 = (0,0,\ldots,0)^T$ is the $n \times 1$ column zero vector.

ILLUSTRATIONS: LINEAR DEPENDENCE AND INDEPENDENCE

Consider the following set $\{x_1, x_2, x_3\}$ of three vectors in $\mathbb{R}^3$:

$$\begin{bmatrix} 3 \\ 1 \\ 9 \end{bmatrix} \quad \begin{bmatrix} 5 \\ 5 \\ 15 \end{bmatrix} \quad \begin{bmatrix} 1 \\ 2 \\ 3 \end{bmatrix} \quad \alpha_1 x_1 + \alpha_2 x_2 + \alpha_3 x_3 = [0]$$
$$\quad\; x_1 \qquad\quad x_2 \qquad\quad x_3$$

We will determine whether this set is linearly independent. Let vector α be $[\alpha_1, \alpha_2, \alpha_3]^T$ and Matrix X be $[x_1, x_2, x_3]$. We determine that vector set $\{x_1, x_2, x_3\}$ is linearly dependent by demonstrating that there exists a vector α that produces $X\alpha = [0]$. By inspection, we find that $\alpha = [1, -1, 2]^T$ is one such vector. Thus, the set $\{x_1, x_2, x_3\}$ is linearly dependent. Also note that any one of these three vectors is a linear combination of the other two.

Vector set $\{y_1, y_2, y_3\}$ below is linearly independent because the only vector satisfying $\alpha^T Y = [0]$ is $\alpha = [0, 0, 0]^T$.[3]

$$\begin{bmatrix} 3 \\ 1 \\ 9 \end{bmatrix} \quad \begin{bmatrix} 5 \\ 5 \\ 15 \end{bmatrix} \quad \begin{bmatrix} 1 \\ 2 \\ 4 \end{bmatrix} \quad \alpha_1 y_1 + \alpha_2 y_2 + \alpha_3 y_3 = [0]$$
$$\quad\; y_1 \qquad\quad y_2 \qquad\quad y_3$$

Furthermore, no vector in set $\{y_1, y_2, y_3\}$ can be defined as a linear combination of the other vectors in set $\{Y\}$. Thus, this set is linearly independent. This means that it is impossible to express any one of the vectors as a linear combination of the other two vectors.

1.3.2.2 *Spanning the Vector Space and the Basis*

A set of m vectors $\{x_1, x_2, \ldots, x_m\}$, where each vector x_i is an n-dimensional vector in $\mathbb{R}^n$, is said to *span* the n-dimensional vector space $\mathbb{R}^n$ if any vector in $\mathbb{R}^n$ can be expressed as a linear combination of the vectors $x_1, x_2, \ldots, x_m$. In other words, for every vector v in $\mathbb{R}^n$, there exist scalars $\alpha_1, \alpha_2, \ldots, \alpha_m$ such that $v = \alpha_1 x_1 + \alpha_2 x_2 + \ldots + \alpha_m x_m$.

If a set of vectors $\{x_1, x_2, \ldots, x_m\}$ is both linearly independent and spans the n-dimensional space $\mathbb{R}^n$, then that set of vectors is called a *basis* for the vector space $\mathbb{R}^n$. However, any basis for $\mathbb{R}^n$ must consist of exactly n vectors. This is because for a set of vectors $\{x_1, x_2, \ldots, x_m\}$ in $\mathbb{R}^n$, if $m < n$, then there are not enough vectors to span $\mathbb{R}^n$. On the other hand, if $m > n$, then it is possible for the set of vectors to span $\mathbb{R}^n$, but there will be too many such vectors for the set to be linearly independent. Thus, any set of $m = n$ linearly independent vectors in $\mathbb{R}^n$ will form a basis for $\mathbb{R}^n$ since any such set will also always span $\mathbb{R}^n$.

ILLUSTRATION: SPANNING THE VECTOR SPACE AND THE BASIS

We return to our illustration above with our linearly independent vector set $\{y_1, y_2, y_3\}$:

$$\begin{bmatrix} 3 \\ 1 \\ 9 \end{bmatrix} \quad \begin{bmatrix} 5 \\ 5 \\ 15 \end{bmatrix} \quad \begin{bmatrix} 1 \\ 2 \\ 4 \end{bmatrix}$$
$$y_1 \qquad y_2 \qquad y_3$$

Since this set is linearly independent, it will form a basis for $\mathbb{R}^3$ if it also spans the three-dimensional space. We will demonstrate that any vector v in $\mathbb{R}^3$ is a linear combination of y_1, y_2, and y_3, thereby demonstrating that vectors y_1, y_2, and y_3 span $\mathbb{R}^3$:

$$v = \alpha_1 \begin{bmatrix} 3 \\ 1 \\ 9 \end{bmatrix} + \alpha_2 \begin{bmatrix} 5 \\ 5 \\ 15 \end{bmatrix} + \alpha_3 \begin{bmatrix} 1 \\ 2 \\ 4 \end{bmatrix}$$

To obtain numerical values for α_1, α_2, and α_3, we combine vectors y_1, y_2, and y_3 into a 3×3 matrix, then invert and multiply by v as follows:[4]

$$\begin{bmatrix} 3 & 5 & 1 \\ 1 & 5 & 2 \\ 9 & 15 & 4 \end{bmatrix} \begin{bmatrix} \alpha_1 \\ \alpha_2 \\ \alpha_3 \end{bmatrix} = \begin{bmatrix} v_1 \\ v_2 \\ v_3 \end{bmatrix}$$
$$Y \qquad\quad \alpha \quad = \quad v$$

$$\begin{bmatrix} \alpha_1 \\ \alpha_2 \\ \alpha_3 \end{bmatrix} = \begin{bmatrix} -1 & -.5 & .5 \\ 1.4 & .3 & -.5 \\ -3 & 0 & 1 \end{bmatrix} \begin{bmatrix} v_1 \\ v_2 \\ v_3 \end{bmatrix}$$
$$\alpha \quad = \quad\qquad Y^{-1} \qquad\quad \times \quad v$$

Thus, we can replicate any vector v with a linear combination of vectors y_1, y_2, and y_3 and some vector α. For example, if $v = [6\ 3\ 1]^T$, then we obtain α as follows:

$$\begin{bmatrix} \alpha_1 \\ \alpha_2 \\ \alpha_3 \end{bmatrix} = \begin{bmatrix} -1 & -.5 & .5 \\ 1.4 & .3 & -.5 \\ -3 & 0 & 1 \end{bmatrix} \begin{bmatrix} 6 \\ 3 \\ 1 \end{bmatrix} = \begin{bmatrix} -7 \\ 8.8 \\ -17 \end{bmatrix}$$
$$\alpha \quad = \quad\qquad Y^{-1} \qquad\quad \times \quad v$$

$$v = -7 \begin{bmatrix} 3 \\ 1 \\ 9 \end{bmatrix} + 8.8 \begin{bmatrix} 5 \\ 5 \\ 15 \end{bmatrix} - 17 \begin{bmatrix} 1 \\ 2 \\ 4 \end{bmatrix}$$

As long as we can invert 3×3 matrix $\mathbf{Y}$, we can replicate any vector in $\mathbb{R}^3$ with some linear combination of vectors $\mathbf{y}_1$, $\mathbf{y}_2$, and $\mathbf{y}_3$ from which coefficients are obtained from vector α.

In a sense, when an $n+1^{st}$ vector is linearly dependent on a set of n other $n \times 1$ vectors, the characteristics or information in the n other $n \times 1$ vectors can be used to replicate the information in the $n+1^{st}$ vector. In a financial sense where elements in a vector represent security payoffs over time or across potential outcomes, the payoff structure of the $n+1^{st}$ security can be replicated with a portfolio comprising the n other $n \times 1$ security vectors. When a set of n payoff vectors spans the n-dimensional outcome or time space, the payoff structure for any other security or portfolio in the same outcome or time space can be replicated with the payoff vectors of the n-security basis. Securities or portfolios whose payoff vectors can be replicated by portfolios of other securities must sell for the same price as those portfolios; otherwise, the *law of one price* is violated.[5]

1.4 REVIEW OF DIFFERENTIAL CALCULUS

The derivative and the integral are the two most essential concepts from calculus. The derivative from calculus can be used to determine rates of change or slopes. They are also useful for finding function maxima and minima. For those functions whose slopes are changing, the derivative is equal to the instantaneous rate of change; that is, the change in y induced by the "tiniest" change in x. Assume that y is given as a function of variable x. If x were to increase by a small (infinitesimal—that is, approaching, though not quite equal to zero) amount Δx, by how much would y change? This rate of change is given by the derivative of y with respect to x, which is defined as follows:

$$\frac{dy}{dx} = f'(x) = \lim_{\Delta x \to 0} \frac{f(x+\Delta x) - f(x)}{\Delta x} \tag{1.1}$$

Consider Figure 1.1, which plots the function $y = 2x - x^2$. Using Eq. (1.1), we will find that dy/dx, the slope of our function is calculated by:

$$\frac{dy}{dx} = f'(x) = \lim_{\Delta x \to 0} \frac{f(x+\Delta x) - f(x)}{\Delta x} = \lim_{\Delta x \to 0} \frac{2(x+\Delta x) - (x+\Delta x)^2 - 2x + x^2}{\Delta x}$$

$$= \lim_{\Delta x \to 0} \frac{2x + 2\Delta x - x^2 - (\Delta x)^2 - 2x\Delta x - 2x + x^2}{\Delta x} = \lim_{\Delta x \to 0} \frac{2\Delta x - (\Delta x)^2 - 2x\Delta x}{\Delta x} = \lim_{\Delta x \to 0} (2 - \Delta x - 2x)$$

$$= 2 - 2x$$

On Figure 1.1, suppose that we start from point $(x_0, y_0) = (0.2, 0.36)$. If the change in x were $\Delta x = .8$, the change in y would be $\Delta y = (1 - .36) = .64$ and the average rate of change would be $\Delta y / \Delta x = .64/.8 = .8$. If the change in x were only $\Delta x = .5$, the change in y would be $\Delta y = 0.55$, and the average rate of change would be $\Delta y / \Delta x = .55/.5 = 1.1$. As the change in x approaches 0 (i.e., $\Delta x \to 0$), the rate of change $\Delta y / \Delta x$ approaches $dy/dx = 1.6$. Thus, when $x_i = .2$, $dy/dx = 1.6$,

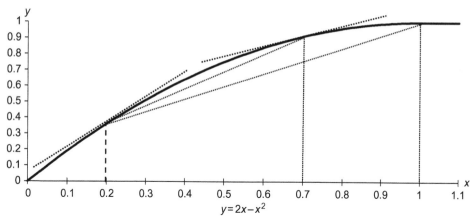

FIGURE 1.1 The derivative of $y = 2x - x^2$. When $x_i = .2$, $dy/dx = 1.6$. As $\Delta x \to 0$, $\Delta y/\Delta x \to dy/dx$. Also, notice that when $x_i = .7$, $dy/dx = .6$.

and an infinitesimal change in x would lead to 1.6 times that rate of change in y. The "point slope" or instantaneous rate of change of $2x - x^2$ is 1.6 when $x_i = .2$. The derivative of y with respect to $x(dy/dx = f'(x))$ can be interpreted to be the instantaneous rate of change in y given an infinitesimal change in x. In addition, notice that the slope (derivative) in Figure 1.1 changes with x. For example, when $x_i = .7$, $dy/dx = .6$. The rate of change of this derivative is the derivative of the derivative function, or the second derivative of the function $f(x)$ is $\frac{d^2y}{dx^2} = f''(x)$. In our example, $f''(x) = -2$. This means that the slope of the tangent line itself is changing at a constant rate of -2. Thus, after each change in x by 1 unit, the value of the slope will decrease by 2 units. This is apparent in Figure 1.1, since as x increases, the slope of the curve decreases.

1.4.1 Essential Rules for Calculating Derivatives

Equation (1.1) provides for a change in y given a very small (infinitesimal) change in x. This definition can be used to derive a number of very useful rules in calculus. A few are discussed below.

1.4.1.1 The Power Rule

One type of function that appears regularly in finance is the polynomial or integer power function. This type of function defines variable y in terms of a coefficient c, variable x, and an exponent n. While the exponents in a polynomial equation are non-negative integers, the rules that we discuss here still apply when the exponents assume negative or non-integer values. Consider a polynomial with a single variable x, a coefficient c, and an exponent n:

$$y = cx^n$$

The derivative of such a function y with respect to x is given by:

$$\frac{dy}{dx} = cnx^{n-1} \tag{1.2}$$

1.4.1.2 *The Sum Rule*

Consider a function that defines variable y in terms of a series of terms or functions involving x:

$$\frac{d}{dx}\left[f(x) + g(x)\right] = \frac{d}{dx}[f(x)] + \frac{d}{dx}[g(x)] \qquad (1.3)$$

The notation $\frac{d}{dx}[f(x)]$ refers to the derivative of the function $f(x)$. In addition, the sum rule applies to any finite sum of terms. For example, consider y as a function of a series of coefficients c_j, variable x, and a series of exponents n_j:

$$y = \sum_{j=1}^{m} c_j \cdot x^{n_j} \qquad (1.4)$$

The derivative of such a function y with respect to x is given by:

$$\frac{dy}{dx} = \sum_{j=1}^{m} c_j \cdot n_j \cdot x^{n_j - 1} \qquad (1.5)$$

That is, simply take the derivative of each term in y with respect to x and sum these derivatives.

1.4.1.3 *The Chain Rule*

Each of the functions discussed in the previous section is written in polynomial form. Other rules can be derived to find derivatives for different types of functions. The *chain rule* is a derivative rule that allows us to differentiate more complex functions of the form:

$$y = f(g(x))$$

where $f(x)$ and $g(x)$ are functions whose derivatives are already known. The chain rule states that:

$$\frac{dy}{dx} = f'(g(x))g'(x) \qquad (1.6)$$

To appreciate when the chain rule is relevant, consider the following two examples. First, consider $y = x^{1/2}$. We obtain the derivative as follows:

$$\frac{dy}{dx} = \frac{1}{2}x^{-1/2}$$

Next, consider the more complicated function $y = (x^3 + 4x - 1)^{1/2}$. We need to use the chain rule to find the derivative of y with respect to x. Observe that if we choose $f(x) = x^{1/2}$ and $g(x) = x^3 + 4x - 1$, then:

$$y = f(g(x)) = (g(x))^{1/2} = (x^3 + 4x - 1)^{1/2}$$

We already know how to find the derivatives: $f'(x) = \frac{1}{2}x^{-\frac{1}{2}}$ and $g'(x) = 3x^2 + 4$. Application of the chain rule to the composite function yields:

$$\frac{dy}{dx} = f'(g(x))g'(x) = \frac{1}{2}\left(x^3 + 4x - 1\right)^{-\frac{1}{2}}(3x^2 + 4)$$

Another way to express the chain rule is to create an intermediate variable, say u, with $u = g(x)$. If $y = f(g(x))$, then $y = f(u)$. With this notation, the chain rule can be expressed as:

$$\frac{dy}{dx} = \frac{dy}{du}\frac{du}{dx}$$

Consider again the example $y = (x^3 + 4x - 1)^{1/2}$. Choose $u = x^3 + 4x - 1$, so that $y = u^{1/2}$. By using the chain rule, we obtain:

$$\frac{dy}{dx} = \frac{dy}{du}\frac{du}{dx} = \frac{1}{2}u^{-1/2}(3x^2 + 4) = \frac{1}{2}(x^3 + 4x - 1)^{-1/2}(3x^2 + 4)$$

Consider one more example where $y = x^3$ and $x = t^2 + 1$ and we wish to find dy/dt. Again, from the chain rule, we have:

$$\frac{dy}{dt} = \frac{dy}{dx}\frac{dx}{dt} = 3x^2(2t) = 3(t^2+1)^2(2t) = 6t(t^2 + 1)^2$$

1.4.1.4 Product and Quotient Rules

The *product rule*, which is applied to a function such as $y = f(x)g(x)$, holds that the derivative of y with respect to x is as follows:

$$\frac{dy}{dx} = f(x)\frac{dg(x)}{dx} + g(x)\frac{df(x)}{dx} \tag{1.7}$$

For example, if $y = (4x + 2)(5x + 1)$ where $f(x)$ is $(4x + 2)$ and $g(x)$ is $(5x + 1)$, the product rule holds that $dy/dx = (4x + 2) \times 5 + (5x + 1) \times 4 = 40x + 14$.

The *quotient rule*, which is applied to a function such as $f(x)/g(x)$, holds that the derivative of y with respect to x is as follows:

$$\frac{dy}{dx} = \left[g(x)\frac{df(x)}{dx} - f(x)\frac{dg(x)}{dx}\right]/g(x)^2 \tag{1.8}$$

For example, if $y = (4x + 2)/5x$ where $f(x)$ is $(4x + 2)$ and $g(x)$ is $5x$, the quotient rule holds that $dy/dx = [(5x \times 4) - 5(4x + 2)]/25x^2 = -2/5x^2$.

The product rule also implies the constant multiple rule:

$$\frac{d}{dx}[cf(x)] = c\frac{d}{dx}[f(x)] \quad \text{(constant multiple rule)} \tag{1.9}$$

1.4.1.5 Exponential and Log Function Rules

Logarithmic and exponential functions and derivatives of these functions are particularly useful in finance for modeling growth. Consider the function $y = e^x$ and its derivative with respect to x:

$$\frac{dy}{dx} = e^x \tag{1.10}$$

Or, more generally, which can be verified with the chain rule:

$$\frac{de^{g(x)}}{dx} = \frac{dg(x)}{dx}e^{g(x)} \tag{1.11}$$

If $y = e^{\ln(x)}$, then, by definition, $y = e^{\ln(x)} = x$, which implies that $de^{\ln(x)}/dx = 1$. Now, consider the following special case of Eq. (1.11):

$$\frac{de^{\ln(x)}}{dx} = \frac{d\ln(x)}{dx} e^{\ln(x)}$$

which implies:

$$1 = \frac{d\ln(x)}{dx} \cdot x$$

$$\frac{d\ln(x)}{dx} = \frac{1}{x} \tag{1.12}$$

Table 1.1 summarizes the rules for finding derivatives covered in Section 1.4.1. We will make regular use of these rules throughout the text.

1.4.2 The Differential

The concept of the differential will be very useful later when we discuss stochastic calculus. The differential of a function can be used to estimate the change of the value of a function $y = f(x)$ resulting from a small change of the x value. Since:

$$f'(x) = \lim_{\Delta x \to 0} \frac{f(x + \Delta x) - f(x)}{\Delta x}$$

then when Δx is small we have:

$$f'(x) \cong \frac{f(x + \Delta x) - f(x)}{\Delta x}$$

The approximation improves as Δx approaches 0. Denote the error in the approximation above by $\epsilon(x, \Delta x)$, so that:

$$\frac{f(x + \Delta x) - f(x)}{\Delta x} = f'(x) + \epsilon(x, \Delta x)$$

TABLE 1.1 Sample Derivative Rules (c and n are Arbitrary Constrants)

1. $\frac{d}{dx}[x^n] = nx^{n-1}$ (power rule)

2. $\frac{d}{dx}[f(x) + g(x)] = \frac{d}{dx}[f(x)] + \frac{d}{dx}[g(x)]$ (sum rule)

3. $\frac{dy}{dx} = \frac{dy}{du}\frac{du}{dx}$ (chain rule)

4. $\frac{d}{dx}[f(x)g(x)] = f(x)\frac{d}{dx}[g(x)] + g(x)\frac{d}{dx}[f(x)]$ (product rule)

5. $\frac{d}{dx}\left[\frac{f(x)}{g(x)}\right] = \frac{g(x)\frac{d}{dx}[f(x)] - f(x)\frac{d}{dx}[g(x)]}{[g(x)]^2}$ (quotient rule)

6. $\frac{d}{dx}[cf(x) = c\frac{d}{dx}[f(x)]$ (constant multiple rule)

7. $\frac{d}{dx}[e^x] = e^x$ (exponential rule)

8. $\frac{d}{dx}[\ln x] = \frac{1}{x}$ (log rule)

Whenever the derivative $f'(x)$ exists, this equality and our definition above for $f'(x)$ imply that $\epsilon(x,\Delta x) \to 0$ as $\Delta x \to 0$. Now, we label the change in y by Δy, so that:

$$\Delta y = f(x + \Delta x) - f(x) = f'(x)\Delta x + \epsilon(x,\Delta x)\Delta x$$

Observe on Figure 1.2 that Δy, the change in y on the curve, can be closely approximated by $f'(x)\Delta x$ when Δx is small. The expression $f'(x)\Delta x$ is the change in y on the tangent line resulting from the change Δx in the value of x. In the case that $\epsilon(x,\Delta x) \to 0$ as $\Delta x \to 0$ (so that the error term is negligible as Δx approaches 0), then one often writes:

$$dy = f'(x)dx$$

where dx has replaced Δx and dy has replaced Δy. The term dy is called the differential of y.

ILLUSTRATION: THE DIFFERENTIAL AND THE ERROR

Reconsider our illustration from earlier with $y = 2x - x^2$, plotted again in Figure 1.2. The differential $dy = (2 - 2x)dx$. Suppose that in this case $x = .6$ and $dx = 0.1$, such that $dy = (2 - 2 \times .6)(.1) = .08$. This tells us that the approximate change in y from $x = .6$ by $\Delta x = .1$ to $x = .7$ will be $\Delta y \approx .08$. The actual change in y can be computed directly since:

$$f(.7) - f(.6) = \left(1.4 - .7^2\right) - \left(1.2 - .6^2\right) = .07$$

The term $\epsilon(x,\Delta x)\Delta x$ itself is the error in using the differential as an approximation to the change in y. More precisely:

$$\epsilon(x,\Delta x)\Delta x = \left[f(x + \Delta x) - f(x)\right] - f'(x)\Delta x$$

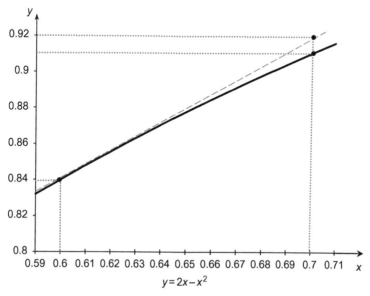

FIGURE 1.2 **The differential.** When $x_0 = .6$, $dy/dx = .8$. As $\Delta x \to 0$, $\Delta y/\Delta x \to dy/dx$. Also, where the tangent dashed line reflects error estimates for y based on error estimates for x and dy, notice that when $x = .7$, $\epsilon \Delta x = .01$.

where Δx is regarded as the same as dx. In our example we have:

$$\epsilon(.6,.1)\Delta x = \left[f(.7) - f(.6)\right] - f'(.6) \times .1 = [.91 - .84] - .08 = .07 - .08 = -.01$$

Observe that the possible error $\epsilon(.6,.1)\Delta x = -.01$ is very small relative to the differential $dy = .08$. Observe that the differential provided a reasonable estimate, and the error term $\epsilon(.6, .1)\Delta x$, because of its size relative to Δx, can be ignored as Δx approaches 0.

1.4.3 Partial Derivatives

If our dependent variable y is a function of multiple independent variables x_j, we can find partial derivatives $\partial y / \partial x_j$ of y with respect to each of our independent variables x_j. For example, in the following, y is a function of x_1 and x_2; function y's partial derivatives with respect to each of its independent variables (while holding the other constant) follow:

$$y = x_1 e^{.05x_2} + .03x_2$$

$$\frac{\partial y}{\partial x_1} = e^{.05x_2}$$

$$\frac{\partial y}{\partial x_2} = .05x_1 e^{.05x_2} + .03$$

1.4.3.1 The Chain Rule for Two Independent Variables

Suppose that $y = f(x)$ and $x = g(t)$. Recall that the chain rule provides:

$$\frac{dy}{dt} = \frac{dy}{dx}\frac{dx}{dt}$$

There is an analogous chain rule for functions of more than one independent variable. Suppose the variable z is a function of the variables x and y, $z = f(x, y)$, and, in turn, each of the variables x and y is a function of the variable t, $x = g(t)$ and $y = h(t)$. This implies that z can be defined as function of the variable t, that is $z = f(g(t), h(t))$.

Now, consider an example where $z = x^2 y + y^3$, $x = t^4$, and $y = 2t$. This implies that $z = (t^4)^2(2t) + (2t)^3 = 2t^9 + 8t^3$. While the derivative $\frac{dz}{dt} = 18t^8 + 24t^2$ can easily be obtained by a direct calculation from this last expression, it can also be found using the chain rule. Since $z = f(x, y)$, $x = g(t)$, and $y = h(t)$, the derivative $\frac{dz}{dt}$ is obtained from the chain rule:

$$\frac{dz}{dt} = \frac{\partial z}{\partial x}\frac{dx}{dt} + \frac{\partial z}{\partial y}\frac{dy}{dt} = 2xy(4t^3) + (x^2 + 3y^2)(2)$$

$$= 2t^4(2t)(4t^3) + \left[\left(t^4\right)^2 + 3(2t)^2\right](2) = 18t^8 + 24t^2$$

Observe that we obtained the same answer earlier by the direct calculation. From the chain rule, we multiply through by dt to derive the total differential:

$$dz = \frac{\partial z}{\partial x}dx + \frac{\partial z}{\partial y}dy$$

We will demonstrate in later chapters that the total differential is a useful tool to find solutions to certain types of differential equations. It is also useful for approximating the change in the variable z (dz) resulting from small changes in the variables x and y (dx and dy).

1.4.4 Taylor Polynomials and Expansions

One can improve on the approximation $\Delta y = f'(x)\Delta x$ by taking into account higher-order derivatives. As long as the function $f(x)$ is differentiable at least n times, the nth-order Taylor polynomial expanded about x_0 is defined by the right side of the following approximation:

$$f(x_0 + \Delta x) \approx f(x_0) + f'(x_0)(\Delta x) + \frac{1}{2!}f''(x_0)(\Delta x)^2 + \frac{1}{3!}f^{(3)}(x_0)(\Delta x)^3$$

$$+ \cdots + \frac{1}{n!}f^{(n)}(x_0)(\Delta x)^n$$

The Taylor polynomial approximation or expansion can be used for finite changes in x. Taylor polynomial expansions are frequently used to evaluate a function $f(x)$ at some point x_1 that differs from an initial point x_0 at which $f(x_0)$ has already been evaluated. That is, the Taylor polynomial can be used to approximate a change in $f(x)$ induced by a change in x. For example, consider the function $f(x) = \ln x$. Choose $x_0 = 1$. We will use the third-order Taylor polynomial to estimate the value of $\ln(1.2)$. Differentiating, we obtain $f'(x) = x^{-1}$, $f''(x) = -x^{-2}$, and $f^{(3)}(x) = 2x^{-3}$.[6] Evaluating at $x_0 = 1$ yields $f(1) = 0$, $f'(1) = 1$, $f''(1) = -1$, and $f^{(3)}(1) = 2$. In this case, $x_1 = 1.2$, such that $\Delta x = 1.2 - 1 = .2$. We obtain our estimate for $f(x_0 + \Delta x)$ as follows:

$$f(x_0 + \Delta x) = \ln(1 + .2) = \ln(1.2) \approx 0 + 1(.2) + \frac{1}{2}(-1)(.2)^2 + \frac{1}{6}(2)(.2)^3 = .182667$$

The actual value of $\ln(1.2)$ is $.18232\ldots$ Observe that the first-order Taylor polynomial would yield the estimate $.2$, and the second-order would give $.18$. In general, the higher the order of the Taylor polynomial, the better the estimate. We are often concerned with changes in the value of the function $f(x)$. Since $\Delta y = f(x + \Delta x) - f(x)$, then after replacing x_0 with x and subtracting $f(x)$, we can express the previous approximation as:

$$\Delta y = f'(x)(\Delta x) + \frac{1}{2!}f''(x)(\Delta x)^2 + \frac{1}{3!}f^{(3)}(x)(\Delta x)^3 + \cdots + \frac{1}{n!}f^{(n)}(x)(\Delta x)^n$$

One can generalize the results above to functions of more than one independent variable. Consider the function $y = f(x,t)$. Define $\Delta y = f(x + \Delta x, t + \Delta t)) - f(x,t)$. This can also be expanded into a two-variable Taylor series where the first few terms out to the second-order derivatives in the expansion take the following form:

$$\Delta y = \frac{\partial f}{\partial x}(x,t)\Delta x + \frac{\partial f}{\partial t}(x,t)\Delta t + \frac{1}{2}\frac{\partial^2 f}{\partial x^2}(x,t)(\Delta x)^2 + \frac{1}{2}\frac{\partial^2 f}{\partial t^2}(x,t)(\Delta t)^2 + \frac{\partial^2 f}{\partial x \partial t}(x,t)\Delta x \Delta t + \cdots$$

For example, consider the function $f(x,t) = 100e^{-x^2+3t}$. In this illustration, we will use a second-order Taylor polynomial to estimate the change $f(.2,.1) - f(0,0)$. So, we must choose $x = 0$, $t = 0$, $\Delta x = .2$ and $\Delta t = .1$. We find first and second derivatives as follows:

$$\frac{\partial f}{\partial x} = -200xe^{-x^2+3t}, \quad \frac{\partial f}{\partial t} = 300e^{-x^2+3t}$$

$$\frac{\partial^2 f}{\partial x^2} = 200(-1 + 2x^2)e^{-x^2+3t}, \quad \frac{\partial^2 f}{\partial x \partial t} = -600xe^{-x^2+3t}, \quad \frac{\partial^2 f}{\partial t^2} = 900e^{-x^2+3t}$$

Evaluating the derivatives at $(x,t) = (0,0)$:

$$\frac{\partial f}{\partial x}(0,0) = 0, \quad \frac{\partial f}{\partial t}(0,0) = 300, \quad \frac{\partial^2 f}{\partial x^2}(0,0) = -200, \frac{\partial^2 f}{\partial x \partial t}(0,0) = 0, \quad \frac{\partial^2 f}{\partial t^2}(0,0) = 900$$

Thus,

$$\Delta f = f(.2,.1) - f(0,0) \cong 0(.2) + 300(.1) + \frac{1}{2}(-200)(.2)^2 + \frac{1}{2}(900)(.1)^2 + (0)(.2)(.1)$$

$$= 30.5$$

1.4.5 Optimization and the Method of Lagrange Multipliers

Differential calculus is particularly useful for determining minima or maxima of functions of many types. In many instances, minima or maxima can be calculated by setting first derivatives with respect to the variable(s) of interest equal to zero (first-order conditions), and then checking second-order conditions (positive second derivative(s) for minima, negative second derivative(s) for maxima).

However, many optimization problems require constraints or limitations on variables. The method of Lagrange multipliers can often enable function optimization in the presence of such constraints. The method of Lagrange multipliers creates a Lagrange function L that supplements the original function $y = f(x)$ to be optimized with an additional expression for each of m relevant constraints. Assume a linear constraint equation of the form $g(x) = c$, with g an $m \times 1$ vector valued function of the vector $x = [x_1, x_2, \ldots, x_n]^T$, which is an $n \times 1$ column vector variable, and c is an $m \times 1$ constant column vector. We will introduce the Lagrange multiplier column vector λ where $\lambda = (\lambda_1, \lambda_2, \ldots, \lambda_m)^T$. The Lagrange function has the form:

$$L = f(x) + \lambda^T(g(x) - c)$$

ILLUSTRATION: LAGRANGE OPTIMIZATION

Suppose that our objective is to minimize the function $y = x_1^2 + 2x_2^2 + .5x_1x_2$ subject to the constraint that $x_1 + .2x_2 = 10$:[7]

$$OBJ : Min\ y = x_1^2 + 2x_2^2 + .5x_1x_2$$
$$s.t. : x_1 + .2x_2 = 10$$

The Lagrange function combines the original function and a revised version of the single constraint as follows:[8]

$$L = x_1^2 + 2x_2^2 + .5x_1x_2 + \lambda(x_1 + .2x_2 - 10)$$

We solve our problem by setting partial derivatives of L with respect to each of our three variables equal to zero. This will result in the following first-order conditions:

$$\frac{\partial L}{\partial x_1} = 2x_1 + .5x_2 + \lambda = 0$$

$$\frac{\partial L}{\partial x_2} = .5x_1 + 4x_2 + .2\lambda = 0$$

$$\frac{\partial L}{\partial \lambda} = x_1 + .2x_2 - 10 = 0$$

This system is structured and solved in matrix format as follows:

$$\begin{bmatrix} 2 & .5 & 1 \\ .5 & 4 & .2 \\ 1 & .2 & 0 \end{bmatrix} \begin{bmatrix} x_1 \\ x_2 \\ \lambda \end{bmatrix} = \begin{bmatrix} 0 \\ 0 \\ 10 \end{bmatrix}$$

$$\begin{bmatrix} x_1 \\ x_2 \\ \lambda \end{bmatrix} = \begin{bmatrix} .010309 & -.05155 & 1.005155 \\ -.05155 & .257732 & -.02577 \\ 1.005155 & -.02577 & -1.99742 \end{bmatrix} \times \begin{bmatrix} 0 \\ 0 \\ 10 \end{bmatrix} = \begin{bmatrix} 10.05155 \\ -.25773 \\ -19.9742 \end{bmatrix}$$

Thus, y is minimized when $x_1 = 10.05155$ and $x_2 = -.25773$. The Lagrange multiplier λ can be interpreted as a sensitivity coefficient that indicates the change in y that would result from a change in the constraint on $x_1 + .2x_2$. If, for example, we were to increase the constraint by 1 from 10 to 11, the value of y would decrease by approximately 19.9742.

1.5 REVIEW OF INTEGRAL CALCULUS

A graphic interpretation of the derivative $f'(x)$ of a function $f(x)$ is that it equals the slope of the curve plotted by that function. A graphic interpretation of the integral of a non-negative function $f(x)$, $\int_a^b f(x)dx$, is that it equals the area under the graph of the function $f(x)$ from $x = a$ to $x = b$, where $\int$ is the *integral sign* and $f(x)$ is the *integrand*. Thus, integrals are useful for finding areas under curves. Integrals can be regarded as the limit of sums involving functions of a continuous variable. Similarly, as we will discuss shortly, they are useful for determining expected values and variances based on continuous probability distributions. As the expectation of a discrete random variable requires summing a discrete countable number of terms, the expectation of a continuous random variable requires integration to handle the continuous (uncountable) number of values of the random variable.

Integral calculus is also useful for analyzing the behavior of variables (such as cash flows) over time. An equation of the form $\frac{dy}{dt} = f(t)$ is known as a differential equation and it might describe the rate of change of the variable y with respect to time t. The solution to this differential equation $y = F(t)$, which is obtained by integration, describes the function y itself over time. For example, $f(t)$ might describe the change in value of the price y of an investment over time (profit) while $F(t)$ provides the actual value of the price.

1.5.1 Antiderivatives

Integrals of many functions can be determined by using the process of *antidifferentiation*, which is the inverse process of differentiation. If $F(x)$ is a function of x whose derivative equals $f(x)$, then $F(x)$ is said to be the antiderivative or integral of $f(x)$, written as follows:

$$F(x) = \int f(x)\, dx \qquad (1.13)$$

The function $F(x)$ has the property that:

$$\frac{dF(x)}{dx} = f(x) \qquad (1.14)$$

One can always add any constant C to the function $F(x)$, where $F(x)$ is any one particular antiderivative of $f(x)$, and it will still be an antiderivative of $f(x)$; that is:

$$\frac{d}{dx}[F(x) + C] = \frac{d}{dx}[F(x)] + \frac{d}{dx}[C] = f(x) + 0 = f(x)$$

Thus, the general form of the *indefinite integral* of $f(x)$ is:

$$\int f(x) dx = F(x) + C$$

where $F(x)$ is one particular antiderivative of $f(x)$. Observe that the indefinite integral of a function is actually a family of functions, since each different choice of the constant C gives a different function.

Suppose, for example, we wished to evaluate $\int 2x\, dx$. We will seek a family of functions for which the derivative is $2x$. Since $\frac{d}{dx}\left[x^2 + C\right] = 2x$, $\int 2x\, dx = x^2 + C$. Using the fact that integrals are the inverse process of differentiation, one can derive integral rules. Table 1.2 provides a short listing of integral rules that will be useful in this book.

TABLE 1.2   Table of Integrals

1. $\int cx^n\, dx = \dfrac{cx^{n+1}}{n+1} + C$ for $n \neq -1$ (power rule)

2. $\int cf(x) dx = c \int f(x) dx$ (constant multiple rule)

3. $\int (f(x) + g(x)) dx = \int f(x) dx + \int g(x) dx$ (sum rule)

4. $\int \dfrac{1}{x} dx = \ln|x| + C$

5. $\int e^{cx}\, dx = \dfrac{1}{c} e^{cx} + C$

Next, suppose that we wished to evaluate $\int \left(\frac{5}{x} + 3e^x + 4x^2 - 6\right) dx$. We will use all five rules in Table 1.2 to evaluate this function, finding that $\int \left(\frac{5}{x} + 3e^x + 4x^2 - 6\right) dx = 5 \ln |x| + 3e^x + \frac{4}{3}x^3 - 6x + C$. Observe that there is only one constant C in the solution. This is sufficient since C can be any arbitrary constant.

1.5.2 Definite Integrals

Using simple rules from geometry, one can find areas of elementary shapes such as squares, rectangles, triangles, and circles. However, if we wish to find the area under the graph of an arbitrary curve, we need a new method. If the values a function $f(x)$ are non-negative so that its graph always lies above the x-axis, then the definite integral of $f(x)$ from $x = a$ to $x = b$ is defined to be the area between the x-axis and its graph from $x = a$ to $x = b$ (see Figure 1.3). For a general function $f(x)$, the definite integral from $x = a$ to $x = b$ equals the area above the x-axis minus the area below the x-axis. The definite integral is denoted by:

$$\int_a^b f(x)\, dx$$

Notice that the notation for this area uses the antiderivative sign. We will show this connection shortly by using the fundamental theorem of calculus.

1.5.2.1 Reimann Sums

The definite integral for any continuous curve can be obtained as a limit of a sum of rectangular areas (or so-called "signed areas" in case that part of the graph of $f(x)$ is below the x-axis). More precisely, consider the graph of a function $f(x)$ on the interval $[a,b]$ of x-values. For the time being, suppose that the function $f(x)$ is non-negative. Divide the interval $[a,b]$ into n subintervals of equal width

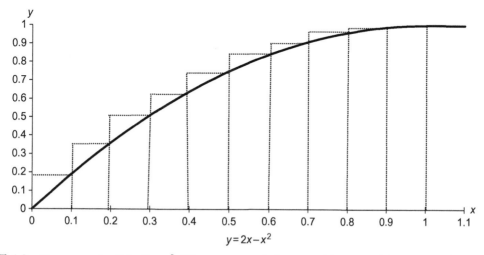

$$y = 2x - x^2$$

FIGURE 1.3   **The area under $f(x) = 2x - x^2$.** When $x_i - x_{i-1} = .1$, the sum of the areas of the 10 rectangles equals 0.715. As the number of rectangles approaches infinity, and their widths approach zero, the sum of their areas will approach 2/3.

$\Delta x = (b-a)/n$. Consider the values of x on the x-axis that are endpoints of the subintervals. They are: $x_0 = a$, $x_1 = a + \Delta x$, $x_2 = a + 2\Delta x$, ..., $x_n = a + n\Delta x = b$. The n subintervals are: $[x_0, x_1]$, $[x_1, x_2]$, ..., $[x_{n-1}, x_n]$. For the ith subinterval $[x_{i-1}, x_i]$, choose a convenient x-value x_i^* in this subinterval; that is, $x_{i-1} \leq x_i^* \leq x_i$. The area between the graph of $f(x)$ and the x-axis can be approximated by the sum of the areas of the n rectangles so that the ith rectangle has height $f(x_i^*)$ and width Δx. Since the area of a rectangle is the product of its height and width, the area of the ith rectangle equals $f(x_i^*)\Delta x$. Thus, the total area from $x = a$ to $x = b$ can be approximated by the sum:

$$\sum_{i=1}^{n} f(x_i^*)\Delta x$$

This sum is known as a *Riemann sum*. To illustrate this Riemann sum, consider the example of estimating the area under the graph of $y = 2x - x^2$ from $x = 0$ to $x = 1$. We choose $n = 10$ so that we are estimating the area under the graph by 10 narrow rectangles (see Figure 1.3). In this case the width of each rectangle is $\Delta x = (1-0)/10 = .1$. The value of $x_i = i/10$. For this example, choose $x_i^* = x_i$. The sum of the areas of the 10 narrow rectangles equals the following Riemann sum, as calculated in Table 1.3:

$$\sum_{i=1}^{10} (2x_i - x_i^2)\Delta x = \sum_{i=1}^{10} \left[\frac{2i}{10} - \left(\frac{i}{10}\right)^2\right].1 = .715$$

Now suppose that we increase the number of rectangles from $n = 10$ an arbitrarily large value of n. For general n, the width of the each rectangle $\Delta x = 1/n$. The ith x-value is $x_i = i/n$. For simplicity, choose $x_i^* = x_i = i/n$. Clearly as n approaches infinity, the Riemann sum should approach

TABLE 1.3 Riemann Sums and Calculating the Area Under $f(x) = 2x - x^2$

i	$f(x_i)$	x_i	Δx	$f(x_i)\Delta x$
1	0.19	.1	0.1	0.019
2	0.36	.2	0.1	0.036
3	0.51	.3	0.1	0.051
4	0.64	.4	0.1	0.064
5	0.75	.5	0.1	0.075
6	0.84	.6	0.1	0.084
7	0.91	.7	0.1	0.091
8	0.96	.8	0.1	0.096
9	0.99	.9	0.1	0.099
10	1	1	0.1	0.100
			$\sum f(x_i)\Delta x =$	**0.715**

the exact area between the curve $y = 2x - x^2$ and the x-axis from $x = 0$ to $x = 1$. So the exact area will equal:

$$\lim_{n \to \infty} \sum_{i=1}^{n} f(x_i)\Delta x = \lim_{n \to \infty} \sum_{i=1}^{n} \left[\frac{2i}{n} - \left(\frac{i}{n}\right)^2 \right] \frac{1}{n}$$

Some algebraic manipulation reveals that the sum on the right equals $\frac{2}{3} + \frac{1}{2n} - \frac{1}{6n^2}$. Thus, the exact area under the curve is 2/3 since $\frac{1}{2n} - \frac{1}{6n^2}$ approaches 0 as n approaches infinity.

Recall that this area is defined to be the definite integral of $f(x) = 2x - x^2$ from $x = 0$ to $x = 1$. This can be expressed as:

$$\text{Area} = \int_0^1 (2x - x^2)\, dx = \frac{2}{3}$$

In general, for any non-negative continuous function $f(x)$, we can express the area as:

$$\int_a^b f(x)\, dx = \lim_{n \to \infty} \sum_{i=1}^{n} f(x_i^*)\Delta x$$

Thus, as the number of subintervals Δx increases, and the widths of each narrow, the area under $f(x)$ approaches the limit of the sum of the rectangular areas. Note that if the graph of $f(x)$ extends below the x-axis, where the values of $f(x)$ are negative, then the terms $f(x_i^*)\Delta x$ are negative, and the terms $f(x_i)\Delta x$ are negative in equation x, such that the contribution to the definite integral will be negative for this portion of the graph. Thus, in general, the definite integral of any continuous function $f(x)$ equals the area of the region above the x-axis minus the area of the region below the x-axis.

It can be challenging or even impossible to compute the right-hand sum of x_i^* and find its limit. The powerful fundamental theorem of calculus often allows us to easily find these areas for a wide range of functions:

Fundamental theorem of calculus: If $f(x)$ is any continuous function on the interval [a,b], then

$$\int_a^b f(x)\, dx = F(x)\Big|_a^b = F(b) - F(a)$$

where $F(x)$ is any particular antiderivative of $f(x)$.

The essential steps of the proof of this theorem can be found in the companion website.

Recall our earlier example, find the area under the graph of $y = 2x - x^2$ from $x = 0$ to $x = 1$. Using the fundamental theorem of calculus, the area equals:

$$\int_0^1 (2x - x^2)\, dx = x^2 - \frac{1}{3}x^3 \Big|_0^1 = \left[1^2 - \frac{1}{3}1^3\right] - \left[0^2 - \frac{1}{3}0^3\right] = \frac{2}{3}$$

The notation $F(x)|_a^b$ is equivalent to $F(b) - F(a)$, but it is useful to use this notation as an intermediate step in evaluating definite integrals, since one first finds the antiderivative $F(x)$ before evaluating $F(x)$ at $x = b$ and $x = a$ and then finally taking their difference. Observe that this is what we did in the example above.

We note that the definite integral is independent of the particular antiderivative that is chosen. To illustrate this point, we recalculate the above definite integral allowing for different choices of the antiderivative:

$$\int_0^1 (2x - x^2)\, dx = x^2 - \frac{1}{3}x^3 + C \bigg|_0^1 = \left[1^2 - \frac{1}{3}1^3 + C \right] - \left[0^2 - \frac{1}{3}0^3 + C \right] = \frac{2}{3}$$

As we shall discuss in Section 4.1.3, the importance of Riemann sums and their limits extends beyond their applications to finding areas under curves. Many continuous valuation models are based on Riemann sums and their limits as the widths of the horizontal intervals approach zero.

1.5.3 Change of Variables Technique to Evaluate Integrals

An important technique for evaluating integrals is a change of variables. This substitution technique can significantly reduce the apparent complexity of many functions. Suppose we wish to integrate some function of the variable t. Suppose that we can choose a new variable x that is a function of t ($x = x(t)$) in just the right way, so that the integral takes the form:

$$\int f(x(t)) \frac{dx}{dt}\, dt$$

In this case, one can change the variable of integration from the variable t to the variable x and integrate the function $f(x)$ to evaluate the integral. Once the integral has been evaluated, one simply substitutes in place of the variable x the function $x(t)$. To express this symbolically:

$$\int f(x(t)) \frac{dx}{dt}\, dt = \int f(x)\, dx = F(x) + C = F(x(t)) + C$$

where $F(x)$ is an antiderivative of $f(x)$.

The proof of this result is quite simple. Suppose $\int f(x) dx = F(x) + C$ is the general antiderivative of $f(x)$. If we differentiate $F(x(t)) + C$ in the variable t, by the chain rule we have:

$$\frac{d}{dt}[F(x(t)) + C] = \frac{d}{dx}[F(x) + C] \frac{dx}{dt} = f(x) \frac{dx}{dt} = f(x(t)) \frac{dx}{dt}$$

Thus, we have proved that the derivative of $F(x(t)) + C$ in the variable t equals $f(x(t)) \frac{dx}{dt}$, thus the antiderivative of $f(x(t)) \frac{dx}{dt}$ in the variable t equals $F(x(t)) + C$, as we wanted to prove.

In order to make use of the technique of change of variables, one needs to be able to find the right choice for the function x in terms of t. This is a matter of practice and being able to visualize the function $x(t)$ in the expression one is attempting to integrate. We should say that the change of variables method only works if the function one is integrating is able to be expressed in the special form $f(x(t)) \frac{dx}{dt}$ and one can find the antiderivative of $f(x)$ in the variable x. We also point out that in calculus textbooks the change of variables method is often called u-substitution. This is because in calculus textbooks the substitution variable is often denoted by u.

ILLUSTRATION: CHANGE OF VARIABLES TECHNIQUE FOR THE INDEFINITE INTEGRAL

Suppose that we seek to evaluate the indefinite integral $\int (t^2 + 1)^3 t \, dt$. First, we notice that the quantity t immediately to the left of dt is almost the derivative of $t^2 + 1$, which is the base of the cubed function in the parentheses. This motivates the attempt to choose $x = t^2 + 1$, such that $\frac{dx}{dt} = 2t$. We now do a little bit of algebra to create the expression that we need:

$$\int (t^2 + 1)^3 t \, dt = \int \frac{1}{2}(t^2 + 1)^3 (2t) \, dt$$

So, in this case $f(x) = \frac{1}{2}x^3$ and $x = t^2 + 1$. Observe that $\frac{dx}{dt} = 2t$ such that $dx = 2t \, dt$. We rewrite as follows:

$$\int \frac{1}{2}(t^2 + 1)^3 (2t) \, dt = \int \frac{1}{2}x^3 \, dx = \frac{1}{8}x^4 + C = \frac{1}{8}(t^2 + 1)^4 + C$$

1.5.3.1 Change of Variables Technique for the Definite Integral

For definite integrals, the change of variables method is the same to determine the antiderivative. The only additional feature is that one can also express the limits of integration in terms of the new variable. If the limits of integration in the variable t are from a to b, then the limits of integration in the variable $x = x(t)$ will be from $x(a)$ to $x(b)$. Thus:

$$\int_a^b f(x(t)) \frac{dx}{dt} \, dt = \int_{x(a)}^{x(b)} f(x) \, dx$$

Now, suppose that we seek to evaluate the indefinite integral $\int_0^1 (t^2 + 1)^3 t \, dt$. We already calculated the integral of this function in the previous example, and now only have only left to evaluate the integral at the endpoints. Since $x = t^2 + 1$, then $x(0) = 1$ and $x(1) = 2$.

$$\int_0^1 (t^2 + 1)^3 t \, dt = \frac{1}{8}x^4 \Big|_{x(0)}^{x(1)} = \frac{1}{8}x^4 \Big|_1^2 = \frac{2^4}{8} - \frac{1^4}{8} = \frac{15}{8}$$

We also could have obtained the solution by expressing the evaluated integral in terms of the original variable t, and then evaluating the integral at the endpoints in terms of t:

$$\int_0^1 (t^2 + 1)^3 t \, dt = \frac{1}{8}(t^2 + 1)^4 \Big|_0^1 = \frac{1}{8}(1^2 + 1)^4 - \frac{1}{8}(0^2 + 1)^4$$

$$= \frac{15}{8}$$

1.6 EXERCISES

1.1. Add the following matrices:

$$
\underbrace{\begin{bmatrix} 2 & 4 & 9 \\ 6 & 4 & 25 \\ 0 & 2 & 11 \end{bmatrix}}_{A} + \underbrace{\begin{bmatrix} 3 & 0 & 6 \\ 2 & 1 & 3 \\ 7 & 0 & 4 \end{bmatrix}}_{B} =
$$

1.2. Subtract **E** from **D**:

$$
\underbrace{\begin{bmatrix} 9 & 4 & 9 \\ 6 & 4 & 8 \\ 5 & 2 & 9 \end{bmatrix}}_{D} - \underbrace{\begin{bmatrix} 5 & 0 & 6 \\ 2 & 1 & 6 \\ 5 & 0 & 9 \end{bmatrix}}_{E} =
$$

1.3. Transpose the following:

a. $\underbrace{\begin{bmatrix} 1 & 8 & 9 \\ 6 & 4 & 25 \\ 3 & 2 & 35 \end{bmatrix}}_{A}$

b. $\underbrace{\begin{bmatrix} 9 \\ 6 \\ 3 \\ 7 \end{bmatrix}}_{y}$

c. $\underbrace{\begin{bmatrix} .09 & .01 & .04 \\ .01 & .16 & .10 \\ .04 & .10 & .64 \end{bmatrix}}_{V}$

1.4. Multiply the following:

$$
\underbrace{\begin{bmatrix} 7 & 4 & 9 \\ 6 & 4 & 12 \\ 3 & 2 & 17 \end{bmatrix}}_{A} \times \underbrace{\begin{bmatrix} 7 & 6 \\ 5 & 1 \\ 9 & 12 \end{bmatrix}}_{B}
$$

1.5. Suppose that $\mathbf{A} = \begin{bmatrix} -2 & 0 \\ 3 & 4 \end{bmatrix}$ and $\mathbf{B} = \begin{bmatrix} 7 & 3 \\ 5 & -1 \end{bmatrix}$. Find the following:

    a. $2\mathbf{A}$
    b. $\mathbf{A}^{\mathsf{T}}$
    c. $\mathbf{A} + \mathbf{B}$
    d. $\mathbf{AB}$
    e. $\mathbf{BA}$

1.6. Invert the following matrices:

    a. $\begin{bmatrix} 8 \end{bmatrix}$

    b. $\begin{bmatrix} 1 & 0 \\ 0 & 1 \end{bmatrix}$

    c. $\begin{bmatrix} 4 & 0 \\ 0 & \frac{1}{2} \end{bmatrix}$

    d. $\begin{bmatrix} 1 & 2 \\ 3 & 4 \end{bmatrix}$

    e. $\begin{bmatrix} .02 & .04 \\ .06 & .08 \end{bmatrix}$

    f. $\begin{bmatrix} -2 & 1 \\ 1.5 & -.5 \end{bmatrix}$

    g. $\begin{bmatrix} \dfrac{100}{3} & -\dfrac{25}{3} \\ -\dfrac{25}{3} & \dfrac{25}{3} \end{bmatrix}$

    h. $\begin{bmatrix} 2 & 0 & 0 \\ 2 & 4 & 0 \\ 4 & 8 & 20 \end{bmatrix}$

1.7. Solve for matrix $\mathbf{X}$ in the matrix equation $\mathbf{AXB} + \mathbf{B} = \mathbf{AB}$. Assume that the inverses of $\mathbf{A}$ and $\mathbf{B}$ exist.

1.8. Solve each of the following for $\mathbf{x}$:

    a. $\underset{\mathbf{C}}{\begin{bmatrix} 100/3 & -25/3 \\ -25/3 & 25/3 \end{bmatrix}} \underset{\mathbf{x}}{\begin{bmatrix} x_1 \\ x_2 \end{bmatrix}} = \underset{\mathbf{s}}{\begin{bmatrix} .01 \\ .11 \end{bmatrix}}$

    b. $\underset{\mathbf{C}}{\begin{bmatrix} .08 & .08 & .1 & 1 \\ .08 & .32 & .2 & 1 \\ .1 & .2 & 0 & 0 \\ 1 & 1 & 0 & 0 \end{bmatrix}} \underset{\mathbf{x}}{\begin{bmatrix} x_1 \\ x_2 \\ x_3 \\ x_4 \end{bmatrix}} = \underset{\mathbf{s}}{\begin{bmatrix} .1 \\ .1 \\ .1 \\ .1 \end{bmatrix}}$

1.9. Use matrices to solve the following system of equations:

$$.02x_1 + .04x_2 = .03$$
$$.06x_1 + .08x_2 = .01$$

1.10. Find the derivative of y with respect to x for the following polynomials:
   a. $y = 7x^4$
   b. $y = 5x^2 - 3x + 2$
   c. $y = -7x^2 + 4x + 5$

1.11. **a.** At what value for x is y minimized in Problem 1.9.b? How do we know that y is not maximized at this point?
   b. At what value for x is y maximized in Problem 1.9.c? How do we know that y is not minimized at this point?

1.12. Suppose the amount of lumber (stumpage value, the value of mature timber before it is cut) that could be produced from the timber in a given forest is a function of time, where s is the amount that can be produced and t is the number of years from today: $s(t) = t^3 - 3t^2 + t + 10$. This function reflects a recent fungus infection in many trees, and this fungus infection is expected to grow.
   a. Find the (instantaneous) rate of change of its stumpage value 1 year from today. Verbally interpret your result.
   b. Find the rate of change of its stumpage value 3 years after today.
   c. Find the average rate of change of stumpage value from year 1 to year 3. Verbally interpret your result.
   d. Suppose that the stumpage value function $s(t)$ reflects the fungus infection and the damage that it is likely to cause over the future. How might this damage be reflected in the value function?

1.13. Find derivatives for y with respect to x for each of the following:
   a. $y = (4x + 2)^3$
   b. $y = (3x^2 + 8)^{1/2}$
   c. $y = 6x(4x^3 + 5x^2 + 3)$
   d. $y = (1.5x - 4)^3(2.5x - 3.5)^4$
   e. $y = 25/x^2$
   f. $y = (6x - 16) \div (10x - 14)$

1.14. Use the following definition of a derivative (a) and the following statement (b) based on the binomial theorem to verify the power rule, also given below (c):
   a. $\dfrac{dy}{dx} = f'(x) = \lim\limits_{\Delta x \to 0} \dfrac{f(x + \Delta x) - f(x)}{\Delta x}$

   b. $(x + \Delta x)^n = \dbinom{n}{0} x^n (\Delta x)^0 + \dbinom{n}{1} x^{n-1}(\Delta x)^1 + \dbinom{n}{2} x^{n-2}(\Delta x)^2 + \dots$

   $$+ \dbinom{n}{n-1} x^1 (\Delta x)^{n-1} + \dbinom{n}{n} x^0 (\Delta x)^n$$

   c. If $y = \sum_{j=1}^{m} c_j \cdot x^{n_j}$, then $\dfrac{dy}{dx} = \sum_{j=1}^{m} c_j \cdot n_j \cdot x^{n_j - 1}$

1.15. Let $y = x^3$ and $x = t^2 + 1$. Use the chain rule to find dy/dt.

1.16. Differentiate each of the following with respect to x:

    a. $y = e^{.05x}$

    b. $y = (e^x)/x$

    c. $y = 5 \ln(x)$

    d. $y = e^x \ln(x)$

    e. $y = x^2 e^x$

    f. $y = \ln(5x^3 + x)$

    g. $y = 5x^3 - 6\sqrt{x} + 2e^x$

    h. $y = x^2 \ln x$

1.17. Here is an exercise unrelated to finance. A square floor is measured to have side length 20 feet, with an error of plus or minus 0.1 feet. Use the differential to estimate the resulting possible error in measuring the area of the floor.

1.18. **a.** Consider the function $y = x^3$. Let $x_0 = 5$. Now, suppose we wish to increase x by $\Delta x = 1$ to $x_1 = 6$. Estimate y_1 based on a third-order Taylor approximation.

    b. How does this approximation compare to an exact solution for y_1? Why?

    c. Estimate y_1 based on a second-order Taylor approximation.

    d. Estimate y_1 based on a first-order Taylor approximation.

    e. Consider the function $y = 10x^3$. Let $x_0 = 2$, and suppose that we wish to increase x by $\Delta x = 3$ to $x_1 = 5$. Use first-, second-, then third-order Taylor polynomial expansions to evaluate y_1.

1.19. Our objective is to find the value for x, which enables us to maximize the function $y = 50x^2 - 10x$ subject to the constraint that $.1x \le 100$. Set up and solve an appropriate Lagrange function for this problem. This exercise is intended to be a somewhat trivial illustration for setting up and solving a Lagrange optimization problem.

1.20. An investor wishes to budget her wealth $w = \$10,000$ in savings so that her spending over 4 years yields the highest level of utility (U, which can be considered to be satisfaction). She has mapped out a utility function that accounts for her consumption (x_t) each year t over the 4-year period:

$$U = 100x_1 + 200x_2 + 250x_3 + 350x_4 - .01x_1^2 - .2x_2^2 - .03x_3^2 - .04x_4^2 - .2x_3x_4$$

Unfortunately, price levels are expected to rise each year such that what \$1 buys in 1 year will cost \$2 in 2 years, \$3 in 3 years, and \$4 in 4 years. Her spending over the 4-year period cannot exceed \$10,000.

    a. If this investor seeks to maximize her total utility over the 4-year period, what are optimal annual consumption levels for each year? Do bear in mind her \$10,000 wealth constraint.

    b. What is the total utility level for the consumer?

1.21. **a.** Find the antiderivative for the function $f(x) = 10x - x^2$.

    b. What is the area under the curve $f(x) = 10x - x^2$ between 0 and 1?

    c. Find the Reimann sum for the function $f(x) = 10x - x^2$ based on five rectangles over the range 0 to 1.

    d. Find the Reimann sum for the function $f(x) = 10x - x^2$ based on ten rectangles over the range 0 to 1.

1.22. Integrate each of the following functions over x:
  a. $f(x) = 0$
  b. $f(x) = 7$
  c. $f(x) = 2x$
  d. $f(x) = 21x^2$
  e. $f(x) = 21x^2 + 5$
  f. $f(x) = e^x$
  g. $f(x) = .5e^{.5x}$
  h. $f(x) = 5^x \ln(5)$
  i. $f(x) = 1/x$
  j. $f(x) = 5/x + 3e^x + 4x^2 - x$

1.23. a. Use the fundamental theorem of integral calculus to find the area between $x = 0$ and $x = 1$ under the function $f(x) = 8x - 9x^2$.
  b. Plot out on an appropriate graph 20 rectangles representing the rectangles for the Reimann sum for this function between $x = 0$ and $x = 1$.

1.24. Consider the function $f(x) = 10x - x^2$. The area under a curve represented by this function over the range from $x = a = 0$ to $x = b = 1$ can be computed with a limit of Riemann sums or through the process of antidifferentiation. Verify that as the number of rectangles used to compute the Riemann sums approaches infinity, and the widths of these rectangles approach zero, the limit of the Riemann sums and antidifferentiation will produce the same area.

1.25. Evaluate $\int_1^3 x^2 \, dx$.

1.26. Suppose that z, y, and x are all functions of t such that:

$$\frac{1}{z}\frac{dz}{dt} = x\frac{dx}{dt} + 5\frac{dy}{dt}$$

Find z in terms of x and y.

NOTES

1. *Normative models*, proposing what "ought to be," are distinguished from *positive models* that predict "what will be."
2. If it is possible to multiply two matrices, they are said to be *conformable* for multiplication. Any matrix can be multiplied by a scalar, where the product is simply each element times the value of the scalar.
3. **Y** is defined similarly to **X**. We can use the Gauss–Jordan elimination procedure to show that $\alpha^T = [0, 0, 0]$ is the only solution to this equation, such that the set is linearly independent.
4. Note here that $\alpha_1 = -v_1 - \frac{1}{2}v_2 + \frac{1}{2}v_3$, $\alpha_2 = \frac{7}{5}v_1 + \frac{3}{10}v_2 - \frac{1}{2}v_3$ and $\alpha_3 = -3v_1 + v_3$.
5. The *law of one price* states that securities or portfolios producing the same payoff structures must sell for the same price. Arbitrage opportunities do not exist when the law of one price holds.
6. $f''(x) = \dfrac{d^2y}{dx^2} = \dfrac{d^2\ln x}{dx^2} = \dfrac{dx^{-1}}{dx} = -x^{-2}$.
7. In many finance problems, we will want to use inequalities as constraints. However, it is convenient to convert them to equalities for Lagrange functions.
8. If the constraint is not binding, λ will equal zero. If the constraint is binding in this example, $x_1 + .2x_2$ will equal 10. Since either the contents within the parentheses or the Lagrange multiplier will equal zero, the numerical value of the function that we have added to our original function to be optimized will be zero. Although the numerical value of our original function is unchanged by the supplement representing the constraint, its derivatives will be affected by the constraint.

2

Probability and Risk

2.1 UNCERTAINTY IN FINANCE

Risk has many definitions; sometimes these meanings are inconsistent or even contradictory, but most definitions are somehow related to uncertainty or undesirable outcomes. Risk can refer to uncertainty, volatility, bad outcomes (e.g., bankruptcy), probability of bad outcomes, extent of bad outcomes, certainty of bad outcomes, etc. Measuring risk is frequently complicated by the difficulty in defining risk, but nonetheless is often accomplished with a variety of tools drawn from probability and statistics. Our usage of the term risk in this book will usually be related to uncertainty, and our measurements will usually pertain to uncertainty or volatility.

2.2 SETS AND MEASURES

We require a set of rules and grammar to effectively communicate with words. Similarly, rules and syntax are needed to communicate ideas in mathematics and probability. Here, we begin to set forth basic rules and syntax for discussions related to probability.

2.2.1 Sets

A *set* A is a collection of well-defined objects called elements such that any given object x either (but not both) belongs to A ($x \in A$) or does not belong to A ($x \notin A$). The *cardinality* of the set is its number of members, which can be either finite or infinite. Set A is a subset of B ($A \subseteq B$) if and only if every element of A is also an element of B; set A is a proper subset of B ($A \subset B$) if and only if every element of A is also an element of B but $A \neq B$. The *union* of sets A and B, $A \cup B$, is the set of all distinct elements that are either in A or B, or both in A and B. The *intersection* of sets A and B, $A \cap B$, is the set of all distinct elements they share in common.

P.M. Knopf & J.L. Teall: Risk Neutral Pricing and Financial Mathematics: A Primer.
DOI: http://dx.doi.org/10.1016/B978-0-12-801534-6.00002-3

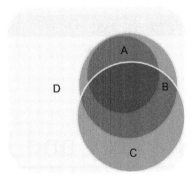

FIGURE 2.1   Venn diagram of sets and subsets.

ILLUSTRATION: TOSS OF TWO DICE

Consider an experiment based on the total of a single toss of two dice. Let D be the set of all possible outcomes, which range from 2 to 12. Let B be the set of outcomes between 5 and 10 inclusive. Let C be the set of all outcomes not less than 9. Let A be the set of odd outcomes between 5 and 9 inclusive. Thus, we have the sets $A = \{5,7,9\}$, $B = \{5,6,7,8,9,10\}$, and $C = \{9,10,11,12\}$. These sets are depicted in Figure 2.1. We can infer the following:

$$A \subset B \subset D$$

Since $\{9, 10\}$ contains the only elements in both sets B and C, we can write:

$$B \cap C = \{9, 10\}$$

That is, the intersection of sets B and C is $\{9, 10\}$. One can also check that the union of sets B and C equals:

$$B \cup C = \{5, 6, 7, 8, 9, 10, 11, 12\}$$

2.2.1.1 Finite, Countable, and Uncountable Sets

To understand the definitions of a measurable space and a probability space, it is necessary to learn the concept of countability. A *finite set* consists of a finite number of elements, meaning that the elements in the set can be numbered from 1 to n for some positive integer n. A finite set is countable. Any finite set A can be expressed in the form $A = \{a_1, a_2, \ldots, a_n\}$. Observe that the number of elements in this set is n. This set can be expressed in a one-to-one correspondence with the set $\{1, 2, \ldots, n\}$ as illustrated below:

$$
\begin{array}{cccccc}
1 & 2 & 3 & & n \\
\downarrow & \downarrow & \downarrow & \cdots & \downarrow \\
a_1 & a_2 & a_3 & & a_n
\end{array}
$$

A one-to-one correspondence between two sets implies that each element in one set can be paired with an element in the second, such that each element in the two sets has exactly one partner.

Denote $N = \{1, 2, 3, \ldots\}$ as the set of the natural numbers; that is, the set of all positive integers. This set is obviously infinite. A *countably infinite* set can be put in a one-to-one correspondence with N. If $A = \{a_1, a_2, a_3, \ldots\}$, then this set is countably infinite since we have the one-to-one correspondence:

$$
\begin{array}{cccc}
1 & 2 & 3 & \cdots \\
\downarrow & \downarrow & \downarrow & \\
a_1 & a_2 & a_3 & \cdots
\end{array}
$$

If set A is finite, a proper subset of A cannot be put in a one-to-one correspondence with the set A. This is not the case with infinite sets. For example, consider the set $A = \{2, 4, 6, \ldots\}$, which is the set of all even positive integers. Although this set is a proper subset of N, it can be put in a one-to-one correspondence with N as follows:

$$
\begin{array}{ccccc}
1 & 2 & 3 & 4 & \cdots \\
\downarrow & \downarrow & \downarrow & \downarrow & \\
2 & 4 & 6 & 8 & \cdots
\end{array}
$$

Thus, A, the set of positive even integers is countably infinite.

Now, consider the set $\mathbb{R}$ of all numbers on the real number line. Every number that is either positive, negative, or zero is a real number. For example, each of the following numbers is real: $3, -4, 7/5, 0, \sqrt{2}, \pi$. Obviously $\mathbb{R}$ is an infinite set; that is, it has an infinite number of elements. However, $\mathbb{R}$ is not countably infinite.[1] One can regard the number of elements in the set $\mathbb{R}$ as a "larger infinity" than the countably infinite number of elements in the set N. The set $\mathbb{R}$ is said to be uncountable.

2.2.2 Measurable Spaces and Measures

A measure is a function that assigns a real number to subsets of a given space. For example, in three-dimensional space, the *Lebesgue measure* of any solid shape (the measurable sets in this space) is simply its volume. In this text, we will discuss a particular type of measure called a probability measure. We will provide a definition of a probability measure shortly.

As an introduction, two necessary properties of a probability measure are:

1. The measure of any subset of the space is non-negative.
2. The measure of the entire space is 1.

A measurable space (Ω, Φ) is a collection of *outcomes* ω comprising the set Ω, referred to as the *universal set* or *sample space*, and a set Φ, a specified collection of subsets of Ω. The set Φ, comprising *events* ϕ, forms an σ-*algebra*, which has the following three properties:

1. $\varnothing \in \Phi$ and $\Omega \in \Phi$ where $\varnothing = \{\ \}$ is the empty set (set containing no elements).
2. If $\phi \in \Phi$, then $\phi^c \in \Phi$ (ϕ^c is the complement of ϕ).
3. If $\phi_i \in \Phi$ for every positive integer i, then $\bigcup\limits_{i=1}^{\infty} \phi_i \in \Phi$.

The σ-algebra contains every countable or finite intersection of sets taken from the σ-algebra.[2] The idea behind the concept of an σ-algebra is to have a sufficient variety of subsets of the sample space, to which we will later assign probabilities. In the case that the sample space Ω is finite or countable, then usually the σ-algebra Φ consists of the power set (set of all possible subsets) of Ω. In the case when Ω is the set of real numbers or an interval of real numbers, then the σ-algebra Φ normally consists of all sets that are generated by countable unions and complements of intervals contained in the sample space Ω. The reason why the σ-algebra is not chosen to be the power set in the case of real numbers is because it is usually impossible to assign probabilities the way we would like to every possible subset of the real numbers.

2.3 PROBABILITY SPACES

A probability space involves assigning or mapping probabilities (numbers between 0 and 1) to the possible outcomes or events that can occur. A probability space (Ω, Φ, P) consists of a measurable space (Ω, Φ) and a probability measure P, to be defined shortly. A *probability space* consists of a sample space Ω, events Φ, and their associated probabilities P. The probability (measure) P is a function that assigns a value to each event ϕ taken from Φ so that the following properties are satisfied:

1. For every event ϕ taken from Φ, $0 \le P(\phi) \le 1$.
2. $P(\Omega) = 1$.
3. For any set of pairwise disjoint (mutually exclusive) events $\{\phi_i\}_{i=1}^{\infty}$, we have
   $P(\bigcup_{i=1}^{\infty} \phi_i) = \sum_{i=1}^{\infty} P(\phi_i)$.[3]

A probability space can be considered to be the triplet (Ω, Φ, P), where again, Ω is the set of outcomes, Φ is an σ-algebra of subsets of Ω, and P assigns probabilities between 0 and 1 to event sets ϕ in Φ.

2.3.1 Physical and Risk-Neutral Probabilities

For financial valuation purposes, it is essential to realize from the outset what is and is not included in the definition for a probability or probability measure. The three properties set forth above essentially say that probabilities associated with events range between 0 and 1, they sum to 1 and that the probability of the union of a set of mutually exclusive events equals the sum of the probabilities for each of the events in the union. Nowhere does this definition of a probability say anything about frequencies or likelihoods of occurrences. This is not part of the general definition that we will use for probability.

However, we often do find it convenient to associate probabilities with frequencies and likelihoods of events. We will refer to these probabilities as physical probabilities. That is, a *physical probability* is a specific type of probability that reflects the frequency or likelihood of an event. While these types of probabilities have a certain intuitive appeal, they are not usually very helpful for financial valuation. First, we often have no meaningful methodology to determine or observe these physical probabilities. Different analysts perceive them differently. Worse, expected values derived from them frequently lead to valuations that lead to price inconsistencies (we will refer to this as arbitrage later in the text). That is, use of physical probabilities simply leads to incorrect market valuations. This inconsistency problem intensifies as investors become increasingly risk averse.

In the next chapter, we will synthesize probabilities that lead to correct and consistent valuations. These synthetic probabilities are known as *risk-neutral probabilities*, which are presumed to be used by risk-neutral investors valuing assets in a consistent manner. Nonetheless, both physical probabilities and risk-neutral probabilities fulfill the conditions set forth above.

ILLUSTRATION: PROBABILITY SPACE

Suppose that a stock will either increase (u) or decrease (d) in each of two periods with equal probabilities. The sample space of outcomes is the set $\Omega = \{uu, ud, du, dd\}$ of all potential outcomes or elementary states ω of the sample space ("world") Ω. An event ϕ is a subset of elementary states ω taken from Ω and is an element of the set of events Φ. In this illustration, we choose Φ to be the power set of Ω. If a set Ω has n elements, its power set has 2^n elements ϕ. Thus, in this case the set of events Φ has $2^4 = 16$ events:

$\varnothing$	$\{dd\}$	$\{ud, du\}$	$\{uu, ud, dd\}$
$\{uu\}$	$\{uu, ud\}$	$\{ud, dd\}$	$\{uu, du, dd\}$
$\{ud\}$	$\{uu, du\}$	$\{du, dd\}$	$\{ud, du, dd\}$
$\{du\}$	$\{uu, dd\}$	$\{uu, ud, du\}$	$\{uu, ud, du, dd\}$

Suppose that physical probabilities associated with each two-period outcome ω (a pair involving u and d) equal 0.25. Physical probabilities associated with each event in the σ-algebra are then:

$P\{\varnothing\} = 0$	$P\{dd\} = 0.25$	$P\{ud, du\} = 0.5$	$P\{uu, ud, dd\} = 0.75$
$P\{uu\} = 0.25$	$P\{uu, ud\} = 0.5$	$P\{ud, dd\} = 0.5$	$P\{uu, du, dd\} = 0.75$
$P\{ud\} = 0.25$	$P\{uu, du\} = 0.5$	$P\{du, dd\} = 0.5$	$P\{ud, du, dd\} = 0.75$
$P\{du\} = 0.25$	$P\{uu, dd\} = 0.5$	$P\{uu, ud, du\} = 0.75$	$P\{uu, ud, du, dd\} = 1$

2.3.2 Random Variables

A *random variable* X defined on a probability space (Ω, Φ, P) is a function from the set Ω of outcomes to $\mathbb{R}$ (the set of real numbers) or a subset of the real numbers. A random variable X is said to be *discrete* if it can assume at most a countable number of possible values: $x_1, x_2, x_3,$ .... For each individual value x_i of the random variable X, we require that the probability $P(X = x_i)$ be defined. See the illustration below. A random variable X is said to be *continuous* if it can assume any value on a continuous range on the real number line. We require that for any interval (a, b) on the real line that the probability $P(a < X < b)$ be defined. We note that the number a can take on the value $-\infty$, and the number b can take on the value ∞.

ILLUSTRATION: DISCRETE RANDOM VARIABLES

Returning to the illustration in Section 2.3.2, the random variable $X(\omega)$ could denote the value of the stock given an outcome ω in the sample space. Suppose that the stock price is defined in the following way: $X(uu) = 30$, $X(ud) = 20$, $X(du) = 20$, and $X(dd) = 10$. Thus, $P(X = 30) = P(\{uu\}) = 0.25$, $P(X = 20) = P(\{ud, du\}) = 0.5$, and $P(X = 10) = P(\{dd\}) = 0.25$. For example, $P(X = 20) = 0.5$ means that there is a 0.5 probability (50% chance) that the stock price will be 20, and this will happen in the event that there is either an increase followed by a decrease (ud) or a decrease followed by an increase (du). This is an example of a discrete random variable since the stock price can only take on discrete values (10, 20, 30). Note that this example fulfills the main requirement for X to be a random variable, since the probability that X equaled each of its values 10, 20, and 30, respectively, was defined: $P(X = 10) = 0.25$, $P(X = 20) = 0.5$, and $P(X = 30) = 0.25$.

2.4 STATISTICS AND METRICS

Suppose that we wish to describe or summarize the characteristics or distribution of a single population of values (or sample drawn from a population). In this section, we discuss various metrics to accomplish this.

2.4.1 Metrics in Discrete Spaces

2.4.1.1 Expected Value, Variance, and Standard Deviation

Two important characteristics of a distribution are central location (measured by mean, median, or mode) and dispersion (measured by range, variance, or standard deviation). The mean is also called the average, and in the setting of probability is often called the expected value. The mean is a measure of location (central tendency) because it is a weighted average of the data points taking into account the frequency or likelihood with which they occur. The *mean* μ of a population is computed by multiplying the possible values x_i from the population by their associated frequencies f_i, summing, and dividing the result by the total number of data values n:

$$\mu = \frac{\Sigma x_i f_i}{n}$$

In the event that we have a random variable X with possible values x_i for $i = 1, \ldots, n$ and associated probabilities P_i, then the *expected value* is:

$$E[X] = \sum_{i=1}^{n} x_i P_i \tag{2.1}$$

The expectation operator $E[*]$ is linear. More precisely, given the random variables $X_1, X_2, \ldots, X_n$, and the constants $c_1, c_2, \ldots, c_n$, we have:

$$E\left[\sum_{j=1}^{n} c_j X_j\right] = \sum_{j=1}^{n} c_j E[X_j] \tag{2.2}$$

Variance σ^2 is a measure of the dispersion (degree of spread) of the values in a data set. In a finance setting, variance is also used as an indicator of risk. *Variance* is defined as the mean of squared deviations of actual data points from the mean or expected value of a data set:

$$\sigma^2 = \frac{\sum_{i=1}^{n}(x_i-\mu)^2}{n} = \frac{\sum_{i=1}^{n}x_i^2}{n} - \mu^2 \tag{2.3}$$

In the event that we have a random variable X, we can express the variance as:

$$\sigma^2 = E[(X-E[X])^2] = E[X^2] - (E[X])^2 \tag{2.4}$$

The variance of a random variable X is also denoted by $Var(X)$. The *standard deviation* σ is simply the square root of variance. The standard deviation is sometimes easier to interpret than variance because its value is expressed in terms of the same units as the data points themselves rather than their squared values.

ILLUSTRATION

Consider again the stock price example in which $P(X = 30) = P(\{uu\}) = 0.25$, $P(X = 20) = P(\{ud, du\}) = 0.5$, and $P(X = 10) = P(\{dd\}) = 0.25$. The expected value of the stock price is:

$$E[X] = \sum x_i P_i = 30(0.25) + 20(0.5) + 10(0.25) = 20$$

The variance of the stock price is:

$$\sigma^2 = E[(X-E[X])^2] = (30-20)^2(0.25) + (20-20)^2(0.5) + (10-20)^2(0.25) = 50$$

2.4.1.2 Co-movement Statistics

A joint probability distribution is concerned with probabilities associated with each possible combination of outcomes drawn from two sets of data. Covariance measures the mutual variability of outcomes selected from each set; that is, covariance measures the variability in one data set relative to variability in the second data set, where variables are selected one at a time from each data set and paired. If large values in one data set seem to be associated with large values in the second data set in terms of their joint probability, the covariance is positive; if large values in the first data set seem to be associated with small values in the second data set, the covariance is negative. If data sets are unrelated, the covariance is zero. The *covariance* between random variables X and Y is measured as follows:

$$\sigma_{X,Y} = E[(X - E[X])(Y - E[Y])] = E[XY] - E[X]E[Y] \tag{2.5}$$

The sign associated with the covariance indicates whether the relationship associated with the random variables is direct (positive sign) or inverse (negative sign). Covariance is sometimes more easily interpreted when it is expressed relative to the standard deviations of each of the random variables. That is, when we divide the covariance by the product of the standard deviations of each of the random variables, we obtain the *coefficient of correlation* $\rho_{X,Y}$ as follows:

$$\rho_{X,Y} = \frac{\sigma_{X,Y}}{\sigma_X \sigma_Y} = \frac{E[XY] - E[X]E[Y]}{\sqrt{E[X^2] - (E[X])^2}\sqrt{E[Y^2] - (E[Y])^2}} \tag{2.6}$$

Correlation coefficients will always range between -1 and $+1$. A correlation coefficient of -1 or $+1$ indicates that the two random variables have a perfect linear relationship (all their values lie on a common line). A correlation coefficient of -1 means that as the X values increase the Y values decrease on the line (negative slope). A correlation coefficient of $+1$ means that as the X values increase the Y values also increase on the line (positive slope). The closer a correlation coefficient is to -1 or $+1$, the stronger is the linear relationship between the two random variables. A correlation coefficient equal to zero implies no linear relationship between the two random variables.

The correlation coefficient may be squared to obtain the coefficient of determination (also referred to as r^2 in some statistics texts and here as ρ^2). The coefficient of determination is the proportion of variability in one random variable that is explained by or associated with variability in the second random variable.

2.4.2 Metrics in Continuous Spaces

A probability density function $p(x)$ is a theoretical model for a frequency distribution. The product $p(x)\,dx$ can be interpreted as the probability that the values of a continuous random variable X lie between x and $(x + dx)$ as $dx \to 0$. The *distribution function* $P(x)$, also called the *cumulative density function*, is defined to be the probability that the values of the random variable X will fall below a given value x; that is, $P(x) = P(X \leq x)$. A density function $p(x)$ can be computed based on a differentiable distribution function $P(x)$ as follows:

$$p(x) = \frac{dP(x)}{dx} = P'(x) \tag{2.7}$$

Conversely, if one is given the density function $p(x)$, the distribution function $P(x)$ is found by integration:

$$P(x) = \int_{-\infty}^{x} p(t)\,dt \tag{2.8}$$

Note that we had to change the variable of integration of the function $p(x)$ from x to t because we are using x as a limit of integration in Eq. (2.8). The mean and variance for a random variable $Y = f(X)$ where f is a function of the random variable X are calculated from the following:

$$E[Y] = \int_{-\infty}^{\infty} f(x)p(x)\,dx \tag{2.9}$$

$$\sigma_Y^2 = E[Y^2] - (E[Y])^2 = \int_{-\infty}^{\infty} (f(x))^2 p(x)\,dx - \left(\int_{-\infty}^{\infty} f(x)p(x)\,dx\right)^2 \tag{2.10}$$

ILLUSTRATION: DISTRIBUTIONS IN A CONTINUOUS SPACE

Consider a random variable X with a very simple density function $p(x) = 0.375x^2$ for $0 \leq x \leq 2$ and 0 elsewhere. From this density function, we can obtain a distribution function by integrating as follows:

$$P(x) = \int_{-\infty}^{x} p(t)\, dt = \int_{0}^{x} 0.375t^2\, dt = 0.125x^3$$

for $0 \leq x \leq 2$, $P(x) = 0$ for $x < 0$, and $P(x) = 1$ for $x > 2$. Note that $P(0) = 0$ and $P(2) = 1$ and that $p(x) \geq 0$ for all x. Now, suppose that the random return on a given stock is $r = f(X) = 0.5X$. We can use a definite integral to compute the cumulative density function (distribution function) to determine the probability that the random variable r is less than some constant c or the probability that the return will fall within a given range. For example, the probability that the stock's return r will be between 0.05 and 0.15 is the same as the probability that X is between 0.1 and 0.3, determined as follows:[4]

$$P(0.05 < r < 0.15) = P(0.1 < X < 0.3) = \int_{0.1}^{0.3} 0.375x^2\, dx = 0.125x^3 \Big|_{0.1}^{0.3}$$

$$= 0.003375 - 0.000125 = 0.00325$$

Thus, there is a 0.00325 probability that X will be between 0.1 and 0.3 and that the stock return will range from 0.05 to 0.15.

2.4.2.1 Expected Value and Variance

The return expected value and variance for our numerical example are calculated as follows:

$$E[r] = \int_{-\infty}^{\infty} f(x)p(x)\, dx = \int_{0}^{2} 0.5xp(x)\, dx = \int_{0}^{2} 0.5x \cdot 0.375x^2\, dx = 0.046875x^4 \Big|_{0}^{2} = 0.75$$

$$\sigma^2 = E[r^2] - (E[r])^2 = \int_{-\infty}^{\infty} (f(x))^2 p(x)\, dx - (0.75)^2$$

$$= \int_{0}^{2} (0.5x)^2 \cdot 0.375x^2\, dx - 0.5625 = \int_{0}^{2} 0.09375x^4\, dx - 0.5625$$

$$= 0.01875x^5 \Big|_{0}^{2} - 0.5625 = 0.6 - 0.5625 = 0.0375$$

Notice the similarity between $E[r] = \int f(x)p(x)\, dx$ for continuous functions and $E[r] = \sum r_i P_i$ that we used earlier for discrete probability functions. Also, notice the similarity between $\sigma^2 = \int f(x)^2 p(x)\, dx - (E[f(X)])^2$ for continuous functions and $\sigma^2 = \sum r_i^2 P_i - (E[r])^2$ that we used earlier for discrete probability functions. That is, expected values are sums of random variable outcomes and their associated probabilities and variances are expected values of squared random variable values, minus the squared value of the expected value.

2.5 CONDITIONAL PROBABILITY

Conditional probability is used to determine the likelihood of a particular event A contingent on the occurrence of another event B. The probability of event A given B is the probability of both A and B divided by the probability of B:

$$P[A|B] = \frac{P[A \cap B]}{P[B]} \tag{2.11}$$

The expected value of a random variable V conditioned on its exceeding some constant C is calculated as follows:

$$E[V|V > C] = \frac{\sum_{V_i > C} V_i P(V = V_i)}{\sum_{V_i > C} P(V = V_i)} \tag{2.12}$$

in the discrete case, and:

$$E[V|V > C] = \frac{\int_C^{\infty} v p(v) dv}{\int_C^{\infty} p(v) dv} \tag{2.13}$$

in the continuous case, with $p(v)$ as the density function for V.

ILLUSTRATION: DRAWING A SPADE

Suppose that we draw a card at random from an ordinary deck of 52 playing cards. What is the probability that a spade is drawn given that the drawn card is black? Assume that spades and clubs are black cards, and that each type comprises 25% of the deck.

The sample space for this draw has 52 outcomes. Let A be the event that a spade is drawn, and let B be the event that a black card is drawn. Since $P[A \cap B] = 13/52 = 1/4$ and $P[B] = 26/52 = 1/2$, $P[A | B] = P[A \cap B]/P[B] = (1/4)/(1/2) = 1/2$. Observe that the probability of obtaining a spade out of all 52 outcomes in the sample space is 1/4. However, if we are given the additional information that the card was black, then we are effectively limiting our sample space to the black cards, and the probability of obtaining a spade is 1/2.

2.5.1 Bayes Theorem

Bayes Theorem is useful for computing the conditional probability of an event B given event A, when we know the value of the reverse conditional probability; that is, the probability of event A given event B. In its simplest form, Bayes theorem states that:[5]

$$P[B|A] = \frac{P[A|B]P[B]}{P[A]} \tag{2.14}$$

ILLUSTRATION: DETECTING ILLEGAL INSIDER TRADING

In US stock markets, insiders are permitted to trade (provided they register their trades), but only when their trades are not motivated by inside information. To evade detection, insiders often engage in activity that leads to particular trading patterns that the New York Stock Exchange (NYSE) uses software systems to detect. However, before implementing a new software system, the NYSE must ensure that the system does not signal an excessive rate of false positives; that is, the new system should not signal too many incidences in which traders are falsely accused of illegal insider trading.

Suppose that 1% of all NYSE trades are known to be motivated by the illegal use of inside information.[6] Further suppose that the NYSE is considering implementation of a new software system for detecting illegal insider trading. In this proposed system, a positive signal from the system has a 90% probability of leading to a conviction for illegal insider trading. That is, 90% of all signals of insider trading are considered to be truthful signals and the other 10% of positive signals are taken to be false based on no conviction. Thus, we assume that a conviction for insider trading only occurs when an illegal act has been committed as alleged. Further, assume that among all trades there is a 5% probability that the system signals positive for an illegal trade; obviously, some proportion of these signals must be false positives. Based on this information, what is the probability that a given positive signal by the proposed system for insider trading actually results in a conviction? Based on your calculation, what is the probability that the system's signal for illegal trading is false?

This problem seems complicated at first glance. However, Bayes rule will simplify its calculation. Let A be the event that a trade tests positive for being initiated illegally and let B be the event that a trade actually is illegally motivated by inside information. Thus, we have the given information that $P[A \mid B] = 0.9$, $P[B] = 0.01$, and $P[A] = 0.05$. We seek to find:

$$P[B|A] = \frac{P[A|B]P[B]}{P[A]} = \frac{0.9(0.01)}{0.05} = 0.18$$

Thus, there is an 18% probability that a trade indicated by the system to be illegal actually was illegal. That is, 82% of signals of illegal trading are actually false. This result may seem surprising at first, because among illegal trades, the test correctly gives a positive signal 90% of the time. Yet only 18% of the trades that test positive on this new system for being motivated by inside information are actually illegally motivated. The intuition is as follows. Even though the signal is positive only 5% of the time among all trades, illegal trades comprise only 1% of the population of all trades. Thus, most of the trades that signal positive for being illegally motivated cannot lead to a conviction.

2.5.2 Independent Random Variables

Events A and B are independent if:

$$P[A \cap B] = P[A]P[B] \tag{2.15}$$

Intuitively, events A and B are independent when the occurrence of one has no impact on the probability of the other. Observe that Eq. (2.15) implies that $P[A] = P[A \mid B]$ and $P[B] = P[B \mid A]$, which is consistent with our intuitive notion of independence.

Discrete random variables X and Y are said to be *independent* if for each possible value of $X = x$ and $Y = y$:

$$P[X = x, Y = y] = P[X = x]P[Y = y] \tag{2.16}$$

Consider the following simple example in which two fair coins are tossed independently of one another. For the first coin, let $X = 1$ if it lands heads and $X = 0$ if it lands tails. For the second coin, let $Y = 1$ if it lands heads and $Y = 0$ if it lands tails. Then:

$$P[X = x, Y = y] = P[X = x]P[Y = y] = \frac{1}{2}\left(\frac{1}{2}\right) = \frac{1}{4}$$

for each possible value of x and y. Thus, X and Y are independent random variables.

Continuous random variables X and Y are said to be *independent* if their joint density function $f(x,y)$ satisfies $f(x,y) = g(x)h(y)$, where $g(x)$ is the density function for X and $h(y)$ is the density function for Y. In other words, as $dx, dy \to 0$, the probability that the random variable X has its values in the interval $x \le X \le x + dx$ and the random variable Y has its values in the interval $y \le Y \le y + dy$ satisfies the condition:

$$\begin{aligned} f(x, y)dx\, dy &= P[x \le X \le x + dx, y \le Y \le y + dy] \\ &= P[x \le X \le x + dx]P[y \le Y \le y + dy] = g(x)h(y)\, dx\, dy \end{aligned} \tag{2.17}$$

where dx and dy are infinitesimals.

ILLUSTRATION

Suppose the random variable X has the density function $g(x) = 0.375x^2$ for $0 \le x \le 2$ and 0 elsewhere. Suppose the random variable Y has the density function $h(y) = 0.1$ for $0 \le y \le 10$ and 0 elsewhere. Assume that the probability that X lies in the interval $[x, x + dx]$ and Y lies in the interval $[y, y + dy]$ satisfies:

$$P[x \le X \le x + dx, y \le Y \le y + dy] = P[x \le X \le x + dx]P[y \le Y \le y + dy]$$

$$= g(x)h(y)\, dx\, dy$$

Thus, the random variables X and Y are independent. If the return on one stock is $r_1 = 0.4X$ and the return on another stock is $r_2 = 0.3Y$, then it is not difficult to show that the returns r_1 and r_2 are also independent random variables. Intuitively, this means that the return on one stock has no effect on the return on the other stock.

2.5.2.1 *Multiple Random Variables*

The notion of independence can be extended to any number of random variables or over numerous units of time when dealing with a stochastic process. For example, the discrete random variables $X_1, X_2, \ldots, X_n$ are said to be *pairwise independent* if:

$$P[X_i = x_i, X_j = x_j] = P[X_i = x_i]P[X_j = x_j] \tag{2.18}$$

for every $i \ne j$. Furthermore, pairwise independence implies by induction that:

$$\begin{aligned} P[X_1 = x_1, X_2 &= x_2, \ldots, X_n = x_n] \\ &= P[X_1 = x_1]P[X_2 = x_2]\ldots P[X_n = x_n] \end{aligned} \tag{2.19}$$

The case of continuous random variables can be extended in an analogous manner. Because of property (2.19), one often simply states that the random variables $X_1, X_2, \ldots, X_n$ are *independent* if condition (2.18) is satisfied. Independent random variables have some useful properties. Suppose, for example, that $X_1, X_2, \ldots, X_n$ are pairwise independent random variables and $c_1, c_2, \ldots, c_n$ and $d_1, d_2, \ldots, d_n$ are constants. The following are three properties that we will use in developing our financial models:

1. $E[X_1, X_2, \ldots, X_n] = E[X_1]E[X_2]\ldots E[X_n]$.
2. $\text{Var}[c_1 X_1 + c_2 X_2 + \cdots + c_n X_n] = c_1^2 \text{Var}[X_1] + c_2^2 \text{Var}[X_2] + \cdots + c_n^2 \text{Var}[X_n]$
3. The random variables $c_1 X_1 + d_1, c_2 X_2 + d_2, \ldots, c_n X_n + d_n$ are pairwise independent.

2.6 DISTRIBUTIONS AND PROBABILITY DENSITY FUNCTIONS

A *probability distribution* is a function that assigns a probability to each possible outcome in an experiment involving chance producing a discrete random variable. In the case of a continuous random variable, a probability density function is used to obtain probabilities of events. Here, we will discuss certain important distributions producing random variables, and examine the probability distribution in the discrete case and the probability density function in the continuous case.

2.6.1 The Binomial Random Variable

Suppose that a random trial (or experiment) with two possible outcomes, one of which is called a "success" and the other a "failure," is performed. We define variable $X = 1$ when the outcome is a success and define $X = 0$ when the outcome is a failure. Suppose that the probability of success is p, with $0 \leq p \leq 1$, such that $P[X = 1] = p$ and $P[X = 0] = 1 - p$. This random variable X is called a *Bernoulli trial* or process. Now suppose that n-independent identically distributed Bernoulli trials are performed. Let X represent the number of successes after n trials. Note that we can express $X = \sum_{i=1}^{n} X_i$ where each X_i is a Bernoulli process. X is a *binomial random variable*. Observe that X is a discrete random variable. The probability of obtaining exactly k successes after n trials equals:

$$P[X = k] = \binom{n}{k} p^k (1-p)^{n-k} \tag{2.20}$$

where $\binom{n}{k} = \dfrac{n!}{k!(n-k)!}$ for $k = 0, 1, \ldots, n$. This follows from the facts that there will be $\binom{n}{k}$ potential outcomes that produce k successes in n trials, and any particular outcome with k successes in n trials is obtained with probability $p^k (1-p)^{n-k}$.

ILLUSTRATION: COIN TOSSING

Suppose five fair coins are tossed. Find the physical probability that exactly three of the coins land on heads. Here "heads" is a "success," and since the coins are fair, $p = \frac{1}{2}$. If X equals the number of heads that are obtained with five tosses, then:

$$P[X = 3] = \binom{5}{3}\left(\frac{1}{2}\right)^3\left(1-\frac{1}{2}\right)^2 = 10\left(\frac{1}{8}\right)\left(\frac{1}{4}\right) = \frac{5}{16}$$

ILLUSTRATION: DK TRADES[7]

Suppose that the probability that any given trade is a DK is 0.02. In a series of 10 trades, find the probability that 2 or more of them are DKs. Assume that among the 10 trades, the event that any one trade is a DK is independent of any other trade being a DK. In this problem, a trade being a DK is a "success." In Bernoulli or binomial processes, "success" simply refers to what we are measuring or counting, not that it is necessarily a beneficial outcome. We wish to find $P[X \geq 2]$. Since $\sum_{i=0}^{10} P[X = i] = 1$ (having accounted for all possible outcomes), the easiest way to calculate $P[X \geq 2]$ is:

$$P[X \geq 2] = 1 - P[X = 0] - P[X = 1] = 1 - \binom{10}{0}(0.02)^0(0.98)^{10} - \binom{10}{1}(0.02)(0.98)^9$$

$$= 1 - (0.98)^{10} - 10(0.02)(0.98)^9 = 0.01618$$

2.6.2 The Uniform Random Variable

Let X be a continuous random variable whose values lie in the interval $a \leq X \leq b$ where a and b are real numbers and $a < b$. X is a *uniform random variable* on the interval $[a,b]$ if the probability that X takes on a range of values in any interval contained inside $[a,b]$ is equal to the probability that X takes on a range of values in any other interval contained inside $[a,b]$ of the same width. The density function for X is equal to:

$$f(x) = \begin{cases} \dfrac{1}{b-a} & \text{for } a \leq x \leq b \\ 0 & \text{otherwise} \end{cases} \tag{2.21}$$

The mean and variance of a uniform distribution on the interval $[a,b]$ are:[8]

$$E[X] = \frac{1}{2}(a + b)$$

$$\sigma_X^2 = \frac{1}{12}(b-a)^2$$

ILLUSTRATION: UNIFORM RANDOM VARIABLE

Suppose that the returns on a series of bonds of varying credit risks in a given year happened to be uniformly distributed on the interval $[0.02, 0.07]$. The probability that the return for a single bond is between 0.03 and 0.05 is calculated from the following density function, where X is the annual return on the bond:

$$f(x) = \begin{cases} \dfrac{1}{0.07 - 0.02} = 20 & \text{for } 0.002 \leq x \leq 0.07 \\ 0 & \text{otherwise} \end{cases}$$

$$P[0.03 < X < 0.05] = \int_{0.03}^{0.05} 20 \, dx = 0.4$$

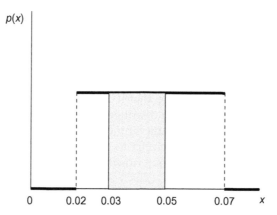

FIGURE 2.2   The uniform density curve.

Thus, we see from this calculation and from Figure 2.2 that there is a probability of 0.4 that our uniformly distributed variable will range from 0.03 to 0.05.

2.6.3 The Normal Random Variable

Many random variables in finance can be approximated with a normal random variable. We say that Y is a normal random variable with parameters mean equal to μ and variance equal to σ^2, and we write $Y \sim N(\mu, \sigma^2)$ if Y has the density function:

$$f(y) = \frac{1}{\sigma\sqrt{2\pi}} e^{\frac{-(y-\mu)^2}{2\sigma^2}} \tag{2.22}$$

The area under the normal density function's graph (the graph is called the normal curve; see Figure 2.3) within a specified range (a,b) is found by evaluating the following definite integral:

$$P(a \le Y \le b) = \frac{1}{\sigma\sqrt{2\pi}} \int_a^b e^{\frac{-(y-\mu)^2}{2\sigma^2}} \, dy \tag{2.23}$$

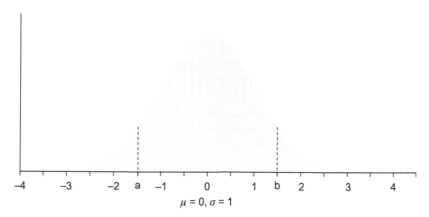

FIGURE 2.3   The standard normal density curve.

The distribution function for Y is:

$$F(y) = P(Y \le y) = \frac{1}{\sigma\sqrt{2\pi}} \int_{-\infty}^{y} e^{\frac{-(t-\mu)^2}{2\sigma^2}} \, dt \qquad (2.24)$$

The standard normal random variable, usually denoted by Z, has mean 0 and variance 1. Thus, we can write $Z \sim N(0,1)$. It is also easy to see that if $Y \sim N(\mu, \sigma^2)$, then $Y = \mu + \sigma Z$. This linear relationship means that every calculation of the distribution function for a normal random variable Y can be reduced to a calculation of the distribution function for the standard normal random variable Z. It is convenient to use the notation $N(z)$ to refer to the distribution function of Z; that is, $N(z) = P(Z \le z)$. The formula that relates the distribution function for Y in terms of that for Z is quite simple:

$$P(Y \le y) = P(\mu + \sigma Z \le y) = P\left(Z \le \frac{y-\mu}{\sigma}\right) = N\left(\frac{y-\mu}{\sigma}\right)$$

2.6.3.1 Calculating Cumulative Normal Density

Unfortunately, no closed-form solution exists for either of the above integrals; they can only be represented as integrals. Thus, in practice, and for textbook tables, the cumulative normal density is standardized as shown immediately above, then $N(z)$ is approximated, typically with a polynomial function such as the following created to resemble the cumulative density:

$$N(z) \approx 1 - \left(\frac{1}{\sqrt{2\pi}}\right) e^{-\frac{z^2}{2}} \left(c_1 k + c_2 k^2 + c_3 k^3 + c_4 k^4 + c_5 k^5\right)$$

where:

$k = 1/(1 + c_0 z)$
$c_0 = 0.2316419$
$c_1 = 0.319381530$
$c_2 = -0.356563782$
$c_3 = 1.781477937$
$c_4 = -1.821255978$
$c_5 = 1.330274429$

2.6.3.2 Linear Combinations of Independent Normal Random Variables

A very useful property of independent normal random variables is that linear combinations of them are still normal random variables. More precisely:

Theorem: If $X_1, X_2, \ldots, X_n$ is a family of pairwise independent random variables, each with a normal distribution so that the mean of X_k is μ_k and its variance is σ_k^2, then for any constants $c_1, c_2, \ldots, c_n$, the random variable defined by:

$$X = \sum_{k=1}^{n} c_k X_k$$

has a normal distribution with mean $\sum_{k=1}^{n} c_k \mu_k$ and variance $\sum_{k=1}^{n} c_k^2 \sigma_k^2$.

2.6.4 The Lognormal Random Variable

The lognormal random variable is often used for modeling stock returns. For example, the multiplicative product of many independent positive random variables such as $(1 + r_t) = S_t/S_{t-1}$ as in $\frac{S_T}{S_0} = \prod_{t=1}^{T}(1 + r_t)$ where T is large can often be modeled as having a lognormal distribution. In such a model, incremental stock returns r_t are arguably independent over time and r_t is never less than -100%. In this case, S_T/S_0 has an approximately lognormal distribution. A random variable Y is said to be lognormal if the logarithm of Y is a normal random variable X. Therefore, $\ln Y = X$ where $X \sim N(\mu, \sigma^2)$, or equivalently $Y = e^X$. Because $-\infty < X < \infty$, $0 < Y < \infty$. Since the density function for X satisfies:

$$\frac{1}{\sigma\sqrt{2\pi}}\int_{-\infty}^{\infty} e^{\frac{-(x-\mu)^2}{2\sigma^2}}\, dx = 1,$$

then by the change of variables, $x = \ln(y)$ and $dx = dy/y$, we obtain:

$$\int_0^{\infty} \frac{1}{\sqrt{2\pi}\sigma y} e^{-\frac{(\ln y - \mu)^2}{2\sigma^2}}\, dy = 1$$

Thus, the density function for the lognormal random variable is:

$$f(y) = \frac{1}{\sqrt{2\pi}\sigma y} e^{-\frac{(\ln y - \mu)^2}{2\sigma^2}} \tag{2.25}$$

A generic version of the lognormal density function for random variable Y is plotted in Figure 2.4.

The lognormal random variable is particularly useful when security returns or interest are compounded over time. First, consider the non-random model based on compound interest. Suppose that a security with value $Y(t)$ at time t pays interest at a rate r compounded continuously. This means that the instantaneous rate of change of $Y(t)$ is equal to $rY(t)$:

$$\frac{dY}{dt} = rY$$

We know from integral calculus that the solution for $Y(t)$ is:

$$Y(t) = Y_0 e^{rt}$$

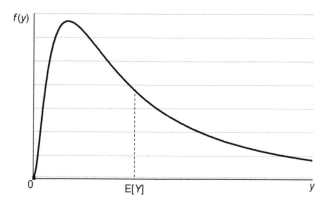

FIGURE 2.4 The lognormal density curve.

where Y_0 is the value of Y at time 0. Observe that $\ln(Y(t)/Y_0) = rt$. Of course, this is a deterministic model. Suppose that we also add a random term of the form $s\sqrt{t}Z$, where Z is the standard normal random variable and $s > 0$ is a constant. The purpose of this random term is to model the unpredictability and volatility of securities. The reason for this particular choice will be discussed later in the book. Our model takes the form $\ln(Y(t)/Y_0) = rt + s\sqrt{t}Z$. If we study the value of the security at a particular fixed time t, and relabel $\mu = rt$ and $\sigma = s\sqrt{t}$, then:

$$\ln\left(\frac{Y(t)}{Y_0}\right) = \mu + \sigma Z$$

But $X = \mu + \sigma Z$ is the normal random variable $X \sim N(\mu, \sigma^2)$. Thus, $Y(t)/Y_0$ is a lognormal random variable for fixed t.

2.6.4.1 The Expected Value of the Lognormal Distribution

An important result that is used regularly in financial modeling, and that will be important in Chapters 5–8, is that if X has a normal distribution with mean μ and variance σ^2, then:

$$E[e^X] = e^{\mu + \frac{1}{2}\sigma^2} \tag{2.26}$$

This, of course, is the expected value of the lognormal random variable. We prove this result as follows: Since X has a normal distribution with mean μ and variance σ^2, we can express X as $X = \mu + \sigma Z$ where Z has a standard normal distribution. Thus:

$$E[e^X] = \frac{1}{\sqrt{2\pi}} \int_{-\infty}^{\infty} e^{\mu + \sigma z} e^{-\frac{1}{2}z^2}\, dz$$

We complete the square in the exponents[9]:

$$\sigma z - \frac{1}{2}z^2 = -\frac{1}{2}(z - \sigma)^2 + \frac{1}{2}\sigma^2$$

and make the change of variables $u = z - \sigma$ in the integral, we then obtain:

$$E[e^X] = e^{\mu + \frac{1}{2}\sigma^2} \frac{1}{\sqrt{2\pi}} \int_{-\infty}^{\infty} e^{-\frac{1}{2}u^2}\, du = e^{\mu + \frac{1}{2}\sigma^2}$$

ILLUSTRATION: RISKY SECURITIES

Suppose that the returns on a security are lognormally distributed with parameters $\mu = 0.08$ and $\sigma = 0.05$. This means that the distribution of price relatives S_t/S_{t-1} (value after one period in proportion to its prior value) is lognormal with the given parameters $\mu = 0.08$ and $\sigma = 0.05$. The logs of the price relatives are distributed normally. If the value of the security was $100 at time 0, we will find the probability that its value after 1 year lies between $95 and $120. First, let S_1 refer to the security's value after 1 year. Given that $\ln(S_1/100) = 0.08 + 0.05Z$, we wish to determine $P(95 < S_1 < 120)$:

$$P(95 < S_1 < 120) = P\left(0.95 < \frac{S_1}{100} < 1.2\right) = P\left(\ln(0.95) < \ln\left(\frac{S_1}{100}\right) < \ln(1.20)\right)$$

$$= P(-0.05129 < 0.08 + 0.05Z < 0.1823) = P\left(\frac{-0.05129 - 0.08}{0.05} < \frac{0.05Z}{0.05} < \frac{0.1823 - 0.08}{0.05}\right)$$

$$= P(-2.626 < Z < 2.046) = N(2.046) - N(-2.626) = 0.97962 - 0.00432$$

$$= 0.9753$$

We can use the standard normal z-table or an appropriate spreadsheet function ($=$NORMDIST $(x,0,1,1)$) to complete this calculation. Note that $\ln(S_1/100) = X \sim N(0.08,0.05^2)$, so that $S_1/100 = e^X$. Using the expectation result derived above (Eq. (2.26)), we can find the one-period expected return for our security as follows:

$$E\left[\frac{S_1}{100} - 1\right] = E[e^X - 1] = e^{\mu + \frac{1}{2}\sigma^2} - 1 = e^{0.08 + \frac{1}{2} \times 0.05^2} - 1 = 0.0846$$

2.6.5 The Poisson Random Variable

Suppose that we are interested in the number of occurrences X of an event over a fixed time period of length t in which X is a Poisson random variable. An example might be one in which a trader might wish to predict the number of 10% price jumps experienced by a stock over a given time period t. We will derive a distribution function for $P[X = k]$. It is assumed that the expected value of the number of occurrences X of the event over a period of time t is known, and we denote this parameter by λt. Variable X will be a *Poisson random variable* with parameter λt if the following three properties hold:

1. The probability is negligible that the event occurs more than once in a sufficiently small interval of length $h \rightarrow 0$.
2. The probability that the event occurs exactly once in a short time interval of length h is approximately equal to λh, such that this probability divided by (λh) approaches one as h approaches 0.
3. The number of times that the event occurs over a given interval is independent of the number of times that the event occurs over any other non-overlapping interval.

2.6.5.1 *Deriving the Poisson Distribution*

We will consider this Poisson distribution in the context of and as a limit of a binomial distribution. Suppose that we divide the time interval $[T,T + t]$ into n subintervals of equal length so that the ith interval takes the form $[T + t(i - 1)/n, T + ti/n]$ for $i = 1, \ldots, n$. By property 1 above, the probability that the event occurs more than once in the ith time interval is negligible if n is large. By property 2 above, the probability that the event occurs in the ith time interval is approximately $p = \lambda t/n$ for $i = 1, \ldots, n$. Thus, the Poisson process follows a Bernoulli process within each of the n time intervals, with probability $p = \lambda t/n$ of success and probability $1 - p = 1 - \lambda t/n$ of failure. Furthermore, by property 3 above, we can regard the Poisson process as approximating n-independent identically distributed Bernoulli processes. Thus, the process follows a binomial distribution with n trials and $p = \lambda t/n$. Referring back to the binomial distribution, we have:

$$P(X = k) \approx \binom{n}{k}(p)^k(1-p)^{n-k} = \binom{n}{k}\left(\frac{\lambda t}{n}\right)^k\left(1 - \frac{\lambda t}{n}\right)^{n-k}$$

$$= \frac{(\lambda t)^k}{n^k}\left(1 - \frac{\lambda t}{n}\right)^n \frac{n(n-1)\ldots(n-k+1)}{k!}\left(1 - \frac{\lambda t}{n}\right)^{-k}$$

$$= \frac{(\lambda t)^k}{k!}\left(1 - \frac{\lambda t}{n}\right)^n \frac{n(n-1)\ldots(n-k+1)}{n^k}\left(1 - \frac{\lambda t}{n}\right)^{-k}$$

$$= \frac{(\lambda t)^k}{k!}\left(1 - \frac{\lambda t}{n}\right)^n\left(\left(1 - \frac{1}{n}\right)\left(1 - \frac{2}{n}\right)\ldots\left(1 - \frac{k-1}{n}\right)\right)\left(1 - \frac{\lambda t}{n}\right)^{-k}$$

Holding k fixed as $n \to \infty$, it follows that:

$$\left(1 - \frac{1}{n}\right)\left(1 - \frac{2}{n}\right) \cdots \left(1 - \frac{k-1}{n}\right)\left(1 - \frac{\lambda t}{n}\right)^{-k} \to 1$$

Recall from calculus that $\left(1 - \frac{\lambda t}{n}\right)^n \to e^{-\lambda t}$ as $n \to \infty$. Thus, the Poisson random variable X with parameter $\lambda t > 0$ is a discrete random variable that takes on values 0, 1, 2, ... with probability:

$$P(X = k) = e^{-\lambda t} \frac{(\lambda t)^k}{k!} \quad \text{for } k = 0, 1, 2, \ldots \tag{2.27}$$

ILLUSTRATION: STOCK PRICE JUMPS

Suppose that a trader needs to predict the probability that a given stock will not experience any 10% price jumps over the course of a month. Normally, the number jumps per year can be modeled as a Poisson variable, with an expected value of $\lambda = 3$; that is, on a monthly basis, $\lambda t = 3/12$:

$$P(X = 0) = P(X = 0) = e^{-0.25}\frac{0.25^0}{0!} = 0.7788$$

ILLUSTRATION: BANK AUDITOR

Suppose the average number of visits from a regulatory bank auditor per year is 1.5. Assume that the number of audits per year is a Poisson random variable. We will find the probability that the number of visits by the auditor to the bank in a 2-year time period is between 3 and 5, inclusive. In this case, $\lambda = 1.5$ and $t = 2$, so that $\lambda t = 3$. Thus, we find the probability of between 3 and 5 visits as follows:

$$P(3 \le X \le 5) = P(X = 3) + P(X = 4) + P(X = 5) = e^{-3}\frac{3^3}{3!} + e^{-3}\frac{3^4}{4!} + e^{-3}\frac{3^5}{5!} = 0.4929$$

2.7 THE CENTRAL LIMIT THEOREM

The central limit theorem states that the distribution of the sample means drawn from a population of independently and identically distributed random variables will approach the normal distribution as the sample size approaches infinity. Regardless of the distribution of the population, as long as observations are independently and identically distributed, the distribution of the sample mean will approach a normal distribution.

More formally, let X_i, $\forall i = 1$ to n, be independent and identically distributed (iid) random variables such that $E[X_i] = \mu$ and $\mathrm{Var}[X_i] = \sigma^2 < \infty$ and:

$$\overline{X} = \sum_{i=1}^{n} \frac{X_i}{n}$$

where $\overline{X}$ is the mean of a sample of X_i of size n. As n approaches ∞, the distribution of $(\overline{X} - \mu)/(\sigma/\sqrt{n})$ approaches the unit normal distribution ($\sim N(0,1)$):

$$\lim_{n \to \infty} P\left[a \leq \frac{\overline{X} - \mu}{\sigma/\sqrt{n}} \leq b\right] = \frac{1}{\sqrt{2\pi}} \int_a^b e^{-x^2/2}\, dx \qquad (2.28)$$

ILLUSTRATION OF THE CENTRAL LIMIT THEOREM

This example will demonstrate the convergence of the sample mean random variable to the normal distribution as the sample size gets large. Here, we will assume that sample values are drawn from a uniform distribution over the range 0, 1. Thus, consider a random variable that is uniformly distributed on the interval $0 \leq X \leq 1$. Its density function is:

$$p(x) = \begin{cases} 1 & \text{for } 0 \leq x \leq 1 \\ 0 & \text{otherwise} \end{cases}$$

and depicted by Figure 2.5.

Now, suppose that we choose a random sample of three values (X_1, X_2, X_3) from this distribution such that each choice is independent of the others. We find the sample mean:

$$\overline{X} = \frac{\sum_{i=1}^3 X_i}{3}$$

The sample mean $\overline{X}$ is itself a random variable with its own distribution function. That is, the mean of a draw of three values will itself produce a random variable. The density function for this particular random variable $\overline{X}$ is rather complex to derive. We present a histogram based on a simulation of this random variable, using a spreadsheet producing 10,000 repeated independently chosen random samples, each of size 3, and tabulating the 10,000 sample means $\overline{X}$. Using a large sample such as 10,000 will ensure that the resulting histogram will tend towards the actual density function for the random variable $\overline{X}$. The histogram of these results is shown in Figure 2.6. Notice that this histogram only slightly resembles one representing a normal distribution.

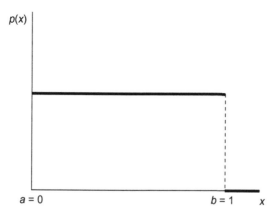

FIGURE 2.5 The uniform density curve.

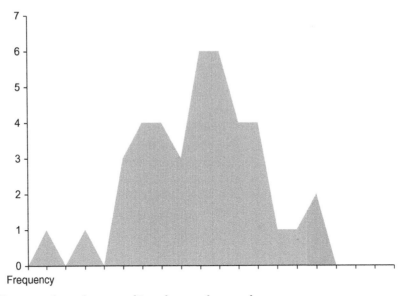

FIGURE 2.6   Histogram of sample means of 3-random number samples.

We would expect this, since the sample size for each $\overline{X}$ numbers only 3, and the original distribution from which the sample is drawn is a uniform distribution, which is quite different in shape from a normal curve.

Next, we draw samples of size 30 from the same uniform distribution such that the 30 values are chosen randomly and independently. The sample mean is computed. If X_1, X_2, ..., X_{30} are the 30 chosen sample values, then the sample mean $\overline{X}$ is:

$$\overline{X} = \frac{\sum_{i=1}^{30} X_i}{30}$$

Since $\overline{X}$ is simply a weighted sum of the individual random variables X_1, X_2, ..., X_{30}, $\overline{X}$ itself is a random variable. Similar to the case when the sample size was 3, we estimated the shape of the density function for the random variable $\overline{X}$ when the sample size is 30. We produced 10,000 repeated independent random samples of size 30 and tabulated their 10,000 sample means. The histogram of these results appears in Figure 2.7. This time the shape of the histogram for the sample mean more closely resembles a normal curve, despite the fact that the original distribution from which the sample was drawn was uniform. This is because the sample size was larger (30). This resemblance to the normal curve is a reflection of the validity of the central limit theorem.

The intuition behind why the distribution of the sample mean random variable should resemble the normal distribution is fairly straightforward. The distribution should be symmetrically shaped, should have the center (mean) of the sample mean distribution equal the mean of the original distribution from which the sample was drawn, and should taper off as the sample mean values deviate from the center. This is because sample values are selected randomly, such that with a large sample size, the values that are picked that are less than the mean (μ) are nearly certain to be offset by values that are larger than μ. Thus, the sample mean $\overline{X}$ has a greater chance of

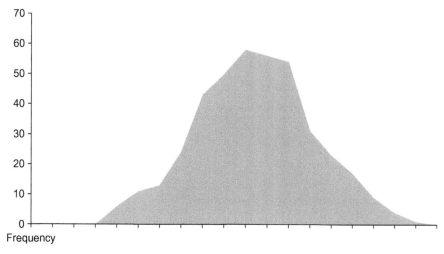

FIGURE 2.7   Histogram of sample means of 30-random number samples.

having its value near μ than far away from μ, which is why the distribution tapers off away from the center μ. The sample mean distribution should be approximately symmetrical because it is just as likely one gets the sample mean to be greater than rather than less than μ.

2.8 JOINT PROBABILITY DISTRIBUTIONS

We have been studying probability from the perspective of a single random variable. Often, one is interested in the probabilistic behavior of two (or more) random variables studied jointly. The joint probability distribution for discrete random variables X and Y is specified by:

$$P(X = x, Y = y) = \text{the probability that } X = x \text{ and } Y = y \tag{2.29}$$

If X and Y are independent random variables, then $P(X = x, Y = y) = P(X = x)P(Y = y)$. If X and Y are continuous random variables, then one specifies the joint density function $p(x,y)$ for X and Y. This density function has the property that $p(x,y)\, dx\, dy$ equals the probability that the values of X are between x and $x + dx$ and the values of Y are between y and $y + dy$. If X and Y are independent random variables, then $p(x,y) = p_X(x)p_Y(y)$ where $p_X(x)$ is the density function for X and $p_Y(x)$ is the density function for Y. Since we still have a probability space, the total probability must sum to 1. In the discrete case, this means:

$$\sum_i \sum_j P(X = x_i)P(Y = y_j) = 1 \tag{2.30}$$

In the continuous case, this is expressed as:

$$\int_{-\infty}^{\infty} \int_{-\infty}^{\infty} p(x, y)\, dx\, dy = 1.$$

It follows from the definition of the joint density function that if the set A is any region in the xy-plane, then the probability that (X,Y) will have its values in the set A equals:

$$\iint_{(x,y)\in A} p(x,y)\,dx\,dy$$

In particular, the joint distribution function for X and Y is:

$$P(x,y) = P(X \le x, Y \le y) = \int_{-\infty}^{y}\int_{-\infty}^{x} p(s,t)\,ds\,dt \tag{2.31}$$

2.8.1 Transaction Execution Delay Illustration

Suppose that X is the amount of time that it takes for a security's transaction to transmit from a hedge fund to a stock exchange. Assume that this time delay has a uniform distribution from 0 to 20 ms. Let Y be the amount of time required for the stock exchange to actually execute the transaction once received. Assume this time has a uniform distribution from 0 to 10 ms. Also assume that the random variables X and Y are independent. Determine the probability that the total transaction completion time (transmission and execution) is less than 10 ms.

To solve this problem, consider that X has a uniform distribution and that its density function is $p_X(x) = 1/20$ for $0 \le x \le 20$ and equals 0 otherwise. Similarly, $p_Y(y) = 1/10$ for $0 \le y \le 10$ and 0 otherwise. Since X and Y are independent, their joint density function is $p(x,y) = (1/20)(1/10) = 1/200$ for $0 \le x \le 20$ and $0 \le y \le 10$, and 0 otherwise. We wish to determine:

$$P(X + Y \le 10) = P(X + Y \le 10, 0 \le Y \le 10) = P(0 \le X \le 10 - Y, 0 \le Y \le 10)$$

$$= \int_{0}^{10}\int_{0}^{10-y} \frac{1}{200}\,dx\,dy = \int_{0}^{10} \frac{x}{200}\Big|_{0}^{10-y}\,dy = \int_{0}^{10}\frac{10-y}{200}\,dy = \frac{1}{20}y - \frac{1}{400}y^2\Big|_{0}^{10} = \frac{1}{4}$$

2.8.2 The Bivariate Normal Distribution

The joint probability distribution for random variables X and Y is said to have a bivariate normal distribution if its density function is given by:[10]

$$f(x,y) = \frac{1}{2\pi\sigma_x\sigma_y\sqrt{1-\rho^2}}e^{\left[-\frac{1}{2(1-\rho^2)}\left[\left(\frac{(x-\mu_x)^2}{\sigma_x^2}\right) + \left(\frac{(y-\mu_y)^2}{\sigma_y^2}\right) - 2\rho\left(\frac{(x-\mu_x)}{\sigma_x}\right)\left(\frac{(y-\mu_y)}{\sigma_y}\right)\right]\right]} \tag{2.32}$$

It can be shown that if X and Y have a bivariate normal distribution, then X and Y are normally distributed random variables, $Y \sim N(\mu_y, \sigma_y^2)$ and $X \sim N(\mu_x, \sigma_x^2)$, with correlation coefficient $\rho = \rho_{x,y}$. However, the converse is not true. That is, if X and Y are normally distributed random variables, $Y \sim N(\mu_y, \sigma_y^2)$ and $X \sim N(\mu_x, \sigma_x^2)$, with correlation coefficient $\rho = \rho_{x,y}$, then it is not necessarily the case that X and Y have a bivariate normal distribution. We will be concerned with the probability that both random variables fall within specified ranges.

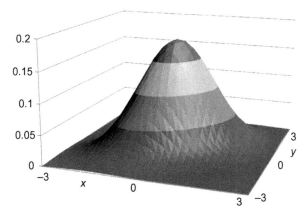

FIGURE 2.8 Bivariate normal density surface: $f(x,y)$; $\rho = 0.5$.

The volume under the bivariate normal density function's graph within specified ranges (a,b) for X and (c,d) for Y is found by evaluating the following:

$$P(a \le X \le b, c \le Y \le d) = \int_c^d \int_a^b f(x,y) \, dx \, dy \tag{2.33}$$

The joint distribution function for X and Y is:

$$F(x,y)$$

$$= \int_{-\infty}^{y} \int_{-\infty}^{x} f(s,t) ds \, dt$$

$$= \int_{-\infty}^{y} \int_{-\infty}^{x} \frac{1}{2\pi\sigma_x\sigma_y\sqrt{1-\rho^2}} e^{\left[-\frac{1}{2(1-\rho^2)}\left[\left(\frac{(s-\mu_x)^2}{\sigma_x^2}\right)+\left(\frac{(t-\mu_y)^2}{\sigma_y^2}\right)-2\rho\left(\frac{(s-\mu_x)}{\sigma_x}\right)\left(\frac{(t-\mu_y)}{\sigma_y}\right)\right]\right]} ds \, dt \tag{2.34}$$

To help visualize the bivariate normal distribution, consider the case involving two standard normal random variables $Z_x = (X - \mu_x)/\sigma_x$ and $Z_y = (Y - \mu_y)/\sigma_y$, in which each has mean 0 and standard deviation 1. Their joint density function is plotted in Figure 2.8, assuming that $\rho = 0.5$. We can write in this special case that our standard normal variables Z_x and Z_y with correlation coefficient ρ have the joint distribution function $M(z_x,z_y; \rho)$:

$$M(z_x, z_y; \rho) = \int_{-\infty}^{z_y} \int_{-\infty}^{z_x} \frac{1}{2\pi\sqrt{1-\rho^2}} e^{\left[-\frac{s^2+t^2-2\rho st}{2(1-\rho^2)}\right]} ds \, dt \tag{2.35}$$

2.8.3 Calculating Cumulative Bivariate Normal Densities

Estimating cumulative probabilities for a bivariate normal distribution with an appropriate polynomial function is rather similar to the process for the univariate normal density, but is somewhat more time consuming. There are a number of spreadsheet-based bivariate distribution calculators online (e.g., on the companion website for this book and at John Hull's website at http://www.rotman.utoronto.ca/~hull/software/bivar.xls), and most statistics packages such as R (which is free online) have features that will perform these computations.

Three inputs will be needed for such a calculator to compute the cumulative probability: the standardized deviate for the first variable X, the standardized deviate for the second variable Y, and the correlation coefficient ρ between the two.

ILLUSTRATION: CALCULATING BIVARIATE NORMAL DENSITIES

Suppose that the log of one stock's price relative $X_t = \ln(S_t/S_{t-1})$ is distributed normally with mean 5% and standard deviation 1%. The log of another stock's price relative $Y_t = \ln(S_t/S_{t-1})$ is distributed normally with mean 6% and standard deviation 2%. Assume that X_t and Y_t have a bivariate normal distribution and that their correlation coefficient is $\rho = 0.5$. Find the probability of the event that the first stock's return is less than 6.5% and the second stock's return is less than 6.6%.

We wish to find the distribution function $P(X \le 6.5, Y \le 6.6)$. Note that:

$$P(X \le 6.5\%, Y \le 6.6\%) = P\left(\frac{X - 5\%}{1\%} \le \frac{6.5\% - 5\%}{1\%}, \quad \frac{Y - 6\%}{2\%} \le \frac{6.6\% - 6\%}{2\%} \right)$$

$$= P(Z_1 \le 1.5, Z_2 \le 0.3)$$

where $Z_1 = (X - 5\%)/1\%$ and $Z_2 = (Y - 6\%)/2\%$ are the standardized normal random variables for X and Y. We also need to know the correlation coefficient for the standardized random variables.

It is a convenient fact (see end-of-chapter exercise 2.20) that the correlation coefficient between any two normal random variables is always equal to the correlation coefficient between their standardized normal random variables. Thus, the correlation coefficient for Z_1 and Z_2 is also 0.5. So, we need to calculate the joint distribution function $M(1.5,0.3;0.5)$. We would find from an appropriate calculator that this cumulative density is approximately 0.603, implying that the probability is 0.603 that the random variables X and Y were both less than 6.5% and 6.6%, respectively. Drezner (1978) provides an appropriate polynomial for this estimation process.

2.9 PORTFOLIO MATHEMATICS

A portfolio is simply a collection of investments held by an investor. It may be reasonable to be concerned with the performance of individual securities only to the extent that their performance affects overall portfolio performance. Thus, the performance of the portfolio is of primary importance. The return of an investor's portfolio is simply a weighted average of the returns of the individual securities that are within it. The expected return of a portfolio may be calculated either as a function of potential portfolio returns and their associated probabilities (as computed in earlier sections) or as a simple weighted average of the expected individual security returns. Generally, the portfolio variance or standard deviation of returns will be less than the weighted average of the individual security variances or standard deviations.

2.9.1 Portfolio Arithmetic

Consider a portfolio of n securities with potential return outcomes R_i for $i = 1, \ldots, n$. Suppose that each of the returns R_i is a random variable associated with the same probability space as we vary i.

Let the possible outcomes in the probability space have probabilities P_k for $k = 1, \ldots, m$. Thus, we can label the possible values of the random variable R_i by $R_{i,k}$ for $k = 1, \ldots, m$ and the probability that the value $R_{i,k}$ occurs equals P_k. This is a very reasonable model, because each different security is subject to the same market forces, and if the market is in a particular state (outcome) with the probability P_k that this state will occur, then this will result in a particular value for the ith security's return of $R_{i,k}$. The expected return on security i and a portfolio p based on m potential return outcomes $R_{i,k}$ and associated probabilities P_k are computed as follows:

$$E[R_i] = \sum_{k=1}^{m} R_{i,k} \cdot P_k \tag{2.36}$$

$$E[R_p] = \sum_{k=1}^{m} R_{p,k} \cdot P_k \tag{2.37}$$

The portfolio return R_p is simply a weighted average of the individual security returns R_i. We denote the weights by w_i, so that:

$$w_i = (\$ \text{ invested in } i)/(\$ \text{ invested in } p)$$

If we express the vector of individual returns by $\mathbf{r} = (R_1, \ldots, R_n)^T$ and the vector of weights by $\mathbf{w} = (w_1, \ldots, w_n)^T$, we can express the portfolio return R_p as follows:

$$R_p = \sum_{i=1}^{n} w_i R_i = \mathbf{w}^T \mathbf{r} \tag{2.38}$$

Taking the expectation of both sides of Eq. (2.38), we find that the expected return of the portfolio in terms of the expected value of each of the individual returns satisfies the equation:

$$E[R_p] = \sum_{i=1}^{n} w_i E[R_i] \tag{2.39}$$

Portfolio return variance is a function of potential portfolio returns and associated probabilities or individual security weights and covariance pairs[11]:

$$\sigma_p^2 = E[(R_p - E[R_p])^2] = \sum_{i=1}^{n} \sum_{j=1}^{n} w_i \cdot w_j \cdot \sigma_{i,j} = \mathbf{w}^T \mathbf{V} \mathbf{w} \tag{2.40}$$

where $\mathbf{V} = (\sigma_{i,j})$ is the $n \times n$ matrix of return covariances.[12]

2.9.2 Optimal Portfolio Selection

The investor's portfolio problem is to select n portfolio weights w_i so as to minimize portfolio risk σ_p^2 subject to some target return r_p as follows:

$$\min_{\mathbf{w}} \sigma_p^2 = \mathbf{w}^T \mathbf{V} \mathbf{w}$$

$$\mathbf{w}^T \mathbf{r} = r_p$$

$$\mathbf{w}^T \boldsymbol{\iota} = 1$$

where $\boldsymbol{\iota} = (1, 1, \ldots, 1)^T$.

This problem is solved with the following Lagrangian:

$$L = \mathbf{w}^T \mathbf{V} \mathbf{w} - \lambda_1(\mathbf{w}^T \mathbf{r} - r_p) - \lambda_2(\mathbf{w}^T \boldsymbol{\iota} - 1)$$

with $n + 2$ first-order conditions as follows[13]:

$$\nabla L(\mathbf{w}) = \mathbf{0}$$

$$\frac{\partial L}{\partial \lambda_1} = -\mathbf{w}^T \mathbf{r} + r_p = 0$$

$$\frac{\partial L}{\partial \lambda_2} = -\mathbf{w}^T \boldsymbol{\iota} + 1 = 0$$

The gradient $\nabla L(\mathbf{w})$ is an $n \times 1$ vector and its ith component will be $2\sigma_{i,1}w_1 + 2\sigma_{i,2}w_2 + \cdots + 2\sigma_{i,n}w_n - \lambda_1 R_i - \lambda_2$ for $i = 1, 2, \ldots, n$.

2.9.2.1 *Portfolio Selection Illustration*

Suppose that an investor wishes to comprise the lowest risk portfolio of three assets. The expected returns of the three assets A, B, and C are 0.1, 0.2, and 0.3, respectively. Their standard deviations are 0.2, 0.3, and 0.4. The returns covariance between assets A and B is 0.03, the returns covariance between assets A and C is 0.04, and the returns covariance between assets B and C is 0.06. What should be the weights of this optimal or lowest risk portfolio if he seeks an expected portfolio return of 15%?

First, since the expected portfolio return is 0.15 and the weights must sum to 1, the LaGrange function is written as either of the following:

$$L = 0.04w_A^2 + 0.09w_B^2 + 0.16w_C^2 + 2 \cdot 0.03w_A w_B + 2 \cdot 0.04w_A w_C + 2 \cdot 0.06w_B w_C$$
$$+ \lambda_1\left(0.15 - 0.1w_A - 0.2w_B - 0.3w_C\right) + \lambda_2(1 - w_A - w_B - w_C)$$

or:

$$L = \begin{bmatrix} w_A & w_B & w_C \end{bmatrix} \begin{bmatrix} 0.04 & 0.03 & 0.04 \\ 0.03 & 0.09 & 0.06 \\ 0.04 & 0.06 & 0.16 \end{bmatrix} \begin{bmatrix} w_A \\ w_B \\ w_C \end{bmatrix}$$

$$- \lambda_1 \begin{bmatrix} w_A & w_B & w_C \end{bmatrix} \begin{bmatrix} 0.1 \\ 0.2 \\ 0.3 \end{bmatrix} + 0.15\lambda_1 - \lambda_2 \begin{bmatrix} w_A & w_B & w_C \end{bmatrix} \begin{bmatrix} 1 \\ 1 \\ 1 \end{bmatrix} + 1\lambda_2$$

First-order conditions are follows, which combines $\nabla L(\mathbf{w}) = \mathbf{0}$, $\frac{\partial L}{\partial \lambda_1} = \mathbf{w}^T \mathbf{r} - r_p = 0$ and $\frac{\partial L}{\partial \lambda_2} = \mathbf{w}^T \boldsymbol{\iota} - 1 = 0$, the latter two of which are slightly rearranged:

$$\underbrace{\begin{bmatrix} 0.08 & 0.06 & 0.08 & -0.1 & -1 \\ 0.06 & 0.18 & 0.12 & -0.2 & -1 \\ 0.08 & 0.12 & 0.32 & -0.3 & -1 \\ -0.1 & -0.2 & -0.3 & 0 & 0 \\ -1 & -1 & -1 & 0 & 0 \end{bmatrix}}_{\mathbf{V}*} \underbrace{\begin{bmatrix} w_A \\ w_B \\ w_C \\ \lambda_1 \\ \lambda_2 \end{bmatrix}}_{\mathbf{w}*} = \underbrace{\begin{bmatrix} 0 \\ 0 \\ 0 \\ -0.15 \\ -1 \end{bmatrix}}_{\mathbf{c}*}$$

We invert Matrix $\mathbf{V^*}$ then solve for Vector $\mathbf{w^*}$. We find the following weights: $w_A = 0.625$, $w_B = 0.25$, and $w_C = 0.125$; our LaGrange multipliers are $\lambda_1 = 0.225$ and $\lambda_2 = 0.0525$. The expected return and standard deviation of portfolio returns are 0.15 and 0.207665, respectively:

$$E[rp] = \mathbf{w}^T\mathbf{r} = \begin{bmatrix} 0.625 & 0.25 & 0.125 \end{bmatrix} \begin{bmatrix} 0.1 \\ 0.2 \\ 0.3 \end{bmatrix} = 0.15$$

$$\sigma_p^2 = \mathbf{w}^T\mathbf{V}\mathbf{w} = \begin{bmatrix} 0.625 & 0.25 & 0.125 \end{bmatrix} \begin{bmatrix} 0.04 & 0.03 & 0.04 \\ 0.03 & 0.09 & 0.06 \\ 0.04 & 0.06 & 0.16 \end{bmatrix} \begin{bmatrix} 0.625 \\ 0.25 \\ 0.125 \end{bmatrix} = 0.043125$$

2.10 EXERCISES

2.1. Suppose we toss a die and observe the outcome. We can obtain an outcome of 1, 2, 3, 4, 5, or 6 of the die. If each outcome is equally likely, we would assign a probability of 1/6 to each of the outcomes.
   a. What is the sample space for numerical values of the roll of the die?
   b. How many events are in Φ?
   c. What is $P(\{2,4,6\})$?

2.2. Suppose we toss three fair coins and observe the result. By a fair coin, we mean that there is a 50−50 chance that the coin comes out either heads or tails.
   a. List the sample space (all possible outcomes) for this experiment.
   b. How many possible events are in the event space for this experiment?
   c. Is {HHH,TTT} an event in this event space? If so, what is its probability?
   d. Is $\varnothing$ an event in this event space? If so, what is its probability?

2.3. Mack Products' management is considering the investment in one of two projects available to the company. The returns on the two projects (A) and (B) are dependent on the sales outcome of the company. Mack management has determined three potential sales outcomes (1), (2), and (3) for the company. The highest potential sales outcome for Mack is outcome (1) or $800,000. If this sales outcome were realized, Project (A) would realize a return outcome of 30%; Project (B) would realize a return of 20%. If outcome (2) were realized, the company's sales level would be $500,000. In this case, project (A) would yield 15%, and Project (B) would yield 13%. The worst outcome (3) will result in a sales level of $400,000, and return levels for Projects (A) and (B) of 1% and 9%, respectively. If each sales outcome has an equal probability of occurring, determine the following for Mack Products:
   a. The probabilities of outcomes (1), (2), and (3).
   b. Its expected sales level.
   c. The variance associated with potential sales levels.
   d. The expected return of Project (A).
   e. The variance of potential returns for Project (A).
   f. The expected return and variance for Project (B).

g. Standard deviations associated with company sales, returns on Project (A) and returns on Project (B).

h. The covariance between company sales and returns on Project (A).

i. The coefficient of correlation between company sales and returns on Project (A).

j. The coefficient of correlation between company sales and returns on Project (B).

k. The coefficient of determination between company sales and returns on Project (B).

2.4. Historical *percentage* returns for McCarthy and Alston companies are listed in the following table along with percentage returns on the market portfolio:

Year	McCarthy	Alston	Market
1988	4	19	15
1989	7	4	10
1990	11	−4	3
1991	4	21	12
1992	5	13	9

Calculate the following based on the preceding table:

a. Mean historical returns for the two companies and the market portfolio.

b. Historical variances associated with McCarthy Company returns and Alston Company returns as well as returns on the market portfolio. Probabilities used for historical mean, variance, and covariance calculations can be assumed to be $1/n$, where n is the relevant number of time periods.

c. The historical covariance and coefficient of correlation between returns of the two securities.

d. The historical covariance and coefficient of correlation between returns of McCarthy Company and returns on the market portfolio.

e. The historical covariance and coefficient of correlation between returns of Alston Company and returns on the market portfolio.

2.5. The following table represents outcome numbers, probabilities, and associated returns for stock A:

Outcome (i)	Return (R_i)	Probability (P_i)	Outcome (i)	Return (R_i)	Probability (P_i)
1	0.05	0.10	6	0.10	0.10
2	0.15	0.10	7	0.15	0.10
3	0.05	0.05	8	0.05	0.10
4	0.15	0.10	9	0.15	?
5	0.15	0.10	10	0.10	0.10

Thus, there are 10 possible return outcomes for Stock A.

a. What is the probability associated with Outcome 9?

b. What is the standard deviation of returns associated with Stock A?

2.6. The Durocher Company management projects a return level of 15% for the upcoming year. Management is uncertain as to what the actual sales level will be; therefore, it associates a standard deviation of 10% with this sales level. Managers assume that sales will be normally distributed. What is the probability that the actual return level will:
   a. fall between 5% and 25%?
   b. fall between 15% and 25%?
   c. exceed 25%?
   d. exceed 30%?

2.7. What would be each of the probabilities in Problem 2.6 if Durocher Company management were certain enough of its forecast to associate a 5% standard deviation with its sales projection?

2.8. Under what circumstances can the coefficient of determination between returns on two securities be negative? How would you interpret a negative coefficient of determination? If there are no circumstances where the coefficient of determination can be negative, describe why.

2.9. Stock A will generate a return of 10% if and only if Stock B yields a return of 15%; Stock B will generate a return of 10% if and only if Stock A yields a return of 20%. There is a 50% probability that Stock A will generate a return of 10% and a 50% probability that it will yield 20%.
   a. What is the standard deviation of returns for Stock A?
   b. What is the covariance of returns between Stocks A and B?

2.10. An investor has the opportunity to purchase a risk-free Treasury bill yielding a return of 10%. He also has the opportunity to purchase a stock that will yield either 7% or 17%. Either outcome is equally likely to occur. Compute the following:
   a. The variance of returns on the stock.
   b. The coefficient of correlation between returns on the stock and returns on the Treasury bill.

2.11. The following daily prices were collected for each of three stocks over a 12-day period.

Company X		Company Y		Company Z	
Date	Price	Date	Price	Date	Price
1/09	50.125	1/09	20.000	1/09	60.375
1/10	50.125	1/10	20.000	1/10	60.500
1/11	50.250	1/11	20.125	1/11	60.250
1/12	50.250	1/12	20.250	1/12	60.125
1/13	50.375	1/13	20.375	1/13	60.000
1/14	50.250	1/14	20.375	1/14	60.125
1/15	52.250	1/15	21.375	1/15	62.625
1/16	52.375	1/16	21.250	1/16	60.750
1/17	52.250	1/17	21.375	1/17	60.750
1/18	52.375	1/18	21.500	1/18	60.875
1/19	52.500	1/19	21.375	1/19	60.875
1/20	52.375	1/20	21.500	2/20	60.875

Based on the data given above, calculate the following:

a. Returns for each day on each of the three stocks. There should be a total of 11 returns for each stock—beginning with the date 1/10.

b. Average daily returns for each of the three stocks.

c. Daily return standard deviations for each of the three stocks. There should be a total of 11 returns for each stock—beginning with the date 1/10.

2.12. Suppose the random variables X and Y are independent, $E[X] = 4$, $E[Y] = 3$, $Var(X) = 0.5$, and $Var(Y) = 1.2$. Find:

a. $E[2X + 5Y]$.

b. $E[XY]$.

c. $Var(2X + 5Y)$.

2.13. Consider the random variable X with the density function $p(x) = 0.5x$ for $0 \le x \le 2$ and 0 elsewhere.

a. Find the distribution function for this density function.

b. What characteristics of this function make it reasonable and meaningful as a distribution function?

c. Suppose that the random return on a given stock is $r = f(X) = 0.8X - 0.2$. Compute the probability that the stock's return r will be between 0.1 and 0.4.

d. What is the expected return on this stock?

e. What is the variance of returns on this stock?

f. What is the expected value of the stock return, conditional on its falling in the range 0.1–0.4?

2.14. Suppose that the terminal cash flow for a given investment is equally likely to assume any value in the range [0, 100] and will not assume any value outside of this range. That is, a uniform density function is associated with the random investment cash flow X. More specifically, the density function for the random variable X is given by $p(x) = 0.01$ for $0 \le x \le 100$ and 0 elsewhere.

a. Find the distribution function $P(x)$ for X.

b. Find the expected value of X conditioned on X being in the range 50–100.

c. Find the expected value of X conditioned on X being in the range 0–50.

d. Find the expected value of X.

e. Find the variance of X.

2.15. Consider the density function $p(x) = 3e^{-3X}$ for $X > 0$ and 0 elsewhere. What is the probability that X is between 0.5 and 1?

2.16. Suppose that X and Y are independent discrete random variables. Verify that $E[XY] = E[X] E[Y]$.

2.17. Suppose that X and Y are independent random variables and a and b are constants. Verify that $Var(aX + bY) = a^2 Var(X) + b^2 Var(Y)$.

2.18. a. In Section 2.6.2, we stated that the mean of the uniform distribution on the interval $[a,b]$ is:

$$E[X] = \frac{1}{2}(a + b)$$

Verify this statement.

b. In Section 2.6.2, we stated that the variance of the uniform distribution on the interval $[a,b]$ is:

$$\sigma_X^2 = \frac{1}{12}(b-a)^2$$

Verify this statement.

2.19. Suppose that the returns on a security priced at $50 at time zero follow a lognormal distribution with parameters $\mu = 0.10$ and $\sigma = 0.5$.

   a. What is the probability that the value of this security after 1 year S_1 will fall between $45 and $60?

   b. What is the expected one-period return for this security?

   c. Provide a non-mathematical intuitive explanation for why the expected return is much higher than μ.

2.20. In Section 2.8.3, we claimed that the correlation coefficient between two normal random variables is equal to the correlation coefficient between their standardized normal random variables. Prove this claim.

2.21. Suppose that a microchip company makes three types of microchips. Let M_1, M_2, and M_3 be the event that a randomly chosen microchip produced by the company is of type 1, 2, and 3, respectively. Let D be the event that a randomly chosen microchip is defective. The company has collected the following data:

	$P[D|M_i]$	$P[M_i]$
type 1	0.001	0.15
type 2	0.003	0.3
type 3	0.004	0.55

As indicated, the left column of numbers lists the probabilities that a microchip is defective given that it is of a certain type (1, 2, or 3). The right column of numbers lists the probabilities that a microchip is of a certain type. Determine the probability that a microchip is of type 1, 2, and 3, respectively, given that it is defective.

2.22. An investor is considering combining Douglas Company and Tilden Company common stock into a portfolio. Fifty percent of the dollar value of the portfolio will be invested in Douglas Company stock; the remainder will be invested in Tilden Company stock. Douglas Company stock has an expected return of 6% and an expected standard deviation of returns of 9%. Tilden Company stock has an expected return of 20% and an expected standard deviation of 30%. The coefficient of correlation between returns of the two securities has been shown to be 0.4. Compute the following for the investor's portfolio:

   a. Expected return.

   b. Expected variance.

   c. Expected standard deviation.

2.23. Work through each of your calculations in Problem 2.22, again assuming the following weights rather than those given originally:

   a. 100% Douglas Company stock; 0% Tilden Company stock

   b. 75% Douglas Company stock; 25% Tilden Company stock

   c. 25% Douglas Company stock; 75% Tilden Company stock

   d. 0% Douglas Company stock; 100% Tilden Company stock

2.24. How do expected return and risk levels change as the portfolio proportions invested in Tilden Company stock increase? Why? Prepare a graph with expected portfolio return on the vertical axis and portfolio standard deviation on the horizontal axis. Plot the expected returns and standard deviations for each of the portfolios whose weights are defined in Problems 2.22 and 2.23. Describe the slope of the curve connecting the points on your graph.

2.25. The common stocks of Landon Company and Burr Company are to be combined into a portfolio. The expected return and standard deviation levels associated with Landon Company stock are 5% and 12%, respectively. The expected return and standard deviation levels for Burr Company stock are 10% and 20%. The portfolio weights will each be 50%. Find the expected return and standard deviation levels of this portfolio if the coefficient of correlation between returns of the two stocks is:

a. 1
b. 0.5
c. 0
d. − 0.5
e. − 1

2.26. An investor is considering combining Securities A and B into an equally weighted portfolio. This investor has determined that there is a 20% chance that the economy will perform very well, resulting in a 30% return for Security A and a 20% return for Security B. The investor estimates that there is a 50% chance that the economy will perform only adequately, resulting in 12% and 10% returns for Securities A and B, respectively. The investor estimates a 30% probability that the economy will perform poorly, resulting in a negative 9% return for Security A and a 0% return for Security B. These estimates are summarized as follows:

Outcome	Probability	R_{ai}	R_{bi}
1	0.20	0.30	0.20
2	0.50	0.12	0.10
3	0.30	−0.09	0

a. What is the portfolio return for each of the potential outcomes?
b. Based on each of the outcome probabilities and potential portfolio returns, what is the expected portfolio return?
c. Based on each of the outcome probabilities and potential portfolio returns, what is the standard deviation associated with portfolio returns?
d. What are the expected returns of each of the two securities?
e. What are the standard deviation levels associated with returns on each of the two securities?
f. What is the coefficient of correlation between returns of the two securities?
g. Based on your answers to part d in this problem, find the expected portfolio return. How does this answer compare to your answer in part b?
h. Based on your answers to parts e and f, what is the expected deviation of portfolio returns? How does this answer compare to your answer in part c?

2.27. Suppose that the number of daily stock price jumps per month for a particular stock of magnitude 7% or higher is Poisson distributed with a mean of 1.25 per month. Find the probability that in a 1 month, there will be for the stock:

    a. at least two price jumps of magnitude 7% or higher

    b. at most two price jumps of magnitude 7% or higher.

2.28. **a.** Suppose that $P[X = k] = e^{-\lambda} \frac{\lambda^k}{k!}$ for $i = 0, 1, \ldots$. Demonstrate that $\sum_{k=0}^{\infty} P[k] = 1$.

    b. Prove that the Poisson distribution with parameter λ has an expected value equal to λ.

2.29. Suppose that the log of one stock's price relative $X_t = \ln(S_t/S_{t-1})$ is distributed normally with mean 5% and standard deviation 10%. The log of a second stock's price relative $Y_t = \ln(S_t/S_{t-1})$ is distributed normally with mean 10% and standard deviation 20%. Assume that X_t and Y_t have a bivariate normal distribution with a correlation coefficient of $\rho = 0.8$.

    a. Find the probability of the event that both stocks' returns are less than 0.

    b. Now, assume that the correlation coefficient is 0. Find the probability of the event that both stocks' returns are less than 0.

    c. Comment on the relationship between the univariate probabilities that X and Y are less than 0 and the probability that both X and Y are less than 0 given a bivariate normal distribution with a correlation coefficient of 0.

2.30. Suppose that an investment's returns are distributed normally with a mean of 4% and a standard deviation of 1.5%. If we consider such an investment that our inside information reveals that the investment will yield at least a 5% return, find the probability that it will give a return of 6% or better.

2.31. Verify that the variance of the random variable (Z^2c) where $Z \sim N(0,1)$ and $c > 0$ equals $2c^2$.

2.32. An investor has combined securities X, Y, and Z into a portfolio. He has invested $1000 in Security X, $2000 into Security Y, and $3000 into Security Z. Security X has an expected return of 10%, Security Y has an expected return of 15%, and security Z has an expected return of 20%. The standard deviations associated with Securities X, Y, and Z are 12%, 18%, and 24%, respectively. The coefficient of correlation between returns on Securities X and Y is 0.8; the correlation coefficient between X and Z returns is 0.7; and the correlation coefficient between Y and Z returns is 0.6. Find the expected return and standard deviation of the resultant portfolio.

2.33. Derive the following result in the chapter:

$$\sigma_p^2 = E[(R_p - E[R_p])^2] = \mathbf{w^T V w}$$

2.34. An investor wishes to combine Stevenson Company stock and Smith Company stock into a riskless portfolio. The standard deviations associated with returns on these stocks are 10% and 18%, respectively. The coefficient of correlation between returns on these two stocks is -1. What must be each of the portfolio weights for the portfolio to be riskless?

2.35. Assume that the coefficient of correlation between returns on all securities equals zero in a given market. There are n securities in this market, all of which have the same standard deviation of returns.

    a. What would be the portfolio return standard deviation if it included all of these n securities in equal investment amounts? Why? (Demonstrate your solution mathematically.)

    b. In the limit as n approaches infinity, what value would the portfolio return standard deviation approach?

2.36. Securities 1, 2, and 3 have expected standard deviations of returns equal to 0, 0.40, and 0.80, respectively. Their expected returns are 0.05, 0.15, and 0.25, respectively. The covariance between returns on securities 2 and 3 is 0. What are the security weights of the optimal portfolio (so as to minimize portfolio risk) with an expected return of 0.2?

NOTES

1. A proof of this result uses the Cantor diagonalization process. Such a proof can be found in a standard textbook on mathematical analysis such as Rudin (1976).

2. This follows from de Morgan's law: $\bigcap_{i=1}^{\infty} \phi_i = (\bigcup_{i=1}^{\infty} A_i^c)^c$.

3. The events $\{\phi_i\}_{i=1}^{\infty}$ are said to be *pairwise disjoint* (mutually exclusive) if for any $i \neq j$, the intersection $\phi_i \cap \phi_j = \varnothing$, the empty set.

4. Stock return is the stock's proportional profit $(S_T - S_0 + \Sigma \mathrm{div}_t)/(TS_0)$, the sum of the gain in stock price plus dividends divided by the product of the number of periods and the initial price.

5. Bayes theorem can be generalized to multiple events. Suppose that events B_1, B_2, ..., B_n form a partition of the sample space S, which means that the events are pairwise disjoint and their union is the sample space S. If A is an event then: $P[B_i|A] = \frac{P[AB_i]P[B_i]}{\sum_{j=1}^{n} P[AB_j]P[B_j]}$

6. All numbers and details provided in this illustration are fictitious.

7. A DK (from the phrase "don't know") is a trade report in the clearing process with discrepancies resulting from recording errors, misunderstanding, and fraud. DKs are sent back to traders to resolve or reconcile the trades.

8. See Exercise 2.18 at the end of the chapter for a verification of the mean and variance formulas for the uniform distribution.

9. Note that: $-\frac{1}{2}(z-\sigma)^2 + \frac{1}{2}\sigma^2 = -\frac{1}{2}z^2 - \frac{1}{2}\sigma^2 + 2\frac{1}{2}\sigma z + \frac{1}{2}\sigma^2 = \sigma z - \frac{1}{2}z^2$ such that $E[e^X] = \frac{1}{\sqrt{2\pi}}\int_{-\infty}^{\infty}$ $e^{\mu+\frac{1}{2}\sigma^2 - \frac{1}{2}(z-\sigma)^2}\,dz = e^{\mu+\frac{1}{2}\sigma^2}$ since $\frac{1}{\sqrt{2\pi}}\int_{-\infty}^{\infty} e^{-\frac{1}{2}u^2}\,du = 1$.

10. This topic is more fully developed with appropriate derivations and proofs in the statistics literature, such as in Hastings (1997).

11. See Exercise 2.33 at the end of the chapter for a derivation of this portfolio variance formula.

12. $\mathbf{V}$ is the security variance–covariance matrix, $\mathbf{w}$ is a vector of portfolio weights, and ι is a vector of 1's.

13. The gradient notation $\nabla L(\mathbf{x})$ refers to a vector of partial derivatives of L with respect to each w_i.

References

Drezner, Z., 1978. Computation of the Bivariate Normal Integral. Math. Comput. 32, 277–279.

Hastings, K.J., 1997. Probability and Statistics. Addison Wesley Longman, Reading, MA.

Rudin, W., 1976. Principles of Mathematical Analysis. third ed. McGraw-Hill, New York, NY.

Discrete Time and State Models

3.1 TIME VALUE

Financial markets enable individuals and institutions to accomplish two important things: allocate capital so that it is placed in its most valuable use over time and allocate risk so that those who are least able to bear it pass it on to those with greater risk-bearing capacity. This chapter is concerned first with allocating capital over time and later with the allocation of risk in financial markets.

Cash flows realized at the present time are worth more to investors than cash flows realized later. The present value concept offers a means to express the value of a one dollar (or other currency unit) cash flow in terms of current cash flows:

$$PV = (1 + r_{0,n})^{-n} = d_n \tag{3.1}$$

where PV is the present value of \$1 paid n equal time periods from now (time 0) with a discount rate (or interest rate or yield) of $r_{0,n}$ (sometimes shortened to r) prevailing from period 0 to period n. The discount function PV, which expresses the present value of \$1 to be received at time n, can be written as d_n. If interest compounds m times per period, present value is computed as follows:

$$PV = \left(1 + \frac{r_{0,n}}{m}\right)^{-mn} \tag{3.2}$$

As $m \to \infty$, compounding occurs continuously, and the present value function becomes:

$$PV = e^{-nr_{0,n}} \tag{3.3}$$

3.1.1 Annuities and Growing Annuities

A series of cash flows CF_t is valued as follows:

$$PV = \sum_{t=1}^{n} \frac{CF_t}{(1 + r_{0,t})^t} = \sum_{t=1}^{n} CF_t d_t \tag{3.4}$$

We will introduce annuity functions shortly to simplify the calculation process when n is large.

P.M. Knopf & J.L. Teall: Risk Neutral Pricing and Financial Mathematics: A Primer.
DOI: http://dx.doi.org/10.1016/B978-0-12-801534-6.00003-5

ILLUSTRATION OF USING CASH FLOWS TO VALUE A BOND

Suppose that a 3-year bond has a coupon rate of 2% and a face value of $1000. The discount rates prevailing from time 0 to time 1, 2, and 3, respectively, are 3%, 3.5%, and 4%, respectively. What is the price of the bond? We first find the discount functions for times 1, 2, and 3 using Eq. (3.1):

$$d_1 = (1+0.03)^{-1} = 0.9709, d_2 = (1+0.035)^{-2} = 0.9335, d_2 = (1+0.04)^{-3} = 0.889$$

Since the coupon rate is 2%, the bond will make a payment of $CF_1 = 0.02(1000) = \$20$ and $CF_2 = 0.02(1000) = \$20$ at times 1 and 2. Since the bond matures at time 3, the bond will pay $CF_3 = 0.02(1000) + 1000 = \$1020$ at time 3. The price of the bond should be the present value of the payments (cash flows) that bond holder has received, which is easily found using Eq. (3.4):

$$PV = 20(0.9709) + 20(0.9335) + 1020(0.889) = \$944.87$$

3.1.1.1 Geometric Series and Expansions

A geometric expansion is an algebraic procedure used to simplify a geometric series. Suppose we wished to solve the following finite geometric series for S:

$$S = c \sum_{t=1}^{n} x^t \tag{3.5}$$

where c is a constant and x is called a ratio. Simplifying the series might save substantial amounts of computation time. Essentially, the geometric expansion is a two-step process. First, divide both sides of the equation by the ratio:

$$Sx^{-1} = c \sum_{t=0}^{n-1} x^t \tag{3.6}$$

Next, to eliminate repetitive terms, subtract the above product from the original equation and simplify:

$$S - Sx^{-1} = cx^n - c$$
$$S = c \frac{x^n - 1}{1 - 1/x} = c \frac{1 - x^n}{1/x - 1} \tag{3.7}$$

for $x \neq 1$.

3.1.1.2 Annuities and Perpetuities

An annuity is defined as a series of identical payments made at equal intervals. We will apply the same discount rate r to all payments CF. From the geometric expansion above (finishing with Eq. 3.7), let $x = (1 + r)^{-1}$, such that the present value of an annuity is:

$$PV_A = \sum_{t=1}^{n} \frac{CF}{(1+r)^t} = CF \frac{1 - (1+r)^{-n}}{(1+r) - 1} = \frac{CF}{r} \left[1 - \frac{1}{(1+r)^n} \right] \tag{3.8}$$

As the value of n approaches infinity in the annuity formula, the value of the right-hand-side term in the brackets $1/(1+r)^n$ approaches zero. Thus, the present value of a perpetual annuity, or perpetuity is:

$$PV_p = \frac{CF}{r} \tag{3.9}$$

ILLUSTRATION OF AN ANNUITY

Find the present value of a 5-year annuity with a monthly discount rate of 0.5% and monthly payments of \$300. In this case, the total number of payments $n = 12(5) = 60$, CF = 300, and $r = 0.005$. The present value will be:

$$PV_A = \frac{300}{0.005} \left[1 - \frac{1}{(1.005)^{60}} \right] = \$15,518$$

Consider that these \$300 monthly payments might be used to pay off a \$15,518 automobile loan.

3.1.1.3 Growing Annuities and Perpetuities

If an annuity's cash flow is CF_1 in year 1, and grows each year at a constant rate g over the prior year's cash flow such that the cash flow in any year t is $CF_{t-1}(1+g)$, the present value growing annuity model can be derived based on Eq. (3.7) as follows:[1]

$$PV_{GA} = \sum_{t=1}^{n} \frac{CF_1(1+g)^{t-1}}{(1+r)^t} = \frac{CF_1}{1+g} \sum_{t=1}^{n} \left(\frac{1+g}{1+r} \right)^t$$

$$= \frac{CF_1}{1+g} \frac{1 - \left(\frac{1+g}{1+r} \right)^n}{\frac{1+r}{1+g} - 1} = \frac{CF_1}{r-g} \left[1 - \frac{(1+g)^n}{(1+r)^n} \right] \tag{3.10}$$

where $r > g$, the growing perpetuity model is simply:

$$PV_{gp} = \frac{CF_1}{r-g} \tag{3.11}$$

3.1.2 Coupon Bonds and Yield Curves

Thus far, we have assumed that discount and interest rates are equal for all periods, along with all bond yields. This means that the *yield curve*, which depicts the yield to maturity of zero-coupon bonds with respect to their terms to maturity, is flat.[2] But, interest rates do change over time, sometimes very predictably, and long-term rates frequently exceed short-term rates. Now, we will allow for yields to vary over terms to maturity, and express long-term interest rates as functions of short-term rates. We will distinguish between yields or rates on instruments originating at time 0 or now (spot rates, on instruments originating now) and yields or rates on instruments originating in the future (forward rates, on instruments to be originated in the future at rates locked in now). More specifically, we will argue that long-term interest rates are a geometric mean of a series of short-term spot rates and forward rates. The compounding effect of interest rates leads to long-term rates being calculated based on geometric rather than arithmetic means. More specifically, the long-term spot rate will be expressed as a geometric mean of short-term spot and forward interest rates.

3.1.2.1 The Term Structure of Interest Rates

The *term structure of interest rates* is concerned with how yields and interest rates vary with respect to dates of maturity. The *pure expectations theory* defines the relationship between long- and short-term interest rates as follows, where $r_{0,n}$ is the rate on an instrument originated at time 0 and repaid at time n:

$$r_{0,n} = \sqrt[n]{\prod_{t=1}^{n}(1 + r_{t-1,t})} - 1 \tag{3.12}$$

ILLUSTRATION: TERM STRUCTURE

Consider an example where the 1-year spot rate $r_{0,1}$ is 2%. Investors are expecting that the 1-year spot rate 1 year from now will increase to 3% such that the 1-year forward rate $r_{1,2}$ on loans originated in 1 year is 3%. Further suppose that investors are expecting that the 1-year spot rate 2 years from now will increase to 5%; thus, the 1-year forward rate $r_{2,3}$ on a loan originated in 2 years is 5%. Based on the pure expectations hypothesis, the 3-year spot rate is calculated as follows:

$$r_{0,3} = \sqrt[n]{\prod_{t=1}^{n}(1 + r_{t-1,t})} - 1 = \sqrt[3]{(1 + r_{0,1})(1 + r_{1,2})(1 + r_{2,3})} - 1 = \sqrt[3]{(1 + 0.02)(1 + 0.03)(1 + 0.05)} - 1 = 0.033258$$

3.1.2.2 Term Structure Estimation with Coupon Bonds

In this section, we will infer points on a yield curve from the prices of bonds trading in the market. For example, if a single 1-year, $1000 face value zero-coupon bond (a bond that makes no coupon payments) trades in the market, with price $P_0 = \$950$, we can infer that $d_1 = 0.95$ and $r_{0,1} = 0.0526$. We will require additional bonds with an appropriate combination of terms to maturity to infer other yield curve points.

Now, consider a slightly more complex situation. Suppose a 2-year, 2% coupon, $1000 face value bond A currently sells for $987. A second bond B, maturing in 2 years, 5% coupon, $1000 face value currently sells for $1045. Now, we let d_1 and d_2 denote the discount functions from time 0 to times 1 and 2, respectively. We can use the present value Eq. (3.4) to construct the system of equations we need to solve for d_1 and d_2. The prices of bonds A and B will satisfy the following system of equations:

$$987 = 20d_1 + 1020d_2$$
$$1045 = 50d_1 + 1050d_2$$

This system can be expressed as the matrix equation:

$$\begin{bmatrix} 987 \\ 1045 \end{bmatrix} = \begin{bmatrix} 20 & 1020 \\ 50 & 1050 \end{bmatrix} \begin{bmatrix} d_1 \\ d_2 \end{bmatrix} \qquad (3.13)$$

$$\mathbf{b_0} \quad = \quad \mathbf{CF} \quad \times \quad \mathbf{d}$$

Note in Eq. (3.13) that $\mathbf{b_0}$ is the vector of bond prices, $\mathbf{CF}$ is the matrix of cash flows at times 1 and 2, and $\mathbf{d}$ is the vector of discount functions. As discussed in Chapter 2, as long as the matrix $\mathbf{CF}$ is invertible, then we can solve for $\mathbf{d}$: $\mathbf{d} = \mathbf{CF}^{-1} \times \mathbf{b_0}$. In this case the discount functions are $d_1 = 0.985$ and $d_2 = 0.948333$, from which it follows that the discount rates are $r_{0,1} = d_1^{-1} - 1 = 0.0152$ and $r_{0,2} = d_2^{-1/2} - 1 = 0.0269$. This illustration can be generalized to any number of time periods.

Now, we will examine T bonds over T years. Matrix $\mathbf{CF}$ below represents cash flows anticipated in each of T years from the T bonds. Bond i cash flows appear in row i of matrix $\mathbf{CF}$; time t cash flows appear in column t. The prices of these bonds are given by vector $\mathbf{b_0}$. We will calculate from these two matrices vector $\mathbf{d}$, which represents discount functions (the value of \$1 to be paid) for each of these T years.

$$\begin{bmatrix} B_{1,0} \\ B_{2,0} \\ \vdots \\ B_{T,0} \end{bmatrix} = \begin{bmatrix} CF_{1,1} & CF_{1,2} & \cdots & CF_{1,T} \\ CF_{2,1} & CF_{2,2} & \cdots & CF_{2,T} \\ \vdots & \vdots & \ddots & \vdots \\ CF_{T,1} & CF_{T,2} & \cdots & CF_{T,T} \end{bmatrix} \begin{bmatrix} d_1 \\ d_2 \\ \vdots \\ d_T \end{bmatrix} \qquad (3.14)$$

$$\mathbf{b_0} \quad = \quad \mathbf{CF} \quad \times \quad \mathbf{d}$$

Double subscripts refer to bond names (e.g., bond 1) and time periods (e.g., time 0). For example, $B_{1,0}$ is the initial market price of bond 1 and $CF_{2,1}$ is the first year cash flow from bond 2. In addition, d_T is the Tth-year discount function. We will solve for the vector $\mathbf{d}$ of discount functions from the vector of prices ($\mathbf{b_0}$) of the T selected already priced bonds and their matrix of cash flows ($\mathbf{CF}$). Once we have found $\mathbf{d}$, we can then obtain interest rates, yield curves, and price any other bond that may enter the market in the T-year time frame. In order to obtain a unique solution for $\mathbf{d}$, we need to have each row vector of the cash flow matrix to be independent of one another. This is required because the inverse $\mathbf{CF}^{-1}$ of the matrix $\mathbf{CF}$ exists if and only if its row vectors are linearly independent. In that case, the solution for the vector of discount functions is given by:

$$\mathbf{d} = \mathbf{CF}^{-1} \times \mathbf{b_0} \qquad (3.15)$$

Any additional bonds entering the market can be priced based on the values of $\mathbf{d}$. The financial interpretation of the cash flow vectors being linearly independent is that the market satisfies the important requirement that bond markets are complete. If we were unable to find bonds that produced a set of T linearly independent cash flows, there would be insufficient information in the bond market in order to uniquely determine the discount functions, and hence the interest rates. Such markets are said to be incomplete.

TABLE 3.1 Coupon Bonds 1, 2, and 3

Bond	Current price	Face value	Coupon rate	Years to maturity
1	992	1000	0.01	2
2	996	1000	0.02	3
3	1025	1000	0.03	3

ILLUSTRATION: COUPON BONDS AND THE YIELD CURVE

Consider an illustration involving three bonds whose current prices, coupon rates, and terms to maturity are given in Table 3.1. Market spot rates will be determined simultaneously. Since d_t is the discount function for time t and equals $1/(1 + r_{0,t})^t$, $r_{0,t}$ is the spot rate or discount rate that equates the present value of a bond with its current price. Thus, using Eq. (3.14), we obtain spot rates as follows:

$$\underbrace{\begin{bmatrix} 10 & 1010 & 0 \\ 20 & 20 & 1020 \\ 30 & 30 & 1030 \end{bmatrix}}_{\mathbf{CF}} \times \underbrace{\begin{bmatrix} d_1 \\ d_2 \\ d_3 \end{bmatrix}}_{\mathbf{d}} = \underbrace{\begin{bmatrix} 992 \\ 996 \\ 1025 \end{bmatrix}}_{\mathbf{b_0}}$$

To solve this system we first invert matrix **CF**, then use this inverse to premultiply vector $\mathbf{b_0}$ to obtain vector **d**:

$$\underbrace{\begin{bmatrix} d_1 \\ d_2 \\ d_3 \end{bmatrix}}_{\mathbf{d}} = \underbrace{\begin{bmatrix} -0.001 & -0.10403 & 0.10302 \\ 0.001 & 0.00103 & -0.00102 \\ 0 & 0.003 & -0.002 \end{bmatrix}}_{\mathbf{CF}^{-1}} \times \underbrace{\begin{bmatrix} 992 \\ 996 \\ 1025 \end{bmatrix}}_{\mathbf{b_0}} = \begin{bmatrix} 0.98962 \\ 0.97238 \\ 0.938 \end{bmatrix}$$

Thus, $d_1 = 0.98962$, $d_2 = 0.97238$, and $d_3 = 0.938$. Since $d_t = 1/(1 + r_{0,t})^t$, $1/d_t = (1 + r_{0,t})^t$, and $r_{0,t} = 1/d^{1/t} - 1$, spot rates are determined as follows:

$$\frac{1}{d_1} - 1 = 0.010489 = r_{0,1}$$

$$\frac{1}{d_2^{\frac{1}{2}}} - 1 = 0.014103 = r_{0,2}$$

$$\frac{1}{d_3^{\frac{1}{3}}} - 1 = 0.021564 = r_{0,3}$$

ILLUSTRATION: PRICING BONDS

In an arbitrage-free market, any other bond with cash flows paid at the ends of some combination of years 1, 2, and 3 must have a market price that is consistent with these three spot rates. For example, a 3-year 4% coupon bond 4 must be valued as follows:

$$B_{4,0} = 40d_1 + 40d_2 + 1040d_3 = \frac{40}{(1 + 0.010489)^1} + \frac{40}{(1 + 0.014103)^2} + \frac{1040}{(1 + 0.021564)^3} = 1054$$

Any other market price for bond 4 will lead to an arbitrage opportunity. We will detail this issue shortly. In addition, prices and cash flows from any combination of three of these bonds above will be consistent with the following forward rates:

$$r_{1,2} = \frac{(1+r_{0,2})^2}{(1+r_{0,1})} - 1 = \frac{(1+0.014103)^2}{(1+0.014089)} - 1 = 0.01773$$

$$r_{2,3} = \frac{(1+r_{0,3})^3}{(1+r_{0,1})(1+r_{1,2})} - 1 = \frac{(1+0.021564)^3}{(1+0.014089)(1+0.01773)} - 1 = 0.036652$$

$$r_{1,3} = \sqrt{\frac{(1+r_{0,3})^3}{(1+r_{0,1})}} - 1 = \sqrt{\frac{(1+0.021564)^3}{(1+0.010489)}} - 1 = 0.027147$$

3.2 DISCRETE TIME MODELS

3.2.1 Arbitrage and No Arbitrage

As we discussed in Chapter 1, *arbitrage* is the simultaneous purchase and sale of instruments producing identical cash flows. The most easily arbitraged financial instruments are those with guaranteed payments or with payments that are perfectly correlated with other instruments. We will focus on riskless bonds here, though the principles that we discuss can be extended to other securities. The key components of arbitrage are that:

1. Arbitrage is riskless. All cash flows, including transactions prices in the market are known.
2. Arbitrage will never produce a negative cash flow in any time period or under any outcome.

An arbitrage opportunity fulfills the above conditions and produces at least one positive cash flow in at least one period and/or outcome. In a perfect market, arbitrage opportunities do not exist since rational and greedy investors will never price securities such that they produce an arbitrage opportunity for a competitor at their own expense. No-arbitrage conditions are used to price securities relative to one another such that they do not produce such an arbitrage opportunity.

3.2.1.1 No-Arbitrage Bond Markets

Consider a frictionless market comprised of n riskless securities i with payoffs $CF_{i,t}$ over $T+1$ periods t, including time $t = 0$ (when investment transactions take place). $CF_{i,0}$ might be taken to be the initial transactions price (purchase or sale price) for security i, which would be a negative value in the event of a purchase. The matrix $\widehat{CF}$ is defined to the matrix of cash flows from time 0 to time T for n securities (where n equals or exceeds T in that market) with payoffs $CF_{i,t}$.[3] Thus, the left column of matrix $\widehat{CF}$ below represents the current transactions prices $CF_{i,0}$ ($-B_{i,0}$ for purchase or $B_{i,0}$ for sale) for the n securities and the first element in vector $\hat{d}$ is 1; the time 0 cash flow is not discounted. The remaining elements in $\hat{d}, d_1, d_2, \ldots, d_T$ are the discount functions that are obtained from Eq. (3.15) for T already priced bonds with linearly independent payoff vectors. Let γ_i represent the commitment made to a given investment i; that is, γ_i is the number of units of a particular security holding. A no-arbitrage (arbitrage-free) market exists where for each and every possible portfolio strategy γ^T:

$$
[\gamma_1 \quad \gamma_2 \quad \cdots \quad \gamma_n]
\begin{bmatrix}
CF_{1,0} & CF_{1,1} & \cdots & CF_{1,T} \\
CF_{2,0} & CF_{2,1} & \cdots & CF_{2,T} \\
\vdots & \vdots & \ddots & \vdots \\
CF_{n,0} & CF_{n,1} & \cdots & CF_{n,T}
\end{bmatrix}
\begin{bmatrix}
1 \\
d_1 \\
d_2 \\
\vdots \\
d_T
\end{bmatrix}
= 0
\tag{3.16}
$$

$$
\gamma^T \qquad \times \qquad \widehat{CF} \qquad \times \qquad \hat{d} \quad = \quad 0
$$

Thus, there is no way to combine securities in a riskless profit-producing portfolio when this equality holds for all possible combinations of γ_i. A nonzero value for this product would imply an arbitrage opportunity (a nonzero NPV portfolio), either through purchases, short sales, or some combination thereof.[4] When no-arbitrage conditions apply in this market, all other securities that might exist in this T-period payoff space can be priced if:

1. n, the number of securities in the market, is no smaller than T, the relevant number of payoff time periods, and
2. there is a subset of n securities whose payoff vectors span the T-period payoff space.

ILLUSTRATION: NO-ARBITRAGE BOND MARKETS

Consider the following three-bond, two-time period scenario represented as follows:

$$
[\gamma_1 \quad \gamma_2 \quad \gamma_3]
\begin{bmatrix}
-950 & 1000 & 0 \\
-900 & 0 & 1000 \\
-1085 & 100 & 1100
\end{bmatrix}
\begin{bmatrix}
1 \\
0.95 \\
0.90
\end{bmatrix}
= 0
$$

$$
\gamma^T \qquad \times \qquad \widehat{CF} \qquad \times \qquad \hat{d} \quad = \quad 0
$$

The matrices represent 1- , 2-, and 3-year zero-coupon instruments that sell for 900, 950, and 1085, respectively. Because $\widehat{CF} \times \hat{d} = [0,0,0]^T$, there is no portfolio strategy γ^T that can produce a nonzero arbitrage profit.[5] More generally, as long as $\gamma^T \times \widehat{CF} \times \hat{d} = 0$, the market is considered to be no-arbitrage. Therefore, no-arbitrage opportunities exist in this simple three-bond market.

3.2.1.2 Pricing Bonds

Now, we use an arbitrage strategy to price a bond directly from bonds whose prices we already know in our no-arbitrage market. In this no-arbitrage market, each bond outside this T-security subset ($i \notin \{1,2,\ldots,T\}$), but whose payoff vector remains in the T-period payoff space, can be priced as a linear combination of the T-security subset. That is, the cash flow structure of bond i can be replicated by some portfolio γ of bonds in subset $\{1,2,\ldots,T\}$ from the market:

$$
\begin{bmatrix}
CF_{i,1} \\
CF_{i,2} \\
\vdots \\
CF_{i,T}
\end{bmatrix}
=
\begin{bmatrix}
CF_{1,1} & CF_{2,1} & \cdots & CF_{T,1} \\
CF_{1,2} & CF_{2,2} & \cdots & CF_{T,2} \\
\vdots & \vdots & \ddots & \vdots \\
CF_{1,T} & CF_{2,T} & \cdots & CF_{T,T}
\end{bmatrix}
\begin{bmatrix}
\gamma_1 \\
\gamma_2 \\
\vdots \\
\gamma_T
\end{bmatrix}
\tag{3.17}
$$

$$
\mathbf{cf_i} \quad = \qquad \qquad CF^T \qquad \qquad \times \qquad \gamma
$$

Note that the vector $\mathbf{cf_i}$ is $T \times 1$ and consists of the cash flows for bond i from time 1 to T. Avoidance of arbitrage opportunities means that if the cash flow structure of bond $i \notin \{1,2,\ldots,T\}$ can be expressed as a linear combination of the cash flows of the bonds in no-arbitrage market subset $\{1,2,\ldots,T\}$, then its price must be a linear combination of the prices of bonds in the no-arbitrage market subset $\{1,2,\ldots,T\}$:

$$
B_{i,0} = \begin{bmatrix} \gamma_1 & \gamma_2 & \cdots & \gamma_T \end{bmatrix} \begin{bmatrix} B_{1,0} \\ B_{2,0} \\ \vdots \\ B_{T,0} \end{bmatrix} \tag{3.18}
$$

$$
B_{i,0} = \gamma^T \times \mathbf{b_0}
$$

where

$$
\gamma = (\mathbf{CF}^T)^{-1} \mathbf{cf_i} \tag{3.19}
$$

We create a portfolio $\gamma^T = \begin{bmatrix} \gamma_1 & \gamma_2 & \cdots & \gamma_T \end{bmatrix}$ of the T already priced bonds so that the payoffs of the priced bonds replicate the payoffs for bond i for each time period. Equations (3.18) and (3.19) merely state that the portfolio γ of bonds $\{1,2,\ldots,T\}$ that replicates bond i payoffs for times 1 through T (rows 1 through T of matrix $\mathbf{CF}^T$ in Eq. (3.17)) should be the same portfolio γ of bonds $\{1,2,\ldots,T\}$ that replicates the time 0 price of bond i. This result is assured by arbitrage-free pricing of bonds.

ILLUSTRATION: PRICING BONDS

Here, we continue our previous illustration with just the 1- and 2-year zero-coupon bonds priced at 900 and 950, respectively. Suppose that a new 2-year bond 3, with a 10% coupon rate, enters the market. We solve for γ to replicate the cash flow structure of bond 3 as follows:

$$
\begin{bmatrix} 100 \\ 1100 \end{bmatrix} = \begin{bmatrix} 1000 & 0 \\ 0 & 1000 \end{bmatrix} \begin{bmatrix} \gamma_1 \\ \gamma_2 \end{bmatrix}
$$

$$
\mathbf{cf_3} = \mathbf{CF}^T \times \gamma
$$

$$
\begin{bmatrix} \gamma_1 \\ \gamma_2 \end{bmatrix} = \begin{bmatrix} 0.001 & 0 \\ 0 & 0.001 \end{bmatrix} \begin{bmatrix} 100 \\ 1100 \end{bmatrix} = \begin{bmatrix} 0.1 \\ 1.1 \end{bmatrix}
$$

$$
\gamma = (\mathbf{CF}^T)^{-1} \times \mathbf{cf_3}
$$

Thus, bond 3 is replicated by a portfolio comprising 0.1 bonds 1 and 1.1 bonds 2. Next, we price bond 3 as follows:

$$
B_{3,0} = \begin{bmatrix} 0.1 & 1.1 \end{bmatrix} \begin{bmatrix} 950 \\ 900 \end{bmatrix} = 1085
$$

$$
B_{3,0} = \gamma^T \times \mathbf{b_0}
$$

3.2.1.3 The Pricing Kernel

The pricing kernel tells the price or value of each dollar to be received in each relevant year t, based on market prices of traded bonds. We define the *pricing kernel* to be vector $\hat{\mathbf{d}}$ such that $\widehat{\mathbf{CF}} \times \hat{\mathbf{d}} = 0$:

$$
\begin{bmatrix}
CF_{1,0} & CF_{1,1} & \cdots & CF_{1,T} \\
CF_{2,0} & CF_{2,1} & \cdots & CF_{2,T} \\
\vdots & \vdots & \ddots & \vdots \\
CF_{n,0} & CF_{n,1} & \cdots & CF_{n,T}
\end{bmatrix}
\begin{bmatrix}
1 \\
d_1 \\
d_2 \\
\vdots \\
d_T
\end{bmatrix}
= 0
\tag{3.20}
$$

$$
\underbrace{\qquad}_{\widehat{\mathbf{CF}}} \quad \times \quad \underbrace{\quad}_{\hat{\mathbf{d}}} \quad = \quad 0
$$

Observe that the ith row of the vector $\widehat{\mathbf{CF}} \times \hat{\mathbf{d}}$ is the result of $\gamma^{\mathsf{T}} \times \widehat{\mathbf{CF}} \times \hat{\mathbf{d}}$ for the case $\gamma^{\mathsf{T}} = (0,\ldots,0,1,0,\ldots,0)$ where the value 1 appears in the ith entry of the vector γ^{T}. This has the effect of assigning a unit purchase to the ith asset.

If asset purchases (sales) are made at time 0, asset purchase (sales) prices are negatives (positives) of the cash flows $CF_{i,0}$ at time 0. All securities with payoff vectors in this space can be priced with this pricing kernel. Elements of the pricing kernel are the market's marginal rates of substitution for cash flows at time 0 relative to those at time t; that is, the pricing kernel consists of discount functions described by the bond market's yield curve.

3.2.2 Arbitrage with Riskless Bonds

Consider a set of riskless bonds with coupon payments and redemption values as provided in Table 3.2, all making payments at year-ends until they mature. The cash flow structure of any 1-, 2-, or 3-year bond (e.g., bond D) added to the market can be replicated with some portfolio of bonds A, B, and C, as long it also makes payments at year-end. For example, assume that there now exists bond D, a 3-year, 20% coupon bond selling in this market for $1360. This bond will make payments of $200 in years 1 and 2 in addition to a $1200 payment in year 3. We will demonstrate that a portfolio of bonds A, B, and C can be comprised to generate the same cash flow series as bond D, and that the price of bond D is inconsistent with a no-arbitrage market.

TABLE 3.2   Coupon Bonds A, B, and C

Bond	Current price	Face value	Coupon rate	Years to maturity
A	1000	1000	0.04	2
B	1055.5	1000	0.06	3
C	889	1000	0	3

3.2.2.1 *Replicating the Future Cash Flow Structure of Bond D*

We will demonstrate that bond D can be replicated by a portfolio of our first three bonds in the following quantities: $\gamma_A = 0$, $\gamma_B = 3\frac{1}{3}$, and $\gamma_C = -2\frac{1}{3}$, where γ_A, γ_B, and γ_C are bond portfolio replicating strategies, which are determined as follows:

$$\begin{bmatrix} 200 \\ 200 \\ 1200 \end{bmatrix} = \begin{bmatrix} 40 & 60 & 0 \\ 1040 & 60 & 0 \\ 0 & 1060 & 1000 \end{bmatrix} \begin{bmatrix} \gamma_A \\ \gamma_B \\ \gamma_C \end{bmatrix}$$

$$\mathbf{cf_d} \qquad = \qquad \mathbf{CF^T} \qquad \qquad \gamma$$

To solve this system we first invert matrix $\mathbf{CF^T}$, then use it to premultiply vector $\mathbf{cf_d}$ to obtain vector γ:

$$\begin{bmatrix} 0 \\ 3.333333 \\ -2.333333 \end{bmatrix} = \begin{bmatrix} \gamma_A \\ \gamma_B \\ \gamma_C \end{bmatrix} = \begin{bmatrix} -0.001 & 0.001 & 0 \\ 0.0173333 & -0.00066670 & 0 \\ -0.018373 & 0.00070667 & 0.001 \end{bmatrix} \begin{bmatrix} 200 \\ 200 \\ 1200 \end{bmatrix}$$

$$\gamma \qquad = \qquad (\mathbf{CF^T})^{-1} \qquad\qquad \mathbf{cf_d}$$

3.2.2.2 *Creating the Arbitrage Portfolio*

Cash flows, starting with time 0 generated by the four-bond arbitrage portfolio, are represented in the following system:

$$\begin{bmatrix} -1000 & -1055.5 & -889 & -1360 \\ 40 & 60 & 0 & 200 \\ 1040 & 60 & 0 & 200 \\ 0 & 1060 & 1000 & 1200 \end{bmatrix} \begin{bmatrix} 0 \\ -3.33333 \\ 2.33333 \\ 1 \end{bmatrix} = \begin{bmatrix} 84 \\ 0 \\ 0 \\ 0 \end{bmatrix}$$

Thus, by buying bonds C and D at quantities $2\frac{1}{3}$ and 1, respectively, and shorting bond B in quantity $3\frac{1}{3}$, the arbitrageur locks in a profit of 84 in time 0, and zero profits in the subsequent 3 years. The remaining 0's in the right-hand vector reaffirm that we have created an arbitrage portfolio with 0 net investment for each of the time periods 1, 2, and 3. The arbitrage portfolio is riskless because all bonds are presumed to be default risk-free and are presumed to be held to maturity, thereby eliminating interest rate and liquidity risk. Obviously, this bond market is not in equilibrium.

ILLUSTRATION: OBTAINING THE PRICING KERNEL

The time 0 value of the portfolio consisting of -3.3333 units of bond B, 2.3333 units of bond C, and one bond D produces a profit equal to 84. Thus, the market could reach equilibrium if the initial price of one or some combination of these bonds changes to reduce the time 0 cash flow of the portfolio by 84. For example, the price of bond D could drop by 84, C could drop by $84/2.33333 = 36$, or B could increase by 25.2 to increase the initial portfolio investment by 84. Instead, combinations of bond prices could change to increase this initial investment by 84. Suppose, for example, that market equilibrium is reached by the market bidding up the price of bond D by 84 to 1444. When the market is in equilibrium, a pricing kernel can be produced. We define the pricing kernel for this market to be vector $\hat{\mathbf{d}}$ such that $\widehat{\mathbf{CF}} \times \hat{\mathbf{d}} = 0$:

$$
\begin{bmatrix}
-1000 & 40 & 1040 & 0 \\
-1055.5 & 60 & 60 & 1060 \\
-889 & 0 & 0 & 1000 \\
-1360 & 200 & 200 & 1200
\end{bmatrix}
\begin{bmatrix}
1 \\
d_1 \\
d_2 \\
d_3
\end{bmatrix}
=
\begin{bmatrix}
0 \\
0 \\
0 \\
0
\end{bmatrix}
$$

$$
\widehat{\mathbf{CF}} \qquad\qquad \times \quad \hat{\mathbf{d}} \quad = \quad 0
$$

This system is rearranged and rewritten to reduce to Eq. (3.14), which is then used to find the discount functions:

$$
\begin{bmatrix}
40 & 1040 & 0 \\
60 & 60 & 1060 \\
0 & 0 & 1000
\end{bmatrix}
\begin{bmatrix}
d_1 \\
d_2 \\
d_3
\end{bmatrix}
=
\begin{bmatrix}
1000 \\
1055.5 \\
889
\end{bmatrix}
$$

$$
\mathbf{CF} \qquad\qquad \times \quad \mathbf{d} \quad = \quad \mathbf{b}_0
$$

$$
\begin{bmatrix}
40 & 1040 & 0 \\
60 & 60 & 1060 \\
0 & 0 & 1000
\end{bmatrix}^{-1}
\begin{bmatrix}
1000 \\
1055.5 \\
889
\end{bmatrix}
=
\begin{bmatrix}
d_1 \\
d_2 \\
d_3
\end{bmatrix}
$$

$$
\begin{bmatrix}
-0.001 & 0.017333 & -0.1837 \\
0.001 & -0.00067 & 0.000707 \\
0 & 0 & 0.001
\end{bmatrix}
\begin{bmatrix}
1000 \\
1055.5 \\
889
\end{bmatrix}
=
\begin{bmatrix}
d_1 \\
d_2 \\
d_3
\end{bmatrix}
=
\begin{bmatrix}
0.96144 \\
0.92456 \\
0.889
\end{bmatrix}
$$

$$
\mathbf{CF}^{-1} \qquad\qquad \times \quad \mathbf{b}_0 \quad = \quad \mathbf{d}
$$

Thus, $\hat{\mathbf{d}} = [1, 0.96144, 0.92456, 0.889]^{\mathsf{T}}$ is the pricing kernel for this illustration. Note that since the payoff structure of any one of these four bonds is a linear combination of the other three, we can obtain the pricing kernel with any combination of three, as long as one of the three remaining payoff structures is not a linear combination of the other two. We can combine bonds A, B, and C as we did here to obtain the pricing kernel, A, B, and D, but not bonds B, C, and D.

3.3 DISCRETE STATE MODELS

So far, we have worked with riskless securities in single state, multiple period frameworks. Now, we will examine multiple state risky markets in a single period.

3.3.1 Outcomes, Payoffs, and Pure Securities

Investors can value securities by valuing them as functions of the known prices of other securities that have already been valued. This process can involve determining the vector space $\mathbb{R}^n$, valuing n

"control" securities with linearly independent payoff vectors, and pricing the payoff vectors of the previously unpriced securities based on linear combinations of prices from the "control" securities. We will initially assume the following for our valuations:

1. There exist n potential *states of nature* (prices) in a one-time period framework.
2. Each security will have exactly one payoff resulting from each potential state of nature.
3. Only one state of nature will occur at the end of the period (states are mutually exclusive) and which state occurs is ex-ante unknown.
4. Each investor's utility or satisfaction is a function only of his level of wealth; the state of nature that is realized is important only to the extent that the investor's wealth is affected (this assumption can often be relaxed).
5. Capital markets are in equilibrium (supply equals demand) for all securities.

3.3.1.1 *Payoff Vectors and Pure Securities*

Suppose that a given economy has n potential states of nature and there exists a security x with a known payoff vector:

$$\mathbf{x} = \begin{bmatrix} x_1 \\ x_2 \\ \vdots \\ x_n \end{bmatrix}$$

Vector $\mathbf{x}$ defines every potential payoff for security x in this n-state world. We will value this security based on known values of other securities existing in this three-state economy.

The first step in the evaluation procedure is to decompose the security into an imaginary portfolio of *pure securities*. Define a pure security (also known as an elementary, primitive, or Arrow–Debreu security) to be an investment that pays \$1 if and only if a particular outcome or state of nature is realized and nothing otherwise. Thus, the payoff vector for a given pure security i in an n-potential outcome economy will comprise n elements:

1. The ith element will equal 1.
2. All other elements will be zero.

The following is the payoff vector of pure security 3 in an n-outcome economy:

$$\mathbf{e}_3 = \begin{bmatrix} 0 \\ 0 \\ 1 \\ \vdots \\ 0 \end{bmatrix}$$

Each security with a payoff vector in this n-dimensional payoff space can be replicated with the n pure securities from this space.

ILLUSTRATION: PAYOFF VECTORS AND PURE SECURITIES

Suppose that an economy has three potential states of nature and there is a security x with the following payoff vector:

$$x = \begin{bmatrix} 8 \\ 4 \\ 1 \end{bmatrix}$$

Vector x defines every potential payoff in this three-state world. Note that by expressing a "real" security payoff vector in terms of the pure securities, it tells us immediately the future value of the security given any particular outcome. In this example, security x will pay 8, 4, or 1, respectively, if and only if outcome 1, 2, or 3, respectively, occurs. We will value this security based on known values of other securities existing in this three-state economy.

The first step in the evaluation procedure is to decompose the security into a portfolio of pure securities. For example, the payoff vector for pure security 2 in a three-potential-outcome world is given as follows:

$$e_2 = \begin{bmatrix} 0 \\ 1 \\ 0 \end{bmatrix}$$

Pure security 2 will pay 1 if outcome 2 is realized; otherwise, it will pay zero. Payoff vectors for three pure securities will span the vector space for a three-outcome economy:

$$e_1 = \begin{bmatrix} 1 \\ 0 \\ 0 \end{bmatrix} \quad e_2 = \begin{bmatrix} 0 \\ 1 \\ 0 \end{bmatrix} \quad e_3 = \begin{bmatrix} 0 \\ 0 \\ 1 \end{bmatrix}$$

The three-dimensional vector space is spanned by three vectors if any vector in that space can be defined in terms of a linear combination of those three vectors. The payoff vector for any security existing in this three-outcome world is a linear combination of the payoff vectors of the three pure securities spanning the three-element vector space. For example, security x is a linear combination of pure securities 1, 2, and 3:

$$\begin{bmatrix} 8 \\ 4 \\ 1 \end{bmatrix} = 8 \cdot \begin{bmatrix} 1 \\ 0 \\ 0 \end{bmatrix} + 4 \cdot \begin{bmatrix} 0 \\ 1 \\ 0 \end{bmatrix} + 1 \cdot \begin{bmatrix} 0 \\ 0 \\ 1 \end{bmatrix}$$

$$x \quad = \quad 8 \times e_1 \quad + \quad 4 \times e_2 \quad + \quad 1 \times e_3$$

We are able to evaluate security x easily if we know values of the three pure securities. Suppose just for the moment that the value of pure security 1 is $\psi_1 = 0.362$, which suggests that an investor is willing to pay 0.362 for a security that pays 1 if and only if outcome 1 is realized. Furthermore, suppose that pure security 2 has a value of 0.523 and pure security 3 has a value of 0.015. The value of security x is determined as follows: $S_x = 8\psi_1 + 4\psi_2 + 1\psi_3 = (8 \times 0.362) + (4 \times 0.523) + (1 \times 0.015) = 5$. Next, we discuss how we value pure securities.

3.3.1.2 *Spanning and Complete Markets*

The n-dimensional payoff space for a market is spanned when the payoff vectors for a set or subset of n securities is linearly independent. Complete capital markets exist when these n securities have known prices. In complete capital markets, the value of any other security, real or synthetic, with a payoff vector in this n-state space is calculated based on values of the original n securities. With the known prices of n traded securities with linearly independent payoff vectors, we can determine values of each of n pure securities.

ILLUSTRATION: SPANNING AND COMPLETE MARKETS

A complete capital market in a three-potential outcome economy occurs when a set of three securities with known prices spans the state space. If the values of each of these three securities are known, the value of any other security in this three-potential state world can be determined based on values of the original three securities. Suppose, for example, securities x, y, and f exist in a three-potential outcome world and have the following payoff vectors:

$$\mathbf{x} = \begin{bmatrix} 8 \\ 4 \\ 1 \end{bmatrix} \quad \mathbf{y} = \begin{bmatrix} 2 \\ 10 \\ 3 \end{bmatrix} \quad \mathbf{f} = \begin{bmatrix} 1 \\ 1 \\ 1 \end{bmatrix}$$

This set of three payoff vectors is linearly independent. Thus, $\mathbf{x}$, $\mathbf{y}$, and $\mathbf{f}$ span the three-dimensional vector space and, if x, y, and f are priced, there exists a complete capital market in this economy. The payoff vector for every other security in this economy is a linear combination of the payoff vectors for securities x, y, and f. This implies that we can decompose the payoff vectors of securities x, y, and f into the following component pure security payoff vectors:

$$\mathbf{e}_1 = \begin{bmatrix} 1 \\ 0 \\ 0 \end{bmatrix} \quad \mathbf{e}_2 = \begin{bmatrix} 0 \\ 1 \\ 0 \end{bmatrix} \quad \mathbf{e}_3 = \begin{bmatrix} 0 \\ 0 \\ 1 \end{bmatrix}$$

where security x replicates a portfolio comprising eight of pure security 1, four of pure security 2, and one of pure security 3. Similarly, security y replicates a portfolio of two of pure security 1, 10 of pure security 2, and so on. Suppose that security x has a market value of 5, security y has a market value of 6, and f has a value of 0.9. In the absence of arbitrage opportunities, portfolios of pure securities replicating the cash flow structures of x, y, and f should also have values of 5, 6, and 0.9:[6]

$$\begin{matrix} \begin{bmatrix} 5 \\ 6 \\ 0.9 \end{bmatrix} & = & \begin{bmatrix} 8 & 4 & 1 \\ 2 & 10 & 3 \\ 1 & 1 & 1 \end{bmatrix} & \begin{bmatrix} \psi_1 \\ \psi_2 \\ \psi_3 \end{bmatrix} \\ \mathbf{S} & = & \mathbf{CF} & \psi \end{matrix} \qquad (3.21)$$

More generally, suppose we are given n priced securities, with future cash flows given by $\mathbf{CF}$ and prices given by vector $\mathbf{S}$. Where the set of security payoff vectors is linearly independent and spans the n-outcome payoff space, we can find the n pure security prices, expressed by the vector ψ, by solving the system:

$$\mathbf{S} = \mathbf{CF} \times \psi \qquad (3.22)$$

The solution is:

$$\psi = \mathbf{CF}^{-1}\mathbf{S} \tag{3.23}$$

There is a very close connection between pricing bonds based on payoffs in the previous section and pricing securities in this section based on pure security payoffs. Observe that Eq. (3.22) is analogous to Eq. (3.14), and Eq. (3.23) is analogous to Eq. (3.15).

For our example in (3.21), we obtain $\psi_1 = 0.362$, $\psi_2 = 0.523$, and $\psi_3 = 0.015$.

Now, consider a fourth security z with the following payoff structure:

$$\mathbf{z} = \begin{bmatrix} 20 \\ 8 \\ 6 \end{bmatrix}$$

The price $S_{z,0}$ of market security z is determined from the prices of our three pure securities:

$$S_{z,0} = \begin{bmatrix} 0.362 & 0.523 & 0.015 \end{bmatrix} \begin{bmatrix} 20 \\ 8 \\ 6 \end{bmatrix} = 11.508$$

$$S_{z,0} = \quad\quad \psi^\mathrm{T} \quad\quad \mathbf{cf_z}$$

Observe that the equation $S_{z,0} = \psi^\mathrm{T}\mathbf{cf}_z$ is analogous to Eq. (3.4) that was used to price bonds.

3.3.2 Arbitrage and No Arbitrage Revisited

As we discussed earlier, arbitrage is the simultaneous and riskless purchase and sale of instruments producing identical cash flows that produce nonnegative cash flows in every state of nature. We will use no-arbitrage conditions to price securities relative to one another such that they do not produce such an arbitrage opportunity. Consider a frictionless market trading n securities i with payoffs $CF_{i,j}$ over n possible future states of nature j (note that our cash flow and kernel matrices may also include time 0 cash flows, security prices). Let γ_i represent the investment commitment made to a given investment i; that is, γ_i is the number of units that could be held by an investor. A no-arbitrage (arbitrage-free) market exists where for each and every possible portfolio strategy γ:

$$\begin{bmatrix} \gamma_1 & \gamma_2 & \cdots & \gamma_n \end{bmatrix} \begin{bmatrix} CF_{1,0} & CF_{1,1} & CF_{1,2} & \cdots & CF_{1,n} \\ CF_{2,0} & CF_{2,1} & CF_{2,2} & \cdots & CF_{2,n} \\ \vdots & \vdots & \vdots & \ddots & \vdots \\ CF_{n,0} & CF_{n,1} & CF_{n,2} & \cdots & CF_{n,n} \end{bmatrix} \begin{bmatrix} 1 \\ \psi_1 \\ \psi_2 \\ \vdots \\ \psi_n \end{bmatrix} = 0$$

$$\gamma^\mathrm{T} \quad\quad \times \quad\quad \widehat{CF} \quad\quad \times \quad\quad \hat{\psi} \quad = \quad 0$$

Observe that this equation is analogous to Eq. (3.16) for bonds. Time 0 cash flows will be negative (positive) if this is when securities are purchased (short-sold). A nonzero value for this product implies an arbitrage opportunity, either through purchases, short sales, or some combination thereof. All securities in the n-state payoff space can be priced in this no-arbitrage market. Thus, in a no-arbitrage market, each security k outside this n-security set ($k \notin \{1, 2, \ldots, n\}$), but whose payoff vector remains in the n-state payoff space, can be priced as a linear combination of the n security set.

That is, if the cash flow structure of security k can be replicated by some specific portfolio γ of securities in subset n from the market:

$$
\begin{bmatrix} CF_{k,1} \\ CF_{k,2} \\ \vdots \\ CF_{k,n} \end{bmatrix} = \begin{bmatrix} CF_{1,1} & CF_{1,2} & \cdots & CF_{1,n} \\ CF_{2,1} & CF_{2,2} & \cdots & CF_{2,n} \\ \vdots & \vdots & \ddots & \vdots \\ CF_{n,1} & CF_{n,2} & \cdots & CF_{n,n} \end{bmatrix}^{T} \begin{bmatrix} \gamma_1 \\ \gamma_2 \\ \vdots \\ \gamma_n \end{bmatrix}
$$

$$
\mathbf{cf_k} \quad = \quad \mathbf{CF^T} \qquad \times \qquad \gamma
$$

(3.24)

Observe that Eq. (3.24) is analogous to Eq. (3.17) for bonds. Then the price of that security $k \notin \{1,2,\ldots,n\}$ in an arbitrage-free market can be expressed as a linear combination of the prices of securities numbered from set $\{1,2,\ldots,n\}$:

$$
S_{k,0} = \begin{bmatrix} \gamma_1 & \gamma_2 & \cdots & \gamma_n \end{bmatrix} \begin{bmatrix} S_{1,0} \\ S_{2,0} \\ \vdots \\ S_{n,0} \end{bmatrix}
$$

(3.25)

This equation is analogous to Eq. (3.18) to price a bond.

3.3.2.1 The Pricing Kernel

Similar to the section on bonds, we define the *pricing kernel* to be vector $\hat{\psi}$ such that $\overline{CF} \times \hat{\psi} = [0]$:

$$
\begin{bmatrix} CF_{1,0} & CF_{1,1} & \cdots & CF_{1,n} \\ CF_{2,0} & CF_{2,1} & \cdots & CF_{2,n} \\ \vdots & \vdots & \ddots & \vdots \\ CF_{n,0} & CF_{n,1} & \cdots & CF_{n,n} \end{bmatrix} \begin{bmatrix} 1 \\ \psi_1 \\ \psi_2 \\ \vdots \\ \psi_n \end{bmatrix} = 0
$$

$$
\widehat{\mathbf{CF}} \qquad \times \qquad \hat{\psi} \quad = \quad 0
$$

(3.26)

If asset purchases (sales) are made at time 0, asset purchase prices are negatives (positives) of the market prices at time 0. The pricing kernel, in effect, provides that the price of a security exactly offsets its future value; security prices are exactly what securities are worth based on pure security prices. All securities with payoff vectors in this state space can be priced with this pricing kernel. Elements of the pricing kernel are pure security prices.

ILLUSTRATION: OBTAINING THE PRICING KERNEL

Consider the three security illustration (securities x, y, and f) in the previous section. When the market is in equilibrium, a pricing kernel for a securities market can be produced. We define the pricing kernel for this market to be vector $\hat{\psi}$ such that $\widehat{CF} \times \hat{\psi} = 0$:

$$
\begin{bmatrix} -5 & 8 & 4 & 1 \\ -6 & 2 & 10 & 3 \\ -0.9 & 1 & 1 & 1 \end{bmatrix} \begin{bmatrix} 1 \\ \psi_1 \\ \psi_2 \\ \psi_3 \end{bmatrix} = \begin{bmatrix} 0 \\ 0 \\ 0 \end{bmatrix}
$$

$$
\widehat{\mathbf{CF}} \qquad \times \qquad \hat{\psi} \quad = \quad 0
$$

As we showed earlier, we can also express this equation in the form of Eq. (3.22):

$$\begin{bmatrix} 8 & 4 & 1 \\ 2 & 10 & 3 \\ 1 & 1 & 1 \end{bmatrix} \begin{bmatrix} \psi_1 \\ \psi_2 \\ \psi_3 \end{bmatrix} = \begin{bmatrix} 5 \\ 6 \\ 0.9 \end{bmatrix}$$

By Eq. (3.23), the solution is:

$$\begin{bmatrix} 8 & 4 & 1 \\ 2 & 10 & 3 \\ 1 & 1 & 1 \end{bmatrix}^{-1} \begin{bmatrix} 5 \\ 6 \\ 0.9 \end{bmatrix} = \begin{bmatrix} \psi_1 \\ \psi_2 \\ \psi_3 \end{bmatrix}$$

$$\begin{bmatrix} 0.1346 & -0.058 & 0.0384 \\ 0.0192 & 0.1346 & -0.423 \\ -0.1538 & -0.0769 & 1.385 \end{bmatrix} \begin{bmatrix} 5 \\ 6 \\ 0.9 \end{bmatrix} = \begin{bmatrix} \psi_1 \\ \psi_2 \\ \psi_3 \end{bmatrix} = \begin{bmatrix} 0.362 \\ 0.523 \\ 0.015 \end{bmatrix}$$

Thus, $\hat{\psi} = [1, 0.362, 0.5223, 0.015]^{\mathsf{T}}$ is the pricing kernel for this illustration.

3.3.3 Synthetic Probabilities

If our analysis includes a riskless asset in the type of no-arbitrage pricing model that we have been using in this section, every security in this no-arbitrage market will have the same expected rate of return as the riskless bond under certain circumstances. We will prove this fact in Chapter 5. An important feature of this type of pricing model in such a market is that it can be used to define "synthetic," "hedging," or "risk-neutral" probabilities q_i. These risk-neutral probabilities do not exist in any sort of realistic sense, nor are they assumed at the start of the modeling process. Instead, they are inferred from market prices of securities and interest rates. These risk-neutral probabilities are essential in that they can be used to calculate the price of any security in the market so that the no-arbitrage nature of the market is maintained. Risk-neutral probabilities have the useful feature that they lead to expected values that are consistent with pricing by investors that are risk neutral, leading to the term risk-neutral pricing. This is important because it means that we do not need to work with unobservable risk premiums when we value securities; in fact, we do not even need to know anything about any investors' risk preferences.

If payoffs for each of the pure securities e_1, e_2, and e_3 are the result of distinct outcomes, which together account for all possible outcomes (thus forming a sample space), then we can view the pricing model in terms of probabilities. The pure security prices ψ_1, ψ_2, ψ_3 are proportional to the market's assessment of the relative likelihood that the outcomes that we number as 1, 2, and 3, respectively, will occur. Consider the example above concerning securities x, y, and f. The pure security prices estimated in this example were estimated to be $\psi_1 = 0.362$, $\psi_2 = 0.523$, and $\psi_3 = 0.015$. This suggests that investors would be willing to pay 0.362 for a security that pays 1 if and only if outcome 1 is realized; they would be willing to pay 0.523 for a security that pays 1 if and only if outcome 2 is realized, and so on. If we assume that investors are risk neutral, we can infer that they believe that outcome 2 is more likely to be realized than outcome 1 and both are more likely than outcome 3. Suppose we create the riskless portfolio with payoff vector

$\mathbf{f}^{\mathbf{T}} = \begin{bmatrix} 1 & 1 & 1 \end{bmatrix}^{\mathbf{T}}$; that is, the portfolio that pays off 1 regardless of which of the three outcomes occurs. From our example, we were given the price that the market gives to this portfolio:

$$\begin{bmatrix} 1 & 1 & 1 \end{bmatrix} \begin{bmatrix} \psi_1 \\ \psi_2 \\ \psi_3 \end{bmatrix} = \psi_1 + \psi_2 + \psi_3 = 0.9$$

However, the pure security prices add to less than one, implying that investors prefer money sooner rather than later. In this example, and as is usually the case, the ψ_i's do not sum to one because of investors' time value for money. Nevertheless, if we accept that ψ_i is proportional to the relative likelihood that outcome i will occur in the case of a risk-neutral investor, then the probability that outcome i occurs is:

$$q_i = \frac{\psi_i}{\displaystyle\sum_{j=1}^{n} \psi_j} \tag{3.27}$$

where ψ_i is the price of pure security i and ψ_j is the price of each of n pure securities j. In our illustration, synthetic probabilities are $q_1 = 0.4022$, $q_2 = 0.5811$, and $q_3 = 0.0167$. These probabilities are referred to as synthetic probabilities because they are constructed from security prices rather than directly from investor assessments of physical probability.

3.3.3.1 Discount Factors

We saw in our example above that our riskless portfolio $\begin{bmatrix} 1 & 1 & 1 \end{bmatrix}$ was priced at 0.9. This is because there is a time gap between the time the portfolio is priced and when it is paid off. We can use the information obtained from pure security prices to infer the market's time value of money in addition to synthetic probabilities. In a one-time period economy, the discount function and riskless rate are obtained from pure security prices as follows:

$$d_1 = \frac{1}{1+r} = \sum_{j=1}^{n} \psi_j \tag{3.28}$$

Time 1 payoffs for a riskless portfolio of pure securities will be 1 in every state of nature; each pure security in the state space will be included in the riskless portfolio. In our example above with three pure securities $\psi_1 = 0.362$, $\psi_2 = 0.523$, and $\psi_3 = 0.015$, d_1 will equal 0.9 and the riskless return r will equal 0.111. Normally, one would expect that pure security prices will sum to less than 1 such that $d_1 < 1$. From Eqs. (3.27) and (3.28), we see that we can also express each synthetic probability in the form:

$$q_i = \frac{\psi_i}{d_1} \tag{3.29}$$

In our illustration, synthetic probabilities are $q_1 = 0.362/0.9 = 0.402$, $q_2 = 0.523/0.9 = 0.581$, and $q_3 = 0.015/0.9 = 0.017$.

3.3.3.2 *The Risk-Neutrality Argument*

In previous sections, we used complete capital markets and no-arbitrage pricing to demonstrate how cash flow structures are replicated and securities are priced relative to other securities. Such replication and pricing are invariant with respect to investor risk preferences. Pure security prices and synthetic probabilities implicitly reflect risk preferences so that such preferences need not be explicitly input into pricing of other securities. Pure security prices and relative pricing relations are enforced by arbitrage. This is the basis of the risk-neutral valuation models. Risk-neutral valuation means that we are able to price securities such that in the risk-neutral probability space (synthetic probability space), risky securities such as stocks will have the same expected return as the return on a riskless asset such as a T-bill. This will be illustrated in the next section.

3.3.4 Binomial Option Pricing: One Time Period

Derivative securities are assets whose values are derived from the performance of other securities, indices, or rates. Stock options are examples of derivative securities. One type of stock option is a *call*, which grants its owner the right (but not the obligation) to purchase shares of an underlying stock at a specified "exercise" price within a given time period (before the expiration date of the call). The expiration payoff of a call is the maximum of either zero or the difference between the stock price S_T at expiration (at time T) and the exercise price X of the call:

$$c_T = \text{Max}[(S_T - X), 0]$$

Consider a one-time-period, two-potential-outcome framework where Company K stock currently sells for $50 per share and a riskless $100 face value T-bill sells for $90. Suppose Company K stock will pay its owner either $20 or $80 in 1 year. A call with an exercise price of $60 underlies K stock shares. This call will be worth either $0 or $20 when it expires, based on the value of the underlying stock. The payoff vectors $\mathbf{k}$ for the stock, the T-bill ($\mathbf{b}$), and the call ($\mathbf{c}$) are given as follows:

$$\mathbf{k} = \begin{bmatrix} 20 \\ 80 \end{bmatrix} \quad \mathbf{b} = \begin{bmatrix} 100 \\ 100 \end{bmatrix} \quad \mathbf{c} = \begin{bmatrix} 0 \\ 20 \end{bmatrix}$$

The current prices of the stock and T-bill are known to be $50 and $90. Since their payoff vectors span the two-outcome space in this two-potential-outcome framework, they form complete capital markets and we can estimate pure security prices as follows:

$$\underset{\mathbf{v}}{\begin{bmatrix} 50 \\ 90 \end{bmatrix}} \underset{=}{=} \underset{\mathbf{CF}}{\begin{bmatrix} 20 & 80 \\ 100 & 100 \end{bmatrix}} \cdot \underset{\psi}{\begin{bmatrix} \psi_1 \\ \psi_2 \end{bmatrix}} \tag{3.30}$$

We can solve this system for ψ_1 and ψ_2 to obtain $\psi_1 = 0.3667$ and $\psi_2 = 0.5333$. The call value is:

$$c_0 = \begin{bmatrix} 0.3667 & 0.5333 \end{bmatrix} \begin{bmatrix} 0 \\ 20 \end{bmatrix} = 10.667$$

The risk-neutral probabilities (synthetic probabilities) are determined as follows:

$$q_1 = \frac{0.3667}{0.3667 + 0.5333} = 0.4074$$

$$q_2 = \frac{0.5333}{0.3667 + 0.5333} = 0.5926$$

Observe that with respect to the risk-neutral probabilities, the expected value of stock K is $E[K] = q_1 K_1 + q_2 K_2 = 0.4074 \times 20 + 0.5926 \times 80 = 55.556$, which gives an expected return of $(55.556 - 50)/50 = 11.11\%$. The return on the riskless T-bill is $(100 - 90)/90 = 11.11\%$. The returns are equal, thus illustrating the risk-neutral nature of the arbitrage-free pricing mechanism.

We can also use the concept of arbitrage to directly value the call. Since the call and bond are priced, and their payoff vectors of the stock and T-bill span the two-outcome space, they form complete capital markets. Thus, a portfolio comprising the stock and T-bill can replicate the payoff structure of the call:

$$\begin{bmatrix} 20 & 100 \\ 80 & 100 \end{bmatrix} \begin{bmatrix} \gamma_q \\ \gamma_b \end{bmatrix} = \begin{bmatrix} 0 \\ 20 \end{bmatrix}$$

$$\begin{bmatrix} -0.016667 & 0.016667 \\ 0.013333 & -0.003333 \end{bmatrix} \begin{bmatrix} 0 \\ 20 \end{bmatrix} = \begin{bmatrix} 0.3333 \\ -0.06667 \end{bmatrix}$$

Thus, the payoff structure of the call is replicated with 0.3333 shares of underlying stock and -0.06667 riskless bonds. This portfolio requires a net investment of $0.3333 \times 50 + (-0.06667) \times 90 = \$10.67$, so the call must be worth $\$10.67$.

3.3.5 Put–Call Parity: One Time Period

A *put* is an option that grants its owner the right to sell the underlying stock at a specified exercise price on or before its expiration date. Consider a European put (European options can be exercised only at expiration) with value p_0 and a European call with value c_0 written on the same underlying stock currently priced at S_0. Both options have exercise prices equal to X and expire at time T. The riskless return rate is r. Since the payoff function of the call at expiration is $c_T = \text{Max}[S_T - X, 0]$ and the payoff function for the put is $p_T = \text{Max}[X - S_T, 0]$, the following n-outcome system describes the pricing of a put:

$$\begin{bmatrix} P_1 \\ P_2 \\ \vdots \\ P_n \end{bmatrix} = -\begin{bmatrix} S_1 \\ S_2 \\ \vdots \\ S_n \end{bmatrix} + \begin{bmatrix} X \\ X \\ \vdots \\ X \end{bmatrix} + \begin{bmatrix} C_1 \\ C_2 \\ \vdots \\ C_n \end{bmatrix} \qquad (3.31)$$

$$\begin{array}{ccccc} \mathbf{p} &=& -\mathbf{s} &+& \mathbf{x} &+& \mathbf{c} \\ \text{Max}[X - S, 0] &=& -S &+& X &+& \text{Max}[S - X, 0] \end{array}$$

This *put–call parity* relation holds regardless of the number of potential outcomes in the state space. The payoff structures of the European call, underlying stock, and bond will always be sufficient to replicate the European put as long as both the put and call expire when the riskless debt matures and the put and call have the same striking price as the face value of the debt. Consider the following numerical example where there are three potential stock prices, 120, 100, and 80 and a 105 exercise price for the options:

$$
\begin{bmatrix} 0 \\ 5 \\ 25 \end{bmatrix} = - \begin{bmatrix} 120 \\ 100 \\ 80 \end{bmatrix} + \begin{bmatrix} 105 \\ 105 \\ 105 \end{bmatrix} + \begin{bmatrix} 15 \\ 0 \\ 0 \end{bmatrix}
$$
$$
\mathbf{p} \quad = \quad -\mathbf{S} \quad + \quad \mathbf{x} \quad + \quad \mathbf{c}
$$

Because the put–call parity relation must hold at option expiry regardless of the underlying stock price, the following put–call parity relation must hold at option expiry date time T:

$$
p_T = -S_T + X + c_T
$$

Similarly, one of the following put–call parity relations must hold at time 0, depending on exactly how bonds are valued relative to the riskless rate (with discrete or continuous compounding):

$$
p_0 = -S_0 + X(1+r)^{-T} + c_0 \quad p_0 = -S_0 + Xe^{-rT} + c_0 \tag{3.32}
$$

3.3.6 Completing the State Space

When a set of priced payoff vectors forms the basis for the n-dimensional state space, any security in that economy can be replicated and priced as a linear combination of those payoff vectors and prices. This means that any security in an n-state economy can be priced if n other securities reflected by a linearly independent set of payoff vectors are priced. Derivative securities have payoff vectors that are contingent on the payoff vectors for other securities; one can define the outcome space relative to the payoff vector for the underlying security. One can create unlimited numbers of different derivative securities such as options on stocks or other existing assets. This is possible because, for example, unlimited numbers of options, all with different exercise prices, can be created and marketed on every underlying security. When this potentially unlimited number of options is marketed and priced, the state space will surely be spanned. That is, with an assured sufficiently large numbered set of linearly independent payoff vectors, the state space can be spanned with the underlying security, a riskless bond, and all the options written on that security. Thus, the bond, the underlying security, and options written on that security can form the basis for the n-potential outcome economy. In fact, the bond can be replaced with another option. Consider a stock that will pay either 20, 40, or 60. Two calls are written on that stock, one with an exercise price of 30 and a second with an exercise price of 50:

$$
\begin{bmatrix} 20 \\ 40 \\ 60 \end{bmatrix} \quad \begin{bmatrix} 0 \\ 10 \\ 30 \end{bmatrix} \quad \begin{bmatrix} 0 \\ 0 \\ 10 \end{bmatrix}
$$
$$
\mathbf{S} \qquad c_{30} \qquad c_{50}
$$

These three securities form the basis for the three-dimensional state space. Thus, their cash flows can be used to replicate any security's cash flow in this three-outcome payoff space. For example, consider a call option with an exercise price of 40. This option with a payoff vector of $[0, 0, 20]^T$ can be replicated with a portfolio with payoff vectors forming the basis as follows:

$$\begin{bmatrix} 0 \\ 0 \\ 20 \end{bmatrix} = \begin{bmatrix} 20 & 0 & 0 \\ 40 & 10 & 0 \\ 60 & 30 & 10 \end{bmatrix} \begin{bmatrix} s \\ c_{30} \\ c_{50} \end{bmatrix}$$

$$c_{40} \quad = \quad \mathbf{CF^T} \quad\quad \boldsymbol{\lambda}$$

Since the cash flow structure of this call can be replicated with the cash flow structures of the three securities in the three-outcome basis, it can be priced as a linear combination of the prices of the three securities forming the basis. Solving for γ where $\gamma = (\mathbf{CF^T})^{-1} \times c_{40}$, leads to $\gamma^T = [0,0 2]$. This means that the call is replicated with 0 shares of stock, 0 calls with exercise price equal to 30, and 2 calls with exercise price equal to 50. Thus, the price of the $X = 40$ call should be twice that of the $X = 50$ call.

More generally, one should always be able to form a basis for an n-state economy with a stock, a riskless bond, and $n - 2$ priced options written on that stock. Any other security (usually other options on that stock) whose payoff vectors are in the same payoff space can be priced as a linear combination of the prices of the securities forming the basis of payoff vectors for that economy. Thus, we should always be able to create complete capital markets by trading and pricing the appropriate number of calls in that economy.

3.4 DISCRETE TIME–SPACE MODELS

We have introduced single-outcome models in multiple period environments and we have introduced uncertainty (multiple states) in single-period environments. Now, we will introduce multiple states into multiple-period environments. However, the methodologies and even subscripting that we used become very cumbersome very quickly. Nevertheless, we do need to be able to understand and work in multiple period frameworks with uncertainty. We offer the material in this section not to suggest that the reader actually use this type of methodology to price securities in uncertain multiple period environments, but just to provide some intuition on how securities evolve in such environments. We will introduce methodologies in Chapter 6 that will be far more useful for pricing securities in these more complex environments.

The full set of pure securities can be synthesized and priced for any time period as long as complete capital markets exist for that period. However, complete capital markets will require that there exist at least as many securities producing a linearly independent set of payoff vectors as there exist outcomes.

ILLUSTRATION: MULTIPLE TIME PERIODS AND STATES

Suppose that there exist three securities x, y, and b_2 (the 2-year riskless bond) in a two-period economy with the following payoff vectors for period 2:

$$\mathbf{x} = \begin{bmatrix} 4 \\ 18 \\ 25 \end{bmatrix} \quad \mathbf{y} = \begin{bmatrix} 5 \\ 7 \\ 10 \end{bmatrix} \quad \mathbf{b_2} = \begin{bmatrix} 1 \\ 1 \\ 1 \end{bmatrix}$$

with prices at time 0: $S_{x,0} = 11$, $S_{y,0} = 5.696$, and $B_{0,2} = 0.907$ for security x, y, and the riskless bond, respectively. The value $B_{0,2}$ refers to the time 0 value of the bond that is worth 1 at time 2. These given prices are depicted in the upper panel of Figure 3.1. In Figure 3.1, the symbols u and d refer to a security price upjump or downjump. Similarly, the notation $\{u,u\}$ refers to consecutive upjumps over two periods while $\{u,d\}$ refers to a price upjump followed by a downjump from times 0 to 2. The notations $\{d,u\}$ and $\{d,d\}$ are defined analogously. The three possible states in this framework ending at time 2 are $\omega_{u,u}$, $\omega_{u \wedge d}$, and $\omega_{d,d}$. Note that there are four outcomes in this

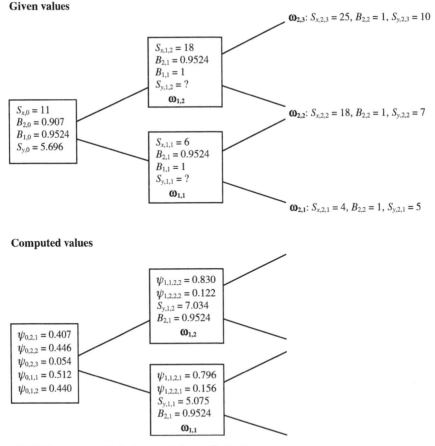

FIGURE 3.1   Multiple states, multiple time periods: an illustration.

recombining binomial framework: $\{u,u\}$, $\{u,d\}$, $\{d,u\}$, and $\{d,d\}$, where $\{\{u,d\}, \{d,u\}\}$ combine and form a single state $\omega_{u \wedge d}$. With three potential states and three securities that span this state space in this binomial framework, we can price the three two-period pure securities at time 0. Time 0 values for each of the three pure securities that pay off in period 2 are obtained from the following:

$$\begin{bmatrix} 11 \\ 5.696 \\ 0.907 \end{bmatrix} = \begin{bmatrix} 4 & 18 & 25 \\ 5 & 7 & 10 \\ 1 & 1 & 1 \end{bmatrix} \begin{bmatrix} \psi_{0,2;d,d} \\ \psi_{0,2;u \wedge d} \\ \psi_{0,2;u,u} \end{bmatrix}$$

Solving this system, we find that our time 0 two-period pure security prices are $\psi_{0,2;d,d} = 0.407$, $\psi_{0,2;u \wedge d} = 0.446$, and $\psi_{0,2;u,u} = 0.054$.[7] These computed prices are depicted in the lower panel of Figure 3.1.

Suppose that none of these securities make a payment at time 1. However, for the purposes of our illustration, we still wish to observe the price evolution of our securities over time. Suppose that the riskless rate will be 0.05 in the first period, which implies that the riskless bond worth 1 at time 2 has a time 1 value of $1/1.05 = 0.9524$. Further suppose that at time 1, security x will be worth either $S_{x,1;d} = 6$ (security x price, time 1 outcome ω_d) or $S_{x,1;u} = 18$, and that these time 1 prices will lead to the time 2 prices of either 4, 18, or 25 in the manner depicted in the upper panel of Figure 3.1. Now for time period 1, we have two priced securities associated with each state, completing our markets in each state for that period, and allowing us to price pure securities and security y in the lower panel of Figure 3.1. In a complete market, arbitrage-free pricing guarantees that any independent set of prices that we pick to accomplish the pricing will result in consistent pricing. Even though we have already priced the securities going from time 0 to 2, if we now price securities going from time 1 to 2 and then from time 0 to 1, the pricing will be consistent. Our pure securities are priced at time 1 by solving the two systems, one for each of two possible states at time 1:[8]

$$\begin{bmatrix} 6 \\ 0.9524 \end{bmatrix} = \begin{bmatrix} 4 & 18 \\ 1 & 1 \end{bmatrix} \cdot \begin{bmatrix} \psi_{1,2;d,d} \\ \psi_{1,2;d,u} \end{bmatrix} \quad \text{State } \omega_d \text{ time 1 prices}$$
$$\mathbf{v}_{1;d} \quad = \quad \mathbf{CF}_{2;d} \quad \cdot \quad \psi_{1,2;d}$$

$$\begin{bmatrix} 18 \\ 0.9524 \end{bmatrix} = \begin{bmatrix} 18 & 25 \\ 1 & 1 \end{bmatrix} \cdot \begin{bmatrix} \psi_{1,2;u,d} \\ \psi_{1,2;u,u} \end{bmatrix} \quad \text{State } \omega_u \text{ time 1 prices}$$
$$\mathbf{v}_{1;u} \quad = \quad \mathbf{CF}_{2;u} \quad \cdot \quad \psi_{1,2;u}$$

We solve these two systems and find that $\psi_{1,2;d,d} = 0.796$, $\psi_{1,2;d,u} = 0.156$, $\psi_{1,2;u,d} = 0.830$, and $\psi_{1,2;u,u} = 0.122$. Thus, contingent on whether state 1 or state 2 is realized, we will know pure security prices for instruments that pay off in period 2. We can also use this information above to price security y at time 1 to be either $0.796 \times 5 + 0.156 \times 7 = 5.075 = S_{Y,1;d}$ or $0.830 \times 7 + 0.122 \times 10 = 7.034 = S_{Y,1;u}$ depending on whether security x decreased (d) or increased (u) in value from time 0 and 1.

Even though none of our priced securities have time 1 payoffs, we can still use our time 1 prices and value pure securities that pay off at time 1. Since only two states are possible at time

1, we can compute two pure security prices based on the time 1 2-year bond value of 0.9524 and the security x value of either 6 or 18:[9]

$$\begin{bmatrix} 11 \\ 0.907 \end{bmatrix} = \begin{bmatrix} 6 & 18 \\ 0.9524 & 0.9524 \end{bmatrix} \cdot \begin{bmatrix} \psi_{0,1;d} \\ \psi_{0,1;u} \end{bmatrix}$$

$$\mathbf{v}_0 \qquad = \qquad \mathbf{CF}_1 \qquad \qquad \psi_{0,1}$$

Thus, time 0 pure security prices are $\psi_{0,1;d} = 0.512$ and $\psi_{0,1;u} = 0.441$. These enable us to price at time 0 any security that pays or has a known value in time 1.

This section serves as an introduction to multiple state–multiple time discrete pricing. In Chapter 5, we will study the pricing mechanism more systematically along with its connection to martingales. We will also generalize these ideas in Chapter 6 to continuous processes. In particular, we will deal with these problems in Chapters 5 and 6 by assuming that our uncertainty can be represented with specific stochastic processes (e.g., binomial or Brownian motion frameworks).

3.5 EXERCISES

3.1. What is the present value of a security to be discounted at a 10% rate promising to pay $10,000 in:
  a. 20 years?
  b. 10 years?
  c. 1 year?
  d. 6 months?
  e. 73 days?
3.2. The Foxx Company is selling preferred stock which is expected to pay a $50 annual dividend per share. What is the present value of dividends associated with each share of stock if the appropriate discount rate were 8% and its life expectancy were infinite?
3.3. The Evers Company is considering the purchase of a machine whose output will result in a $10,000 cash flow next year. This cash flow is projected to grow at the annual 10% rate of inflation over each of the next 10 years. What will be the cash flow generated by this machine in:
  a. its second year of operation?
  b. its third year of operation?
  c. its fifth year of operation?
  d. its tenth year of operation?
3.4. The Wagner Company is considering the purchase of an asset that will result in a $5000 cash flow in its first year of operation. Annual cash flows are projected to grow at the 10% annual rate of inflation in subsequent years. The life expectancy of this asset is 7 years, and the appropriate discount rate for all cash flows is 12%. What is the maximum price Wagner should be willing to pay for this asset?
3.5. What is the present value of a stock whose $100 dividend payment next year is projected to grow at an annual rate of 5%? Assume an infinite life expectancy and a 12% discount rate.

3.6. What would be the present value of $10,000 to be received in 20 years if the appropriate discount rate of 10% were compounded:
  a. annually?
  b. monthly?
  c. daily?
  d. continuously?

3.7. a. What would be the present value of a 30-year annuity if the $1000 periodic cash flow were paid monthly? Assume a discount rate of 10% per year.
  b. Should an investor be willing to pay $100,000 for this annuity?
  c. What would be the highest applicable discount rate for an investor to be willing to pay $100,000 for this annuity?

3.8. a. Suppose that a $1000 face value bond will make a single interest payment at an annual rate of 5%. Suppose this bond is currently selling for 102 (actually meaning 102% of its face value, or 1020) and that it matures in 1 year when its coupon payment is made. What is the 1-year spot rate?
  b. Now, drawing on your results from part a of this problem, consider a second $1000 face value 2-year bond making interest payments at an annual rate of 5%. Suppose this bond is currently selling for 101.75 (meaning 101.75% or 1017.5) and that it matures in 2 years when its second coupon payment is made. What is the 2-year spot rate implied by this bond, considering the 1-year spot rate?
  c. What is the 3-year spot rate $y_{0,3}$ implied by the 3-year 5% coupon bond priced at 101.5 based on parts a and b of this question?
     In parts d through g, assume that the pure expectations theory applies.
  d. What is the 1-year forward rate on a loan originating in 1 year?
  e. What is the 1-year forward rate on a loan originating in 2 years?
  f. What is the 2-year forward rate on a loan originating in 1 year?
  g. Map out the yield curve based on your answers from the preceding parts of this problem.

3.9. a. Suppose that there are three bonds whose characteristics are given in the table below. Determine spot rates for years 1, 2, and 3.

Bond	Current Price	Face Value	Coupon Rate	Years to Maturity
A	947.376	1000	0.05	2
B	904.438	1000	0.06	3
C	981	1000	0.09	3

  b. Calculate $r_{1,2}$, $r_{2,3}$, and $r_{1,3}$.
  c. What would be the market value of a 3-year 2% bond with face value 1000 in this market?

3.10. Consider the following four bonds:

F	n	c	P_0
1000	1	0.01	1005
1000	2	0.05	1040
1000	3	0.04	1020
1000	4	0.04	990

Based on the cash flows and prices associated with these bonds, determine the following:
a. Spot rates $r_{0,n}$ for each of 4 years 1–4. These are interest rates on loans originating at time 0.
b. Forward rates $r_{1,t}$ for each of three periods beginning with year 1. These are interest rates on loans originating at time 1.
c. Forward rates $r_{2,t}$ for each of two periods beginning with year 2. These are interest rates on loans originating at time 2.
d. Forward rates $r_{3,n}$ for the period beginning with year 3. These are the interest rates on loans originating at time 3.

3.11. There are two 3-year bonds with face values equaling $1000. The coupon rate of bond A is 0.05 and 0.08 for bond B. A third bond C also exists, with a maturity of 2 years. Bond C also has a face value of $1000; it has a coupon rate of 11%. The prices of the three bonds are $878.9172, $955.4787, and $1055.419, respectively. Find a portfolio of bonds A, B, and C that would replicate the cash flow structure of bond D, which has a face value of $1000, a maturity of 3 years, and a coupon rate of 3%.

3.12. Suppose that securities x and y exist in a two-potential outcome world and have the following payoff vectors:

$$x = \begin{bmatrix} 8 \\ 4 \end{bmatrix} \quad y = \begin{bmatrix} 2 \\ 10 \end{bmatrix}$$

Suppose that security x has a market value of 5 and security y has a market value of 8.
a. Is the set of two payoff vectors linearly independent?
b. Do these two payoff vectors span the two-dimensional space?
c. Is there a complete capital market in this economy?
d. What are the prices of pure securities in this economy?
e. What is the price of a third security z with the following payoff structure:

$$z = \begin{bmatrix} 20 \\ 8 \end{bmatrix}$$

f. Calculate risk-neutral probabilities for each of the two states.

3.13. Security A will pay 5 in outcome 1, 7 in outcome 2, and 9 in outcome 3. Security B will pay 2 in outcome 1, 4 in outcome 2, and 8 in outcome 3. Security C will pay 9 in outcome 1, 1 in outcome 2, and 3 in outcome 3. Both securities A and C currently sell for $5 and security B currently sells for $3. What would be the value of security D, which will pay 1 in each of the three outcomes?

3.14. Rollins Company stock currently sells for $12 per share and is expected to be worth either $10 or $16 in 1 year. The current riskless return rate is 0.125. What would be the value of a 1-year call with an exercise price of $8?

3.15. Verify that in a one-time-period binomial economy with a riskless return rate equal to r and a stock currently selling for S_0 that the synthetic (risk-neutral) probability for upward price movement in this economy is characterized by $(1 + r - d)/(u - d)$ where u and d are multiplicative upward and downward stock price movements, respectively, meaning that the stock price will equal either $S_1 = uS_0$ or $S_1 = dS_1$.

3.16. In a one-time-period binomial economy with a riskless return rate equal to 10%, a stock currently sells for $50. Its potential terminal prices are either $80 or $40. What are the synthetic (risk-neutral) probabilities for this economy?

3.17. Harper Company stock currently sells for $14 per share and is expected to be worth either $10, $16, or $25 in 1 year. The current riskless return rate is 0.125. A 1-year call with an exercise price of $15 currently sells for $3.
  a. What would be the value of a 1-year call with an exercise price of $9?
  b. What are the synthetic probabilities in this economy?

3.18. Buford Company stock currently sells for $24 per share and is expected to be worth either $20 or $32 in 1 year. The current riskless return rate is 0.125. What would be the value of a 1-year call with an exercise price of $16?

3.19. Robinson Company stock currently sells for $20 per share and will pay off either $15 or $25 in 1 year. A 1-year call with an exercise price equal to $18 has been written on this stock. This call sells for $7.
  a. What is the value of a 1-year call with an exercise price equal to $22?
  b. What is the riskless return rate for this economy?
  c. What is the value of a 1-year put that can be exercised for $40?

3.20. Consider a two 6-month time period framework depicted in the Figure 3.2 in which Company J stock currently sells for $50 per share and a riskless $100 face value T-bill currently sells for $90 ($d_2 = 0.9$). Assume that the discount rate is the same for each 6-month period. There are two possible outcomes for each 6-month period contingent on the outcomes that preceded them. At the end of the year (i.e., two 6-month periods), Company J stock will pay its owner either $34.722, $50, or $72 per share. Note that these payments are $(1/1.2^2) \times 50$, $(1.2/1.2) \times 50$, and $1.2^2 \times 50$; that is, the stock price will either decrease by 30.56%, stay the same, or increase by 44% over the course of the year. We will assume that proportional increases and decreases in the stock's price are the same in each 6-month interval during the year; that is, each is by the factor $\sqrt{1.44} = 1.2$. The stock will make no payments to shareholders (no dividends) at the end of 6 months. A 1-year call with an exercise price of $60 trades on J stock shares. This call will be worth either $0 or $12 when it expires, based on the value of the underlying stock.

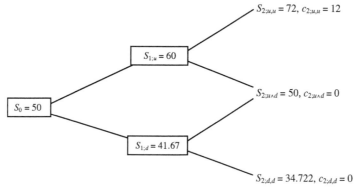

$S_0 = 50$ → $S_{1;u} = 60$ → $S_{2;u,u} = 72$, $c_{2;u,u} = 12$
$S_{1;u} = 60$ → $S_{2;u \wedge d} = 50$, $c_{2;u \wedge d} = 0$
$S_{1;d} = 41.67$ → $S_{2;u \wedge d} = 50$, $c_{2;u \wedge d} = 0$
$S_{1;d} = 41.67$ → $S_{2;d,d} = 34.722$, $c_{2;d,d} = 0$

FIGURE 3.2 Figure for problem 20.

a. Write 1-year payoff vectors for the stock, T-bill, and the call.

b. Assume that risk-neutral probabilities associated with upjumps are constant over the two 6-month intervals during the year in this framework. Are markets complete in this four-outcome two-period market? Why or why not?

c. Assume that risk-neutral probabilities associated with downjumps are constant over the two 6-month intervals during the year as are riskless rates in this framework. What is the pure security price associated with each upjump? What is the pure security price associated with each downjump?

d. From your answers to a, b, and c above, find risk-neutral probabilities for the first 6-month interval, again assuming that the 6-month riskless interest rate is $r = 0.0541$. That is, find $q_{0,1;d}$ and $q_{0,1;u}$.

e. What are the three two-period pure security prices $\psi_{0,2;d,d}$, $\psi_{0,2;u \wedge d}$, and $\psi_{0,2;u,u}$?

f. What is the current value of the call?

NOTES

1. The ratio here is $x = (1 + g)/(1 + r)$ where $g \neq r$.
2. A zero-coupon bond, also known as a pure discount bond or strip, makes no explicit interest payments, but is purchased at a discount from its face or maturity value.
3. Recall from Section 3.1 that we worked with a matrix $\mathbf{CF}$ that was $T \times T$, consisting of cash flows for T bonds from time 1 to T. Matrix $\widehat{\mathbf{CF}}$ is $n \times (T + 1)$, consisting of cash flows for n bonds, $n \geq T$, from time 0 to T. Similarly, vector $\mathbf{d}$ is $T \times 1$ and consists of discount functions for times 1 to T. Vector $\hat{\mathbf{d}}$ is $(T + 1) \times 1$ and consists of discount functions from time 0 to T.
4. The net present value (NPV) of an investment is the discounted future value of that investment minus the initial cash flow (CF_0, usually negative) associated with its purchase price P_0. Zero NPV implies that the investment purchase price P_0 is exactly equal to the present value of the investment's future cash flows. Nonzero NPVs are inconsistent with market equilibrium.
5. We will refer to $\hat{\mathbf{d}}$ as the pricing kernel when $\widehat{\mathbf{CF}} \times \hat{\mathbf{d}} = [0]$.
6. Upper case $\mathbf{S}$ is used to denote this vector of security prices so as not to confuse with a lower case $\mathbf{s}$ used for states.
7. $\psi_{0,2;u,u}$ reads: the time 0 price of pure security for state $\omega_{u,u}$, paying off at time 2. In general when pricing pure securities, the first subscript refers to the time the security is being priced, the second subscript is when the pure security is paid off, and after the semicolon is the state or outcome(s). $S_{x,2;u,u}$ is the price of security x at time 2 if it is in state $\omega_{u,u}$.
8. $\psi_{1,2;d,u}$ reads: the time 1 price of pure security for state $\omega_{u \wedge d}$, paying off at time 2, contingent on realizing state ω_d at time 1. *Note*: Remember that $u \wedge d$ includes both $\{u,d\}$ and $\{d,u\}$.
9. $\psi_{0,1;d}$ reads: the time 0 price of pure security for state ω_d, paying off at time 1.

Continuous Time and State Models

4.1 SINGLE PAYMENT MODEL

In Chapter 3, we derived the continuous time present value for a single future payment:

$$PV = FVe^{-rT} \tag{4.1}$$

where r is the constant discount rate. Any asset making a single payment in the future can be valued with this model, as long as a continuous time model is appropriate for this purpose. For example, a T-year riskless zero-coupon bond with face value F is valued as $B_0 = PV_B = Fe^{-rT}$.

4.1.1 Forward Contracts

As we discussed in Chapters 1 and 2, a forward contract represents an agreement that specifies the delivery of a given quantity of an asset at a specific future date for a given price. In the absence of risk, asset storage costs, dividend, and market inefficiencies, pricing the forward contract relative to the asset spot price can be straightforward in a riskless or risk-neutral environment:

$$F_T = S_0 e^{rT}$$

where F_T is the forward price of the asset to be delivered at time T, the contract settlement date, and S_0 is the spot price (time 0 price) of the asset. In effect, the cash flow structure of the forward contract to purchase an asset at time T at price F_T is identical to that of borrowing S_0 dollars at time 0 and repaying $S_0 e^{rT}$ at time T.

A forward contract is struck before any cash or assets change hands. By convention, the price f of a forward contract is zero, which, in our model above, would equate the forward (delivery) price of the asset to the value of the asset:

$$f = 0 = F_T - S_0 e^{rT}$$

In a risk-neutral environment, the price f of a forward contract is still zero, which, in our model above, will equate the forward price of the asset to the time 0 expected future value of the asset:

$$f = 0 = F_T - E_0[S_T]$$

P.M. Knopf & J.L. Teall: Risk Neutral Pricing and Financial Mathematics: A Primer.
DOI: http://dx.doi.org/10.1016/B978-0-12-801534-6.00004-7

In other words, no cash changes hands at time 0 when the forward contract is originated at time 0. The contracting parties merely agree that the holder of the short position will deliver the specified asset at time T for the forward price F_T to the holder of the long position.

SIMPLE ILLUSTRATION: FORWARD CONTRACT

Suppose, for example, that the spot price for one ounce of gold is $S_0 = \$1152.95$. If the 2-year interest rate is 2%, we calculate the 2-year forward price for gold as follows:

$$F_T = S_0 e^{rT}$$
$$\$1200 = \$1152.95 e^{0.02 \times 2}$$

In effect, the counterparty taking the long position in gold agrees to purchase one ounce for $1200 at time T from the counterparty taking the short position in gold. If the counterparty with the short position in gold currently owns one ounce of gold, the forward contract provides her the riskless opportunity (and obligation) to sell her gold for $1200 in 1 year, consistent with the absence of arbitrage opportunities given the spot price of $1152.95 and the riskless rate equal to 0.02.

Thus far, we have calculated the forward price from the spot price. We could just as easily have calculated the spot price from the forward price. That is, if the forward price were $1200, then the spot price would be $1152.95 given the riskless rate of 0.02 and the absence of arbitrage opportunities. Both scenarios are consistent with the statement that we made above:

$$f = 0 = \$1200 - \$1152.95 e^{0.02 \times 2}$$

Again, no money changes hands at time 0 when the counterparties enter into the forward contract. At time $T = 2$, the long counterparty receives one ounce of gold and the short counterparty receives $1200. These contract terms hold in the absence of credit (default) risk, though other risks and complications might affect the relationship between the spot and forward prices.

4.1.2 Forward Market Complications

4.1.2.1 Dividends

In our models above, we avoided complications such as asset storage costs, dividends, and other income. For example, consider an equity index forward contract in which the underlying stocks pay dividends. These dividends, when paid (technically, on ex-dividend dates), reduce stock prices by amounts roughly comparable to dividend amounts. Here, we propose simple adjustments for such complications. First, consider the effect of income spun off by the asset prior to time T, where δ is the instantaneous periodic income divided by the asset value:

$$F_T = S_0 e^{(r-\delta)T} \tag{4.2}$$

This type of dividend yield adjustment tends to be most appropriate when the index comprises a larger number of stocks whose ex-dividend dates are spread out over the course of time. The equity dividend payout ratio δ reflects the proportion of equity value that the long position in the forward contract will not accrue from the equity waiting delivery at time T. The dividend yield δ reflects asset value that will not be captured by the counterparty with the long position in the forward contract.

4.1.2.2 Foreign Exchange

In a sense, an FX (foreign exchange) forward contract has a similar complication. An FX contract provides for one counterparty to purchase one currency from a second counterparty in exchange for a second currency. An FX contract provides for one counterparty to be long in one currency (e.g., buy Chinese yuan) and short in the other (e.g., sell dollars), with the second counterparty taking opposite positions in the two currencies. Each counterparty can earn interest on the currency that he shorts (perhaps the domestic currency from his perspective) until delivering it at time T. Suppose that $r(d)$ is the riskless rate in the country issuing the currency associated with the short position in the contract and that $r(f)$ is the interest rate associated with the long position in the contract (the foreign currency):

$$F_T = S_0 e^{(r(d) - r(f))T}$$

In this scenario, the long position in a currency forward precludes the counterparty from investing the longed currency at the rate $r(f)$ until time T, but provides the opportunity for him to invest in the shorted currency at rate $r(d)$ until time T.

Suppose, for example, one counterparty to a 2-year FX forward contract longs $CNY1$ (1 Chinese yuan) and shorts $USD0.16$ (0.16 US dollars). This counterparty agrees to purchase 1 yuan for USD0.16 and can invest the $USD0.16$ at rate $r(d)$ until time $T = 2$. However, this counterparty loses the opportunity to invest $CNY1$ at rate $r(f)$ until time $T = 2$. If the US interest rate is $r(d) = 0.02$ and the Chinese rate is $r(f) = 0.04$, the spot rate $S_0 = USD0.1665$ is consistent with our pricing relation above:

$$0.16 = 0.1665 e^{(0.02 - 0.04) \times 2}$$

4.1.2.3 Carry Costs

Similarly, an asset on which storage or transportation costs are incurred, e.g., a commodity such as natural gas, leads to a negative cash flow from the asset. The forward price F_T of an asset with carry cost κ (expressed as a proportion κ of asset value) is calculated as follows:

$$F_T = S_0 e^{(r + \kappa)T}$$

Note that this "negative income" involves a simple sign change from our dividend illustration. The counterparty with the long position in gas does not incur the cost of storing the gas. Not incurring storage costs until time T increases the forward price of the gas to its prospective buyer.

4.1.3 Pricing a Zero-Coupon Bond with Continuous Deterministic Interest Rates

Since future interest rates are normally unknown, most models for interest rates are probabilistic in nature. Nevertheless, a deterministic model can serve as an excellent introduction to the more sophisticated models that we will cover extensively later in this text. In this section, we will assume that the interest rate $r(t)$ is a known real-valued continuous function of time t. The rate, while not stochastic, may or may not change with time.

As we mentioned in Section 1.5.2, the importance of Riemann sums and their limits extends beyond their applications to finding areas under curves. Many continuous valuation models are derived by first approximating equations using discrete models resulting in Riemann sums. Then, taking the limit as n approaches infinity, we derive an appropriate continuous valuation model. Here we will see this principle illustrated to price a zero-coupon bond.

As we reviewed in Chapter 3, the price of a zero-coupon bond is simply the present value of its face value at maturity. Let F denote the face value of the bond, and assume that it is purchased at time 0 and matures at time T. First, we approximate this continuous model with a discrete model in which interest is accumulated and compounded over a total of n equal intervals of time from 0 to T. This means that $\Delta t = (T - 0)/n = T/n$. The n time intervals are $[t_{i-1}, t_i]$ for $i = 1, 2, \ldots, n$ with $t_0 = 0$ and $t_n = T$. Consider the last time interval $[t_{n-1}, t_n]$. The interest rate over this time period is $r(t_{n-1})$. Since $t_n - t_{n-1} = \Delta t$, then the amount of interest paid on one unit of money over the time interval $[t_{n-1}, t_n]$ equals $r(t_{n-1})\Delta t$. If we were to regard time t_{n-1} to be the present time, then the present value of the bond equals $F(1 + r(t_{n-1})\Delta t)^{-1}$. Next, consider the time interval $[t_{n-2}, t_{n-1}]$. If we regard time t_{n-2} as the present time, and repeat the same argument, then the present value of the bond at time t_{n-2} equals $F(1 + r(t_{n-2})\Delta t)^{-1}(1 + r(t_{n-1})\Delta t)^{-1}$. Continuing this process all the way back to time 0 and choosing Δt to be very small, we see that we can approximate the present value of the continuously modeled bond by the equation:

$$B_0 \approx F(1 + r(t_0)\Delta t)^{-1}(1 + r(t_1)\Delta t)^{-1}(1 + r(t_2)\Delta t)^{-1} \cdots (1 + r(t_n)\Delta t)^{-1}$$

Next, take the log of both sides of the equation above:

$$\ln B_0 \approx \ln F - \ln(1 + r(t_0)\Delta t) - \ln(1 + r(t_1)\Delta t) - \cdots - \ln(1 + r(t_n)\Delta t)$$

To estimate the log terms on the right-hand side of the equation, we find the second-order Taylor polynomial of the function $f(x) = \ln(1 + x)$ about $x = 0$. Taking derivatives, we calculate that: $f'(x) = (1 + x)^{-1}$, $f''(x) = -(1 + x)^{-2}$, and so $f(0) = 0$, $f'(0) = 1$, $f''(0) = -1$. The function $\ln(1 + x)$ can be estimated by its second-order Taylor polynomial about 0:

$$\ln(1 + x) \approx 0 + (1)(x - 0) + \frac{1}{2}(-1)(x - 0)^2 = x - \frac{1}{2}x^2$$

This can be used to approximate the log terms in the estimate for $\ln B_0$:

$$\ln(1 + r(t_{i-1})\Delta t) \approx r(t_{i-1})\Delta t - \frac{1}{2}[r(t_{i-1})]^2(\Delta t)^2$$

As $\Delta t \to 0$, the term $\frac{1}{2}[r(t_{i-1})]^2(\Delta t)^2$ becomes negligible, resulting in the approximation[1]:

$$\ln(1 + r(t_{i-1})\Delta t) \approx r(t_{i-1})\Delta t$$

Using this approximation in the estimate for $\ln B_0$ gives:

$$\ln B_0 \approx \ln F - r(t_0)\Delta t - r(t_1)\Delta t - \cdots - r(t_{n-1})\Delta t = \ln F - \sum_{i=1}^{n} r(t_{i-1})\Delta t$$

Exponentiating both sides of the equation results in:

$$B_0 = Fe^{\left(-\sum_{i=1}^{n} r(t_{i-1})\Delta t\right)}$$

In the limit, as n approaches infinity, we obtain the present value of the bond:

$$B_0 = Fe^{\left(\lim\limits_{n \to \infty} - \sum\limits_{i=1}^{n} r(t_{i-1})\Delta t\right)}$$

Notice that on the right-hand side of the equation, the expression in the exponent is a limit of Riemann sums with $t_i^* = t_{i-1}$. By the definition of the definite integral:

$$B_0 = Fe^{\left(-\int_0^T r(t)\mathrm{d}t\right)} \tag{4.3}$$

Equation (4.3) gives the price of the zero-coupon bond.

ILLUSTRATION: PRICING A ZERO-COUPON BOND WITH A DETERMINISTIC CONTINUOUS RATE

Here, we will find the price of a 10-year zero-coupon bond with a face value of $100 where the interest rate is given by $r(t) = 0.05 - 0.001t$. Notice that the interest rate declines proportionally over time in this simple model of the term structure. Using Eq. (4.3), we can easily find the price of the bond:

$$B_0 = 100e^{\left(-\int_0^{10}(0.05-0.001t)\mathrm{d}t\right)} = 100e^{-\left(0.05t - 0.0005t^2|_0^{10}\right)} = 100e^{-0.45} = \$63.76$$

4.2 CONTINUOUS TIME MULTIPAYMENT MODELS

In this and the following sections, we will examine riskless securities whose prices evolve continuously over time. We will also examine interest and exchange rates that evolve continuously over time. In later chapters, we will examine these processes under various types of uncertainty.

4.2.1 Differential Equations in Financial Modeling: An Introduction

Financial economists and practitioners are often concerned with the development or change of a variable or asset over time. A *differential equation* can be structured to model the change (evolution or direction) of an asset's price over time. From this equation, a second equation (*solution*) might be derived to describe the asset's value (state or path) at a given point in time. The asset price is the dependent variable, which is a function of the independent variable (time). More generally, one is interested in determining the solution for a dependent variable as a function of one or more independent variables. Often, the relationship between the dependent variable in terms of its independent variables is described by a differential equation. A differential equation is defined to be an equation that relates the dependent variable and one or more of its derivatives. A differential equation is defined to be an equation that relates the dependent variable with one or more of its derivatives. The solution to a differential equation is an explicit function that, when substituted for the dependent variable in the differential equation, leads to an identity. The following is a simple differential equation along with its solution involving dependent variable x and independent variable t:

$$\frac{dx}{dt} = t \tag{4.4}$$

$$x = \frac{1}{2}t^2 + C \tag{4.5}$$

where C is a constant. We verify the solution to differential equation (4.4) by noting that it represents the derivative of x with respect to t in its solution (Eq. (4.5)). Equation (4.4) represents the change in variable x over time ($dx = t\, dt$). Note that this rate of change increases as t increases. Equation (4.5) represents the state or value of x at a given point in time t. Because Eq. (4.4) concerns only the first derivative of the function of t, it is referred to as a *first-order differential equation*.

4.2.1.1 Separable Differential Equations

A differential equation is said to be *separable* if it can be rewritten in the form $g(x)\, dx = f(t)\, dt$. A separable differential equation written in this form can be solved by the following:

$$\int g(x)\, dx = \int f(t)\, dt \tag{4.6}$$

The following is an example of a separable differential equation:

$$\frac{dx}{dt} = tx^2 - 2tx$$

We separate as follows:

$$\frac{dx}{dt} = t(x^2 - 2x)$$

$$\frac{dx}{(x^2 - 2x)} = t\, dt \text{ or } \frac{1}{(x^2 - 2x)}dx = t\, dt$$

$$\int \frac{1}{(x^2 - 2x)}dx = \int t\, dt$$

4.2.1.2 Growth Models

Consider the following example of a separable differential equation:

$$\frac{dx}{dt} = rx \tag{4.7}$$

To solve this equation, we first separate the variables as follows:

$$\frac{1}{x}dx = r\, dt$$

Next we integrate both sides and to obtain a *general solution* for x:

$$\int \frac{1}{x}dx = \int r\,dt$$

$$\ln|x| + C_1 = rt + C_2$$
$$\ln|x| = rt + C_2 - C_1$$

When we integrate both sides of the equation, one obtains arbitrary constants C_1 and C_2. Since these constants are arbitrary, we can define $C = C_1 - C_2$, which is still an arbitrary constant. Thus, whenever we integrate both sides of an equation, we only need to add an arbitrary constant to one side of the equation. So, we have in this case:

$$\ln|x| = rt + C$$
$$e^{\ln|x|} = e^{rt} \times e^C$$
$$|x| = e^{rt} \times e^C \tag{4.8}$$
$$x = \pm e^C e^{rt}$$
$$x = Ke^{rt} \quad \text{where } K = \pm e^C$$

The constant K can assume any value. Thus, the general solution (or family of solutions) for our differential equation involves a constant that can assume any value. A *particular solution* results when K assumes a specific value. Stating the value of the function at a given moment of time is known as an *initial condition*. In our example, we have $x(0) = x_0$. The initial condition will determine the constant K, and we will then be able to write down the unique solution that satisfies both the differential equation and the initial condition. In this case, one particular solution for x would be $x = x_0 e^{rt}$, where x_0 is the value of x when $t = 0$. In any case, this type of differential equation is typical of those used for modeling growth.

4.2.1.3 Security Returns in Continuous Time

Continuous time and continuous space models involve securities whose values evolve continuously over time (their prices can be observed at every instant) and can take on any real number value. Suppose that the evolution of a stock's price is modeled by the following separable differential equation:

$$\frac{dS_t}{dt} = \mu S_t \tag{4.9}$$

The term μ represents the security's drift or mean instantaneous rate of return. This differential equation is identical to Eq. (4.7) with $x = S_t$ and $r = \mu$. By Eq. (4.8), the solution is:

$$S_t = Ke^{\mu t} \tag{4.10}$$

Equation (4.10) represents a general solution to our differential equation (4.9). Since K is equal to the stock's price S_0 at time zero (K can equal any constant), the particular solution to Eq. (4.9) would be:

$$S_t = S_0 e^{\mu t} \tag{4.11}$$

Differential equations such as Eq. (4.9) are useful in the modeling of security prices and are adaptable to the modeling of stochastic (random) return processes.

ILLUSTRATION: DOUBLING AN INVESTMENT AMOUNT

Next, consider a security with value S_t in time t generating returns on a continuous basis such that the security's price doubles every 7 years. Suppose that the value of this security after 10 years was $50. What would have been the initial value S_0 of this security?

Equation (4.9) models this security price, which can also be depicted by the security's return generating process:

$$\frac{dS_t}{S_t} = \mu \, dt$$

The solution to this equation is obtained with Eq. (4.11). If we substitute $t = 7$ into the solution, we obtain:

$$S_7 = S_0 e^{7\mu} = 2S_0$$

Thus, $\mu = \ln(2) \div 7 = 0.09902$. With this result, we can easily solve for the security's initial value:

$$S_0 = S_{10}e^{-10 \times 0.09902} = \$50e^{-10 \times 0.09902} = \$18.58$$

ILLUSTRATION: MEAN REVERTING INTEREST RATES

While interest rates vary over time, they tend to be more likely to increase when they are "low" and decrease when they are high; that is, they tend to drift or revert to some long-term mean rate (long-term mean rate, not long-term rate). Suppose that μ represents the long-term mean interest rate, r_t the short-term rate at a particular time, and λ an interest rate adjustment mechanism (also known as a "pullback factor"):

$$dr = \lambda(\mu - r_t)dt \tag{4.12}$$

We will divide both sides by $(\mu - r_t)$ to separate and integrate:

$$\int \frac{dr}{(\mu - r_t)} = \int \lambda \, dt$$

Since $\mu - r_t = -(r_t - \mu)$, then:

$$\int \frac{1}{(r_t - \mu)} dr = -\int \lambda \, dt$$

$$\ln|r_t - \mu| = -\lambda t + C \tag{4.13}$$

$$|r_t - \mu| = e^{-\lambda t}e^C$$

$$r_t - \mu = \pm e^C e^{-\lambda t}$$

Define the constant $K = \pm e^{C}$ so that K can be any constant (positive or negative). Substituting K into Eq. (4.13) and solving for r_t gives the solution:

$$r_t = \mu + Ke^{-\lambda t} \tag{4.14}$$

Suppose, for example, the interest rate 2 months ago was 18%, and currently it is 16.5% as it drifts back to the long-term mean rate of 7%. How long will it take for the interest rate to drop below 10%?[2] First, we use the long-term drift to obtain $K = 11\%$:

$$r_0 = 18\% = 7\% + Ke^{-\lambda \times 0}$$

Next, we solve for the "pullback factor" λ by using the interest rate drift over the past 2 months:

$$r_2 = 16.5\% = 7\% + 11\% \times e^{-2\lambda}$$

$$\frac{16.5\% - 7\%}{11\%} = e^{-2\lambda} = 0.86363636$$

$$\lambda = \frac{\ln(0.86363636)}{-2} = 0.0733017$$

Finally, we solve for the time (in months) for the interest rate to drop below 10%:

$$r_t = 10\% = 7\% + 11\% \times e^{-0.0733017t}$$

$$t = \frac{1}{-0.0733017}\ln\frac{3}{11} = 17.7251$$

We see now that it will take 17.7251 months from the start of the process for the interest rate to drift below 10% or 15.7251 months from now.

4.2.2 Annuities and Growing Annuities

We discussed annuities and perpetuities in discrete time in the previous chapter. Now, we discuss them in continuous time with continuous compounding of discount, interest, and growth rates. Note similarities in the expressions that we derive to those in Chapter 3.

4.2.2.1 Annuities and Perpetuities

Consider an investment that leaks cash flows continuously. The present value of cash flows leaked during any infinitesimal interval $[t, t + dt]$ is:

$$PV[t, t + dt] = PV[0, t + dt] - PV[0, t] \tag{4.15}$$

where $PV[t, t + dt]$ equals the present value of cash flows leaked over the interval $[t, t + dt]$. Equation (4.1) describes the infinitesimal change in the investment value, which is more commonly written as $dPV[0,t]$. The amount of leaked cash flow in any infinitesimal time interval dt is $f(t)\, dt = CF_t dt$. The present value of this leakage equals $f(t)e^{-rt}dt$. Thus, the present value of leaked cash flows received over the infinitesimal interval dt is:

$$dPV[0, t] = PV[0, t + dt] - PV[0, t] = f(t)e^{-rt}\, dt \tag{4.16}$$

To find the present value of a sum received over a finite interval beginning with $t = 0$ and ending with $t = T$, we apply the definite integral as follows:

$$PV[0, T] = \int_0^T f(t)e^{-rt}\, dt \tag{4.17}$$

In the case that the cash flow rate is constant $f(t) = CF_t = CF$, then the integral can be evaluated:

$$PV[0, T] = CF \int_0^T e^{-rt}\, dt = \frac{CF}{r}(1 - e^{-rt}) \tag{4.18}$$

More generally, the annuity could be deferred to time $t_0 > 0$, in which case its present value is calculated as follows:

$$PV[t_0, T] = CF \int_{t_0}^T e^{-rt}\, dt = \frac{CF}{r}(e^{-rt_0} - e^{-rT}) \tag{4.19}$$

4.2.2.2 Perpetuities

As T approaches ∞, the annuity and deferred annuity values approach that of a continuous perpetuity (an annuity whose cash flows never cease) or deferred perpetuity:

$$PV[0, \infty] = \int_0^\infty CFe^{-rt}\, dt = \frac{CF}{r} \tag{4.20}$$

$$PV[t_0, \infty] = \int_{t_0}^\infty CFe^{-rt}\, dt = \frac{CF}{r}(1 - e^{rt_0}) \tag{4.21}$$

ILLUSTRATION: CONTINUOUS ANNUITY

An investor's brokerage account continuously credits dividends and interest at a rate of $10,000 per year. The credits appear in equal installments during each interval of time (day or smaller time period dt) such that the installments can be modeled as though they are continuous. If these credits are discounted at an annual rate of $r = 5\%$, what is the present value of the cash flow stream over a 1-year period?

The amount of credits received by the investor during any infinitesimal interval dt equals $f(t)\, dt = \$10,000\, dt$. The present value of this sum credited at time t equals $f(t)e^{-rt}dt = \$10,000e^{-0.05t}dt$. To find the present value of the sum received over the year beginning with $t = 0$, solve Eq. (4.17):

$$PV[0, 1] = \int_0^1 f(t)e^{-rt}\, dt = \int_0^1 10,000e^{-0.05t}\, dt = \$10,000 \left[-\frac{e^{-0.05t}}{0.05} \right] \Big|_0^1$$

$$= \frac{\$10,000}{0.05}(1 - e^{-0.05}) = \$9754.12$$

ILLUSTRATION: CONTINUOUS DIVIDEND STREAMS

Many companies make dividend payments on a quarterly basis. However, payment calendars vary from firm to firm. An index simulating a portfolio of a large number of stocks paying dividends will probably reflect dividend payments scattered throughout the year. For example, a portfolio of securities constructed to replicate the S&P 500 Index would probably receive dividends from companies on close to a continuous basis (500 stocks, each four dividend payments per over 365 days).

Suppose that a fund receives dividends on a continuous basis at a rate of $25,000 per year starting at time $t = 0$. What is the value of the dividend stream received by this fund over a 10-year period if continuously discounted at a 4% rate? The present value of all dividends received in any infinitesimal interval $[t, t + dt]$ is as follows:

$$PV[t, t + dt] = PV[0, t + dt] - PV[0, t]$$

The sum of dividend payments over any infinitesimal time interval dt equals $f(t)\, dt = 25,000\, dt$. The present value of this payment equals $f(t)e^{-rt}\, dt = 25,000e^{-0.04t}\, dt$. To find the present value of the sum received over a finite interval beginning with $t = 0$, one can apply the definite integral as follows:

$$PV[0, 10] = \int_0^{10} f(t)e^{-rt}\, dt = \int_0^{10} 25,000e^{-0.04t}\, dt = \$25,000 \left[-\frac{e^{-0.04t}}{0.04} \right] \Big|_0^{10}$$

$$= \frac{\$25,000}{0.04}(1 - e^{-0.4}) = \$206,050$$

4.2.2.3 Growing Annuities

Now, consider a growing series of continuous cash flows, which can be derived with a growing annuity formula. First, we structure and combine the growth and discount functions to be integrated and obtain:

$$PV[t, t + dt] = f(t)e^{-rt}\, dt = CF_0 e^{gt} e^{-rt}\, dt = CF_0 e^{(g-r)t}\, dt$$

$$PV[0, T] = CF_0 \int_0^T e^{(g-r)t}\, dt \tag{4.22}$$

Now, suppose that our dividend stream is growing at a continuously compounded rate of 3% per year with an initial annual payment rate of $25,000 and is discounted at a 4% rate per year. Its present value over 10 years is:

$$PV[0, 10] = 25,000 \int_0^{10} e^{(0.03-0.04)t}\, dt = \$25,000 \left[-\frac{e^{(0.03-0.04)t}}{0.04 - 0.03} \right] \Big|_0^{10}$$

$$= \frac{\$25,000}{0.01}\left(1 - e^{(0.03-0.04) \times 10}\right) = \$237,906.50$$

4.2.3 Duration and Convexity in Continuous Time

Bonds are normally subject to three types of risk:

- *Default risk*: The risk that the issuer does not fulfill its obligations
- *Transactions risk*: The risk that the bond cannot be quickly and inexpensively sold in secondary markets
- *Interest rate risk*: The risk that interest rate changes on new instruments affect existing bond values.

Default risk and transactions risk tend to be somewhat negligible (this might be an exaggeration in some instances) for most US Treasury issues. Hence, we will focus on interest rate risk in this section. In order to estimate the change in the bond price resulting from a sudden change in the interest rate r, we can regard the bond price B_0 (say, at time 0) as a function of the interest rate: $B_0(r)$. Suppose the interest rate were to instantaneously change from r to $r + dr$, and that we are still at time 0. We can estimate the value of the resulting bond price at time 0 by its second-degree Taylor polynomial:

$$B_0(r + dr) \approx B_0(r) + \frac{dB_0}{dr} dr + \frac{1}{2} \frac{d^2 B_0}{d^2 r} (dr)^2 \tag{4.23}$$

We could also write the differential and divide both sides by B_0 as follows:

$$\frac{dB_0}{B_0} \approx \frac{1}{B_0} \frac{dB_0}{dr} dr + \frac{1}{2B_0} \frac{d^2 B_0}{d^2 r} (dr)^2 \tag{4.24}$$

Duration can be defined at the proportional bond price sensitivity to changes in interest rates and convexity as one-half of the second derivative divided by B_0 as follows:

$$\text{Dur} = \frac{1}{B_0} \frac{dB_0}{dr} \tag{4.25}$$

and

$$\text{Conv} = \frac{1}{2B_0} \frac{d^2 B_0}{dr^2} \tag{4.26}$$

From Eq. (4.23), one can see that duration provides a linear approximation to the proportional shift in the bond price per unit shift in the interest rate. Convexity, obtained from the change in duration resulting from a change in interest rates, captures some portion of the error associated with the linear approximation of the nonlinear bond price.[3]

Thus, by Eq. (4.23), the bond price after a change of dr in the interest rate can be approximated in terms of duration and convexity in the form:

$$B_0(r + dr) \approx B_0(r) + \text{Dur} \times B_0 \, dr + \text{Conv} \times B_0 (dr)^2 \tag{4.27}$$

We can also study the effects on the bond price as one moves forward in time. The bond price can be regarded as a function of the interest rate r and time t: $B_t(r)$. Suppose, over the time interval $[t, t + dt]$, the interest rate changes from r to $r + dr$. Then we can estimate the change in the bond price by

$$dB_t(r) \approx \text{Dur} \times B_t \, dr + \text{Conv} \times B_t (dr)^2 + \frac{\partial B}{\partial t} dt \tag{4.28}$$

with the understanding that the definitions of duration and convexity in Eqs. (4.25) and (4.26) are replaced with partial derivatives in r. This relationship between the sensitivity of bond prices to interest rate shifts will be important in Chapter 8 when we discuss interest rate processes.

4.2.3.1 Zero-Coupon Instruments

Suppose that a T-year zero-coupon bond with face value $F = B_T$ has a price path, so that the price B_t at time t satisfies the following differential with a given current interest rate r_0:

$$\frac{dB_t}{B_t} = r_0\, dt \qquad (4.29)$$

Solving in a manner similar to earlier calculations, we obtain the following:

$$B_t = B_T e^{r_0(t-T)} = F e^{r_0(t-T)}$$

At time 0, we have:

$$B_0 = F e^{-r_0 T}$$

Assume that the current rate r_0 applies to all time periods before T (flat yield curve). Suppose the interest rate (or continuous yield to maturity) r_0 applicable to all time periods (parallel yield curve shift) prior to T were to suddenly shift by dr at time 0. Let $B_0(r_0)$ denotes the price of the bond at time 0 with interest rate r_0. We can estimate $B_0(r_0 + dr)$ using Eq. (4.27) above. First, we calculate the duration and the convexity:

$$\text{Dur} = \frac{1}{B_0} F e^{-r_0 T}(-T) = -T$$

and

$$\text{Conv} = \frac{1}{2B_0} F e^{-r_0 T}(-T)(-T) = \frac{1}{2}T^2$$

Observe that for a zero-coupon bond that is modeled by Eq. (4.29), the duration equals the negative of its remaining life before maturity. Equation (4.27) gives the estimate for the price after the shift dr:

$$B_0(r_0 + dr) \approx B_0\left[1 - T\,dr + \frac{1}{2}T^2(dr)^2\right]$$

4.2.3.2 Coupon Instruments

Coupon bonds make payments at regular intervals in addition to a balloon payment when the bond matures at time T. When yields are continuously compounded, the time zero value of a bond paying interest at rate c (not to be confused with the constant of integration that we discussed earlier) on face value F is given by

$$B_0 = F e^{-r_0 T} + \sum_{t=1}^{T} cF e^{-r_0 t}$$

This expression reflects both the principle and all coupon payments made from time 0 until the bond matures at time T. The duration of a bond reflects the proportional change in the value of a bond given a unit change in the interest rate:[4]

$$\text{Dur} = \frac{1}{B_0}\frac{dB_0}{dr}$$

We find the derivative of the bond's value with respect to interest rates as follows:

$$\frac{dB_0}{dr} = -T\text{FV}e^{-r_0 T} - \sum_{t=1}^{T} t\text{cFV}e^{-r_0 t}$$

Now substitute the derivative and the current bond values into the duration formula to obtain:

$$\text{Dur} = \frac{1}{B_0}\frac{dB_0}{dr} = \frac{-T\text{FV}e^{-r_0 T} - \sum_{t=1}^{T} t\text{cFV}e^{-r_0 t}}{\text{FV}e^{-r_0 T} + \sum_{t=1}^{T} c\text{FV}e^{-r_0 t}} \tag{4.30}$$

We observe here that duration can be interpreted as a weighted average maturity of payment obligations associated with the bond. Similarly, convexity is computed as follows:

$$\text{Conv} = \frac{1}{2B_0}\frac{d^2 B_0}{dr^2} = \frac{1}{2}\frac{T^2\text{FV}e^{-r_0 T} + \sum_{t=1}^{T} t^2 c\text{FV}e^{-r_0 t}}{\text{FV}e^{-r_0 T} + \sum_{t=1}^{T} c\text{FV}e^{-r_0 t}} \tag{4.31}$$

Again, convexity captures some portion of the error associated with the linear approximation of the nonlinear function:

$$\text{Dur} = \frac{dB_t}{dr}\frac{1+r}{B_0}$$

ILLUSTRATION: DURATION AND CONVEXITY

Consider a 3-year 10% coupon bond with face value equal to 1000. The current yield to maturity is 10%. Coupon payments are made at the end of each year. The current market value, duration, and convexity for this bond are computed as follows:

$$B_0 = 1000e^{-0.1 \times 3} + 100e^{-0.1 \times 3} + 100e^{-0.1 \times 2} + 100e^{-0.1 \times 1} = 987.2569$$

$$\text{Dur} = \frac{-3 \times 1000e^{-0.1 \times 3} - 3 \times 100e^{-0.1 \times 3} - 2 \times 100e^{-0.1 \times 2} - 1 \times 100e^{-0.1 \times 1}}{987.2569} = -2.733$$

$$\text{Conv} = \frac{1}{2}\frac{9 \times 1000e^{-0.1 \times 3} + 9 \times 100e^{-0.1 \times 3} + 4 \times 100e^{-0.1 \times 2} + 1 \times 100e^{-0.1 \times 1}}{987.2569} = 3.926$$

Notice that the bond's duration is simply a weighted average (based on discounted cash flows) of the dates associated with bond payments. Bond convexity is a similar weighted average. Now, suppose that interest rates increased by 2% to 0.12. We can use duration and convexity in first- and second-order Taylor expansions to approximate the resulting bond price. First- and second-order approximations are as follows:

$$B_3(0.12) \approx 987.2569 \times [1 - 0.02(2.733)] = 933.27$$

$$B_3(0.12) \approx 987.2569 \times \left[1 - 0.02(2.733) + (0.02)^2(3.976)\right] = 934.83$$

The actual bond value under the new interest rate at 0.12 is 934.7988. The second-order Taylor series approximation provides a reasonably good estimate of this value.

4.2.4 The Yield Curve in Continuous Time

As we discussed in Chapter 3, the yield curve depicts varying spot rates over terms to maturity. Spot rates are interest rates on loans originated at time 0 (e.g., now). The yield curve is typically constructed from the yields of benchmark, highly liquid (where possible), default-free fixed income, and/or zero-coupon instruments. Debt instruments, including Treasury bills and Treasury bonds issued by national governments, are often used as benchmarks since they are generally considered to be the least likely to default and have the highest liquidity in most economies. We calculate spot rates in continuous time from cash flows yielded by bonds (i.e., coupon payments and principal repayments) and their market prices as we did in Chapter 3 for discrete time. The term structure is crucial because yields on securities with fixed maturities are the backbone for obtaining discount rates and riskless return rates needed for valuing bonds, stocks, derivatives, and other instruments.

We start by modifying a few of our definitions from Chapter 3 and earlier in this chapter to accommodate zero-coupon bonds in continuous time. We will make extensive use of these modified definitions later, particularly in Chapter 8. For now, we will focus on zero-coupon bonds in arbitrage-free markets. Given a market spot rate $r_{0,T}$, the current market price of a T-year zero-coupon bond with face value $F = B(T,T) = B_T = 1$ is[5]:

$$B_0 = B(0, T) = e^{-r_{0,T} T} \tag{4.32}$$

Given the absence of coupon payments and arbitrage opportunities, the yield to maturity $r_{0,T}$ of the T-year zero-coupon bond described above equals the T-year spot rate:

$$r_{0,T} = \frac{-\ln(B(0, T))}{T} \tag{4.33}$$

Yield to maturity is simply the discount rate that equates the market price of the zero-coupon bond equal to the present value of its principal or face value. The various spot and forward rates are determined in the marketplace from varying market prices of bonds. In the absence of arbitrage, the yield to maturity of the zero-coupon bond will equal the spot rate. Now, we will divide the lifetime of the bond $(0, T)$ into two periods, $(0, t)$ and (t, T), so that we can distinguish between a short-term spot rate $r_{0,t}$, a long-term spot rate $r_{0,T}$, and a forward rate $r_{t,T}$. Based on the long-term spot rate $r_{0,T}$, we see that the relationship among the market price $B(0, T)$ of the T-year one dollar face value zero-coupon bond, the short-term spot rate $r_{0,t}$, and the short-term forward rate $r_{t,T}$ is:

$$B_0 = B(0, T) = e^{-r_{0,T} T} = e^{-r_{0,t} t} e^{-r_{t,T}(T-t)} \tag{4.34}$$

assuming that we can infer rates from prices. Notice that we have assumed that $r_{0,T} \times T = (r_{0,t} \times t) + (r_{t,T} \times (T - t))$, which is a statement of the pure expectations theory of interest rates, modified from the discrete case presented in Chapter 3. A forward rate $f(t, T) = r_{t,T}$ is the rate agreed to (implied agreement in this case) at time 0 on a zero-coupon bond originating at time t and maturing at time T:

$$f(t, T) = \frac{\ln(B(0, t)) - \ln(B(0, T))}{T - t} = r_{t,T} \tag{4.35}$$

Now, we will allow for additional time periods in our framework. In a similar vein, we can state that the time 0 instantaneous forward rate at time t for the bond described above is:

$$f(t, t) = \lim_{dt \to 0} \frac{\ln(B(0, t)) - \ln(B(0, t + dt))}{dt} \tag{4.36}$$

where $T = t + dt \to t$ in Eq. (4.35) in order to obtain Eq. (4.36). This instantaneous rate applies to a loan that is instantly repaid after its origination. Though the instantaneous rate is unobservable in actual financial markets, it might be proxied by a forward overnight rate. If the "instantaneous loan" is extended now, $f(t,t) = r_0$ is an instantaneous spot rate. In this same framework, we rewrite our bond pricing equation (4.34) in a continuous market with instantaneous spot rate r_0 and instantaneous forward rates $f(s,s)$ for all periods $0 < s \le T$ in continuous time:

$$B_0 = B(0, T) = e^{-r_{0,T}T} = e^{-\int_0^T f(s,s)ds} \tag{4.37}$$

which, in arbitrage-free markets, produces the T-year spot rate $r_{0,T}$:

$$r_{0,T} = \frac{\int_0^T f(s, s)\, ds}{T} = \frac{-\ln(B(0, T))}{T}$$

Observe that the T-year spot rate is the average (mean) of the instantaneous forward rates from time 0 to T.

4.2.5 Term Structure Theories

As we discussed earlier, the term structure is concerned with how interest rates, for example, how a given long-term spot rate, varies with respect to term to maturity. However, this relationship between rate or rates and term to maturity itself shifts over time. Term structure theories are concerned with how and why these term structure shifts occur. Why does the yield curve take one particular shape at one point in time and a very different shape a year later? There are a variety of term structure theories, and they should not be understood to be mutually exclusive. The theories that we focus on in this book are largely based on or derived from the *expectations theory*, meaning that forward rates of interest are functions of expected future short-term interest rates. This implies that spot rates are functions of expected short-term interest rates. More specifically, the *pure expectations theory* states that forward rates of interest are equal to expected future short-term spot rates, implying that spot rates are the average of expected instantaneous spot and forward interest rates r_s prevailing over time 0 to T as in Eq. (4.33) above:

$$r_{0,T} = \frac{-\ln(B(0,T))}{T} = \frac{-\ln\left[E\left[Fe^{\left(-\int_0^T r_s\, ds\right)}\middle| r_0\right]\right]}{T}$$

where the face value of the T-year zero-coupon bond is F. Essentially, the pure expectations theory holds that an investor should be indifferent between holding a long-term bond to its maturity and continuously rolling over short-term (instantaneous maturity) bonds sequentially to the same termination date. Notice that the above equation describing the term structure is based on the expected geometric mean of expected instantaneous spot rates over time to the maturity date T.

However, we generally observe that yield curves are upward sloping ("normal" yield curves, as in Figure 4.1) far more often than downward sloping (when yield curves are said to be inverted), suggesting an upward bias in the curve. Taken to an extreme, the pure expectations theory would imply that in the long run, interest rates tend towards infinity since the yield curve tends to be upward sloping. But, we know that spot interest rates do not approach infinity over time. Alternatively, this upward bias in the yield curve might be explained by risk aversion; risk premiums in longer loans lead to a *biased expectations theory*. These risk premiums result from unknown selling prices for bonds sold prior to their maturities and from associated reinvestment risks (unknown returns from selling bonds prior to maturity then reinvesting the proceeds). Biased expectation theories account for these risk premiums as follows:

$$r_{0,T} = \frac{-\ln(B(0,T))}{T} = \frac{-\ln\left[E\left[Fe^{\left(-\int_0^T r_s\, ds + f_T\right)}\middle| r_0\right]\right]}{T}$$

where f_T is a function of the volatility σ_s associated with these risks and the market price of risk Θ_s over the times s from 0 to T. The terms f_T, σ_s, and Θ_s are used rather loosely here, and can be defined and used more precisely as in Chapter 8. This risk adjustment can lead to an upward bias in the yield

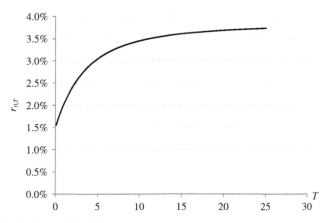

FIGURE 4.1 Upward sloping (normal) yield curve.

curve as we would anticipate. The buy and hold strategy associated with a long-term bond will produce higher returns than the short-term rollover strategies when yield curves are upward sloping.

The *preferred habitat theory* (Modigliani and Sutch, 1966) is a generalization of the biased expectations theory and shares similarities to other theories, allowing borrowers and lenders to have any combinations of preferred time horizons, which, in turn, could allow the market price of risk to be either positive or negative. Essentially, the preferred habitat theory maintains that some borrowers and lenders prefer to hold or issue long-term instruments while others prefer to hold or issue short-term instruments. The balance of these two types of participants determines the shape of the yield curve.

The *liquidity preference theory* (Hicks, 1946) can also explain the upward bias in the yield curve. This theory maintains that lenders generally prefer to make loans for short periods and borrowers prefer to borrow for longer periods at fixed rates (consider, for example, commercial bank customers). This leads to an excess of supply for long-term instruments, raising their rates, and excess demand for short-term instruments, decreasing their rates. These supply/demand imbalances lead to long-term rates generally exceeding short-term rates. As with the biased expectation theory, the liquidity preference theory can account for these risk premiums as follows:

$$r_{0,T} = \frac{-\ln(B(0,T))}{T} = \frac{-\ln\left[E\left[Fe^{\left(-\int_0^T r_s \, ds + LP_T\right)}|r_0\right]\right]}{T}$$

where LP_T is simply a liquidity premium. The difference between the two theories, when there is a difference, is that the liquidity preference theory specifies the particular reason for the higher long-term rates.

The *market segmentation theory* (Culbertson, 1957) focuses on fundamental supply and demand factors underlying debt instruments. Macroeconomic variables, business cycle issues, government policy, liquidity needs, and risk aversion all play roles in the supply and demand of debt instruments, where instruments of different terms are taken to be nonsubstitutable. It is not necessary for risk premiums to monotonically rise with terms to maturity. This theory is useful in explaining why segments of the yield curve fail to shift in unison, but fails to explain why they often move together.

While these term structure theories generally offer explanations as to how the yield curve is shaped at any point in time, they do not adequately explain how the yield curve itself evolves over time. In Chapter 8, we will discuss term structure models, which do seek to explain the evolution of yield curves over time.

4.3 CONTINUOUS STATE MODELS

In the previous chapter, we worked with securities in single-period economies with multiple states. Thus far in this chapter, we have worked only with single-state models that imply certainty. If only a single outcome is possible, the economy is certain. Multiple states imply uncertainty, but pricing is still possible in complete markets. In this chapter, we will seek to price securities relative to others in economies with infinitely many states, though we will still rely on the same risk-neutral pricing methodology as in the previous chapter. First, we will start with a few preliminaries leading to the distribution of states.

4.3.1 Option Pricing: The Elements

In this section, we will value options in single time-period frameworks with multiple potential outcomes in continuous outcome space. We will construct very simple probability distributions to compute expected values of options as functions of conditional expected stock values. Notice that we will work with risk-neutral q probabilities.

4.3.1.1 Expected Values of European Options

The expected future value of a European call is equal to its expected value conditional on its exercise at time T: $E[(S_T - X)|S_T > X]$ multiplied by the probability that it will be exercised $P[S_T > X]$. We will define $q[S_T]$ to be the hedging probability density function S_T.[6] If the range of potential stock prices is continuous, this expected value is written as follows:

$$E[c_T] = E\left[\text{Max}(S_T - X), 0\right] = \int_X^\infty (S_T - X)q[S_T]\, dS_T$$

where $q(S_T)$ is the density associated with a stock priced at S_T at time T. Note that the call's value is zero when $S_T < X$. The probability that the call will be exercised is:

$$P[S_T > X] = \int_X^\infty q[S_T]\, dS_T$$

The expected value of the stock and call given that the stock price exceeds the call exercise price is:

$$E[S_T | S_T > X] = \frac{\int_X^\infty S_T q[S_T] dS_T}{\int_X^\infty q[S_T] dS_T}$$

$$E[c_T | S_T > X] = \frac{E[c_T]}{P[S_T > X]} = \frac{\int_X^\infty (S_T - X)q[S_T]\, dS_T}{\int_X^\infty q[S_T]\, dS_T}$$

These conditional expected values do not account for stock prices when $S_T < X$. The expected value of the call is simply the product of its conditional expected value and the probability that it is exercised.

ILLUSTRATION: CALL OPTIONS AND UNIFORMLY DISTRIBUTED STOCK PRICES

Suppose that a stock's price at time T is expected to be uniformly distributed over the integer values that range from 1 to 100; that is:

$$q(x) = \begin{cases} \dfrac{1}{100}, & 0 \le x \le 100 \\ 0, & \text{elsewhere} \end{cases}$$

Further suppose that a call option with exercise price $X = 60$ trades on this stock with the following terminal payoff function:

$$c_T = \text{Max}[S_T - X, 0]$$

The probability that the call will be exercised is computed as follows:

$$Q[S_T > 60] = \int_{60}^{100} \frac{1}{100} dS_T = \frac{1}{100} S_T \Big|_{60}^{100} = 1 - 0.6 = 0.4$$

The expected value of the stock, contingent on its price exceeding 60, is:

$$E[S_T | S_T > 60] = \frac{\int_{60}^{100} \frac{S_T}{100} dS_T}{\int_{60}^{100} \frac{1}{100} dS_T} = \frac{\frac{1}{200} S_T^2 \Big|_{60}^{100}}{0.4} = \frac{32}{0.4} = 80$$

The expected value of the call, contingent on it being exercised, is:

$$E[c_T | S_T > X] = \frac{\int_X^\infty (S_T - X) q[S_T] dS_T}{\int_X^\infty q[S_T] dS_T} = \frac{\int_{60}^{100} \frac{S_T - 60}{100} dS_T}{\int_{60}^{100} \frac{1}{100} dS_T} = \frac{\frac{1}{200} S_T^2 - 0.6 S_T \Big|_{60}^{100}}{0.4}$$

$$= \frac{(50 - 60) - (18 - 36)}{0.4} = \frac{8}{0.4} = 20$$

The expected value of the call at time T is:

$$E[c_T] = \int_{60}^{100} (S_T - 60) \frac{1}{100} dS_T = \frac{1}{200} S_T^2 - 0.6 S_T \Big|_{60}^{100} = (50 - 60) - (18 - 36) = 8$$

The present value of this call would simply be its discounted value:

$$c_0 = \int_X^\infty (S_T - X) q[S_T] dS_T e^{-rT} = 8e^{-rT}$$

4.4 EXERCISES

4.1. Find the value of a bond that has a face value of $1000 that matures in 1 year. The monthly rate at which interest accrues at time t (in months) after origination is determined by the following simple model: $r(t) = 0.007 - 0.00003t^2$.

4.2. Which of the following are separable differential equations?

a. $\dfrac{dy}{dt} = ty^2 - 2t^2 y$

b. $\dfrac{dy}{dt} = ty + t$

c. $\dfrac{dy}{dt} = y + t$

4.3. Solve the following initial value problem:

$$\frac{dy}{dt} = yt, \quad y_0 = 100$$

4.4. a. Solve the following equation assuming that $B_0 = 1$. This equation is used for pricing riskless bonds:

$$\frac{dB_t}{dt} = r_t B_t$$

b. Solve the above equation assuming that $B_4 = 1$.

4.5. Suppose that a statistical time series analysis revealed that an investment grows at a rate proportional to the square root of its current value with constant of proportionality k:

$$\frac{dV}{dt} = k\sqrt{V} \text{ and } V(0) = V_0$$

Find the value of this investment as a function of time.

4.6. Suppose that the following separable differential equation reflects the continuous growth in a stock's price over time:

$$\frac{dS}{dt} = 0.01S$$

a. Solve this differential equation.

b. Suppose that $S_0 = 50$. Write the particular solution to this differential equation.

4.7. A T-year zero-coupon bond with face value $B_T = F$ has a price path given by the following differential equation:

$$\frac{dB_t}{B_t} = r_0 \, dt$$

Provide a general and a financially appropriate particular solution to this differential equation.

4.8. Suppose that a stock's price grows over time at rate μ as the firm produces profit, but the firm also pays a continuous dividend at a rate δ.

a. Write a differential equation to model the stock price over time.

b. What is the solution to this differential equation.

c. Suppose that $S_0 = 50$, $\mu = 0.05$, and $\delta = 0.02$. Write the particular solution to this differential equation.

4.9. Suppose that a particular process S_t satisfies the differential $dS_t = \mu(M - S_t)\, dt$ and initial value S_0 with $0 < S_0 < M$. Find the solution for S_t that is valid as long as $0 < S_t < M$.

4.10. Suppose a particular contract calls for accruals on an account to be credited such that over the time interval $[t, t + dt]$, where t is time in years, the amount that is credited equals $\$(10,000 + 500t)dt$. If these credits are discounted at an annual rate of 4%, what is the present value of the account after T years?

4.11. The People's Republic of Chrystal seeks to maintain a target exchange rate of $\mu = PRC6$ relative to the US dollar. Should the actual exchange rate r differ from its target rate, it will tend to drift towards the target rate as per the following function where λ and e^K are constants:

$$|\mu - r| = e^{\lambda t} e^K$$

Suppose the exchange rate 3 days ago was 6.5 and is currently 6.4. How long will it take for the rate to drop below 6.1?

4.12. Suppose that short-term interest rates follow the following mean reverting process:

$$dr = \lambda(\mu - r_t)dt$$

The long-term mean interest rate μ equals 0.06, the short-term rate r_0 currently equals 0.02, and the "pullback factor" λ equals 0.4.
a. What will be the short-term rate r_1 in 1 year?
b. What will be the short-term rate r_2 in 2 years?

4.13. Here, we introduce the logistic growth function. Suppose that μ represents the long-term mean interest rate, r the rate at a particular time, and λ an interest rate adjustment mechanism:

$$dr = \lambda r(\mu - r)dt$$

1. Is this differential equation separable? If so, demonstrate.
2. Solve this differential equation.
3. Suppose that the People's Republic of Chrystal seeks to maintain a target exchange rate of $\mu = PRC6$ relative to the US dollar, the adjustment mechanism λ in its logistic growth function is known to be 0.007 and the current exchange rate is 5. Find an equation that provides the exchange rate for any time t.
4. What is the exchange rate 10 days from now?

4.14. A pension fund collects $800,000 in dividends per year from its various securities. The dividends are received in equal installments during each interval (day or dt) during the year such that they can be modeled as being continuous. If dividends are to be discounted at an annual rate of 4%, what would be the present value of the dividend stream over the next 10 years?

4.15. A dividend stream with an initial annual rate of $10,000 will grow at a continuously compounded rate of 3% per year. Cash flows are discounted at 5%. Find the present value of this stream over its first 2 years.

4.16. An individual wishes to retire with $1,000,000 invested at an annual interest rate of 4%. She plans to withdraw $50,000 per year for living expenses after retirement. Assume that she will make withdrawals continuously throughout each year.
a. Let PV designate the present value of the account, FV_t designate the value of the account at time t, PMT designate the payment made from the account during each year, and r designate the interest rate. Devise an appropriate differential equation describing the rate of change in the retiree's account during any given time period.
b. Solve the differential equation to find the balance of the account at any time T.

c. Based on the solution to the relevant differential equation and the numbers given in the example, how much money would the retiree have in her account after 8 years?

d. If the retiree continues to spend $50,000 per year, at what point (in how many years) will she run out of money?

4.17. Work through each of the parts of problem 4.16 assuming that the investor wished for her withdrawals to start at an annual rate of $100,000 but increase at an annual rate of 2% to cover anticipated inflation.

4.18. Suppose that the time t price of a zero-coupon bond is given by the following:

$$B_t \approx Ve^{-r_0(T-t)}$$

Further assume that duration (which equals modified duration when yields are continuous) is generally given by the formula:

$$\text{Dur} = \frac{dB_t}{dr}\frac{1}{B_t}$$

Demonstrate that duration equals $-(T - t)$.

4.19. Consider a 5-year 12% coupon bond with face value equal to 1000. The current yield to maturity and market rate are 10%. Coupon payments are made at the end of each year.

a. Calculate the current market value, duration, and convexity for this bond.

b. Suppose that interest rates increased by 3% to 0.13. Use duration and convexity calculations to approximate the resulting bond price.

c. Calculate the actual bond value under the new market interest rate.

4.20. Suppose that the current market yield to maturity on a 7-year riskless zero-coupon bond is 5% and the current market yield to maturity on a 10-year riskless zero-coupon bond is 6%. Calculate the following:

a. The value of a 7-year $F = \$1$ face value bond.

b. The value of a 10-year $F = \$1$ face value bond.

c. The 3-year forward rate on a bond originating in 7 years.

4.21. Suppose that a stock's price at time T is expected to be uniformly distributed over integers in the range 10 to 20, that is:

$$q(x) = \begin{cases} \dfrac{1}{10}, & 10 \le x \le 20 \\ 0, & \text{elsewhere} \end{cases}$$

Further suppose that a put option with exercise price $X = 16$ trades on this stock with the following terminal payoff function:

$$p_T = \text{Max}[X - S_T, 0]$$

a. Calculate the probability that the put will be exercised.

b. Calculate the expected value of the stock's price contingent on it exceeding the exercise price of the put.

c. Calculate the conditional expected value of the put contingent on it being exercised.

d. Calculate the expected future value of the put.

e. If $T = 2$ and $r = 0.1$, what is the current value of the put? Assume that the stock price after the first period does not affect the parameters of the distribution of stock prices for the second period.

NOTES

1. When Δt is very small $\ln(1 + r(t_{i-1})\Delta t) \approx r(t_{i-1})\Delta t$. As $\Delta t \to 0$, this approximation improves, as long as r has a continuous second derivative.

2. Note the similarity of this mean reverting interest structure to Newton's law of cooling. Also, note that the rate will never actually revert all the way back to its long-term mean.

3. Note that for bonds with continuous yields, Macaulay's duration formula equals modified duration. Do not confuse these formulas for duration and convexity with those that apply to discrete yields.

4. Sometimes duration is defined as the proportional change in bond value induced by a proportional change in interest rates.

5. B_T is simply short-hand for $B(T,T)$, the value of a T-year bond when it matures at time T. B_0 is short-hand for $B(0,T)$, the time zero value of a T-year bond.

6. A hedging probability, as discussed in Chapter 3, is the probability assigned by a risk-neutral investor who values an asset at the prevailing market price.

References

Culbertson, J.M., 1957. The term structure of interest rates. Q. J. Econ. 71, 485–517.

Hicks, J.R., 1946. Value and Capital. 2nd ed Oxford University Press, London.

Modigliani, F., Sutch, R., 1966. Innovations in interest rate policy. Am. Econ. Rev. 56, 178–197.

An Introduction to Stochastic Processes and Applications

5.1 RANDOM WALKS AND MARTINGALES

In the previous two chapters, we priced securities in the marketplace as functions of pure security prices, synthetic probabilities, interest rates, and discount functions. Here, we consider the functions that drive the random natures of one or more of these values or functions.

5.1.1 Stochastic Processes: A Brief Introduction

A *stochastic process* is a sequence of random variables X_t defined on a common probability space $(\Omega, \mathcal{F}, \mathbb{P})$ and indexed by time t. In other words, a stochastic process is a random series of values X_t sequenced over time. The values of $X_t(\omega)$ as t varies define one particular sample path of the process associated with the state or outcome $\omega \in \Omega$. The terms $X(\omega, t)$ and $X_t(\omega)$ are used synonymously here.

A *discrete time process* is defined for a countable set of time periods. This is distinguished from a *continuous time process* that is defined over an interval of the real line that consists of an infinite number of times. The *state space* X is the set of all possible values of the stochastic process $\{X_t\}$:

$$X = \{X_t(\omega) \text{ for some } \omega \in \Omega \text{ and some } t\}$$

The state space can be discrete (countable) or continuous. For example, if a bond price changes in increments of eighths or sixteenths, the state space for prices of the bond is said to be discrete. The state space for prices is continuous if prices can assume any real value.

The value of a real-world security is its value at any particular time, which depends on all past and present information. As we know, the set of possible events that determine the value of a random variable X_t is called its σ-algebra, which we will denote by $\mathcal{F}_t$. This motivates the following definition: A sequence of *filtrations* $\mathcal{F}_t$ for an σ-algebra $\mathcal{F}$ is a sequence of σ-algebras with the property:[1]

$$\mathcal{F}_0 \subset \mathcal{F}_1 \subset \mathcal{F}_2 \subset \cdots \subset \mathcal{F}$$

P.M. Knopf & J.L. Teall: Risk Neutral Pricing and Financial Mathematics: A Primer.
DOI: http://dx.doi.org/10.1016/B978-0-12-801534-6.00005-9

for discrete processes. Typically, the σ-algebras $\mathscr{F}_t$ are getting larger as they evolve in time t. Similarly, continuous processes will have the following property:

$$\bigcup_{s<t} \mathscr{F}_s \subset \mathscr{F}_t \subset \mathscr{F}$$

for all $0 \le s \le t$

A stochastic process X_t is said to be *adapted* to a filtration $\{\mathscr{F}_t\}$ if every measurable event for the random variable X_t is in the σ-algebra $\mathscr{F}_t$. Filtrations arise when securities are modeled by stochastic processes, because as time passes, the number of possibilities for the history of the security grows. In effect, a filtration represents the increasing stream of information (history) concerning the process.

ILLUSTRATION: FILTRATIONS IN A TWO-TIME PERIOD RANDOM WALK

Suppose a stock has an initial price of X_0 at time 0, which, at time 1, will either increase or decrease by 1. Independently, the same potential price changes will occur at time 2. All of the possible distinct price two-period change outcomes for the stock to time 2 are $(1,1)$, $(1,-1)$, $(-1,1)$, and $(-1,-1)$. For example, $(1,-1)$ will mean the stock's price increased by 1 at time 1 and decreased by 1 at time 2. The sample space for this process is $\Omega = \{(1,1), (1,-1), (-1,1), (-1,-1)\}$.

Actually, stochastic process is defined for all times $t = 0, 1, 2, 3, \ldots$. The example above is just for illustrative purposes. Suppose that the investor sells her stock at time 2, then keeps the proceeds in cash. In this scenario, for each time period $t = 3, 4, \ldots$, the value of the stock $X_t = X_2$. This would then conform to the proper definition of a stochastic process. The σ-algebras $\mathscr{F}_t = \mathscr{F}_2$ for all $t = 3, 4, \ldots$.

The price of the stock X_t for $t = 0, 1,$ and 2 is adapted to a filtration that we will now describe. At time $t = 0$, the stock price is known. Since there has been no increase or decrease in the price at time 0, the time 0 σ-algebra consists simply of:

$$\mathscr{F}_0 = \{\varnothing, \Omega\}$$

At time 1, we have acquired information regarding whether the price increased or decreased by 1. At this time, the sets $\{(1,1), (1,-1)\}$ and $\{(-1,1), (-1,-1)\}$ are added to the σ-algebra such that the σ-algebra at time 1 becomes:

$$\mathscr{F}_1 = \{\varnothing, \{(1,1), (1,-1)\}, \{(-1,1), (-1,-1)\}, \Omega\}$$

We cannot decouple the outcome $(1,1)$ from the outcome $(1,-1)$ in the σ-algebra $\mathscr{F}_1$ because we are unable to distinguish these two outcomes at time 1. That is, at time 1, the outcome for time 2 has yet not occurred and is still unknown; therefore, it is not part of the time 1 information set.

At time 2, each of the separate outcomes $(1,1)$, $(1,-1)$, $(-1,1)$, $(-1,1)$ are possible. Thus, $\mathscr{F}_2$ consists of the power set of $\Omega = \{(1,1), (1,-1), (-1,1), (-1,-1)\}$:

$$\mathscr{F}_2 = \begin{matrix}
\{\varnothing & \{(-1,-1)\} & \{(1,-1),(-1,1)\} & \{(1,1),(1,-1),(-1,-1)\} \\
\{(1,1)\} & \{(1,1),(1,-1)\} & \{(1,-1),(-1,-1)\} & \{(1,1),(-1,1),(-1,-1)\} \\
\{(1,-1)\} & \{(1,1),(-1,1)\} & \{(-1,1),(-1,-1)\} & \{(1,-1),(-1,1),(-1,-1)\} \\
\{(-1,1)\} & \{(1,1),(-1,-1)\} & \{(1,1),(1,-1),(-1,1)\} & \Omega\}
\end{matrix}$$

Note that there are $2^4 = 16$ members of this σ-algebra. Observe that $\mathscr{F}_0 \subset \mathscr{F}_1 \subset \mathscr{F}_2$ and define $\mathscr{F} = \mathscr{F}_2$. Thus, $\{\mathscr{F}_0, \mathscr{F}_1, \mathscr{F}_2\}$ is the filtration for this two-period stock process.

5.1.2 Random Walks and Martingales

A *discrete Markov process*, often called a *discrete random walk*, is a noncontinuous stochastic process in which the probability evolves to a given state at time t depends only on its immediately prior state at time $t - 1$ and not on the remainder of its history. A Markov process is characterized by the condition:

$$P(X_t | X_{t-1}, X_{t-2}, \ldots, X_0) = P(X_t | X_{t-1}) \tag{5.1}$$

The following provides an interpretation of this condition. Given the entire history of the process X_t from its start to time $t - 1$; namely, X_0, X_1, ..., X_{t-1}, the probability that the process will be in state X_t at the next moment time t depends only on its present state X_{t-1}. This property can be described as saying that a Markov process is memoryless. It should also be noted that the notation in Eq. (5.1) is an abbreviation for writing:

$$P(X_t = b | X_0, X_1 = a_1, X_2 = a_2, \ldots, X_{t-1} = a_{t-1}) = P(X_t = b | X_{t-1} = a_{t-1})$$

That is, the probability that X_t will take on a particular value, say b, given the values of X_s up to time $s = t - 1$ equals the probability that X_t takes on the value b just given the value X_{t-1}. The probability that the process will be in state X_t at the next moment in time t depends only on its present state X_{t-1}.

It can be proven that condition (5.1) implies the more general condition:

$$P(X_t | X_0, X_1, \ldots, X_s) = P(X_t | X_s) \forall s < t \tag{5.2}$$

This characterization of a Markov process can be generalized further to the condition:

$$P(X_t | \mathscr{F}_s) = P(X_t | X_s) \ \forall s < t \tag{5.3}$$

Recall that $\mathscr{F}_s$ is an σ-algebra of sets whose elements are drawn from all possible outcomes up to time s. Thus, if $\mathscr{F}_s$ is given, then we know which particular history of outcomes occurred to time s. This information determines the values of X_s, X_{s-1}, ..., X_0, so that condition (5.3) implies condition (5.2). Condition (5.3) is in some sense the most natural of the three conditions in pricing securities in the market. That is, we typically wish to project some future price X_t based on some history to time s prior to time t. The σ-algebra $\mathscr{F}_s$ denotes all possible events that can occur in the market up to time s. Any security X_s that we wish to price in this market will be a random variable with $\mathscr{F}_s$ as its σ-algebra. These concepts will be illustrated shortly when we consider a family of Markov processes modeling security prices.

Let X_t be a stochastic process over continuous time t, and random variables X_t be continuous. We define a *continuous Markov process* by the condition:

$$P(X_t \in A | \mathscr{F}_s) = P(X_t \in A | X_s) \forall s < t \tag{5.4}$$

where A is any measurable set on the real line. In Eq. (5.4), the parameters s and t take on nonnegative real numbers. Note that it is not sufficient to define the probabilities for only single outcomes X_t as we did for the case when X_t is discrete, since the probability of any given single outcome is normally equal to 0 for continuous random variables X_t. Similar to the discrete case, given $\mathscr{F}_s$, one knows X_i for all $0 \le i \le s$. Thus, condition (5.4) implies:

$$P(X_t \in A|X_i, 0 \le i \le s) = P(X_t \in A|X_s) \,\forall s < t$$

Observe that Eq. (5.4) is analogous to Eq. (5.3). In fact, if we choose the set $A = \{a\}$ for a particular value a of X_t and let the parameters s and t in Eq. (5.4) take on nonnegative integer values rather than continuous values, then Eq. (5.4) is exactly Eq. (5.3).

ILLUSTRATION: THE RANDOM WALK

Consider the following type of random walk.[2] A person starts a walk at a certain integer position X_0 along an x-axis ($\ldots, -3, -2, -1, 0, 1, 2, 3, \ldots$) and at every moment ($t = 0, 1, 2, \ldots$), she chooses to take either one step to the right with probability p ($0 < p < 1$) or one step to the left with probability $1 - p$, such that each step to the right or left is independent of prior steps. Let Z_i be the random variable that equals 1 if the step is to the right at time i and -1 if the step is to the left at time i. It is obvious that if each step is taken independently of any previous steps, then $P(Z_s = a, Z_t = b) = P(Z_s = a)P(Z_t = b)$ when $s \ne t$. Thus, the random variables $\{Z_t\}$ are pairwise independent. Let X_t denote the position of the person on the x-axis after t steps have been taken. Note that Z_t equals the increments (differences) $Z_t = X_t - X_{t-1}$ in the position of the person from time $t - 1$ to time t. Thus, X_t can be expressed as $X_t = X_0 + Z_1 + Z_2 + \cdots + Z_t$. When the increments $\{Z_t\}$ are pairwise independent, we will prove below that:

$$P(X_t|X_0, X_1, \ldots, X_{t-1}) = P(X_t|X_{t-1})$$

This result is intuitively clear. If each step is independent of the previous steps, then the probability that the person will be at a certain position X_t at time t is only dependent upon her position X_{t-1} at the previous step. The rest of the past history of her walk is superfluous information. Thus, this random walk is a Markov process. Random walks are of particular interest since they can serve as models of security prices.

5.1.2.2 Markov Processes and Independent Increments

Consider the increments (differences) of a discrete stochastic process from time $t - 1$ to time t: $Z_t = X_t - X_{t-1}$. One can express:

$$X_t = X_0 + Z_1 + Z_2 + \cdots + Z_t$$

Many discrete stochastic processes are those in which the random increments Z_t are independent random variables. If $\{Z_t\}$ are pairwise independent random variables, then X_t is a Markov process. We verify this as follows:[3]

$$P(X_t = b|X_0, X_1 = a_1, \ldots, X_{t-2} = a_{t-2}, X_{t-1} = a_{t-1})$$

$$= P(Z_t = b - a_{t-1}|Z_1 = a_1 - X_0, \ldots, Z_{t-2} = a_{t-2} - a_{t-3}, Z_{t-1} = a_{t-1} - a_{t-2})$$

$$= \frac{P(Z_t = b - a_{t-1}, Z_1 = a_1 - X_0, \ldots, Z_{t-2} = a_{t-2} - a_{t-3}, \ldots, Z_{t-1} = a_{t-1} - a_{t-2})}{P(Z_1 = a_1 - X_0, \ldots, Z_{t-2} = a_{t-2} - a_{t-3}, Z_{t-1} = a_{t-1} - a_{t-2})}$$

$$= \frac{P(Z_t = b - a_{t-1})P(Z_1 = a_1 - a_2)\cdots P(Z_{t-2} = a_{t-2} - a_{t-3})P(Z_{t-1} = a_{t-1} - a_{t-2})}{P(Z_1 = a_1 - a_2)\cdots P(Z_{t-2} = a_{t-2} - a_{t-3})P(Z_{t-1} = a_{t-1} - a_{t-2})}$$

$$= P(Z_t = b - a_{t-1}) = P(Z_t + a_{t-1} = b) = P(X_t = b|X_{t-1} = a_{t-1})$$

Because the top line equals the right-hand value in the bottom line, this process must be Markov. Although independent stochastic increments are sufficient to characterize a process as being Markov, the converse it not true.[4]

ILLUSTRATION: FAMILIES OF MARKOV PROCESSES

Consider the following simple model for the market, developed in terms of filtrations. At each time t, there are only two possible market states: upswing (u) or downswing (d). The probability that u occurs at time t equals p with $0 < p < 1$, and d occurs with probability $1 - p$. We study the market for times $t = 0, 1, 2, 3, \ldots$. The sample space Ω consists of every outcome $\omega \in \Omega$ of the form $\omega = (\omega_0, \omega_1, \omega_2, \ldots)$ where each ω_i is either u or d. We assume that the following probability is satisfied:[5]

$$P(\omega_1, \omega_2, \ldots, \omega_t) = P(\omega_1)P(\omega_2) \cdots P(\omega_t)$$

Suppose, for example, the probability that the market states are u, d, and u at times 1, 2, and 3, respectively, is $P(u\,du) = p(1 - p)p$. The σ-algebra $\mathcal{F}$ consists of every possible measurable subset of Ω. Define the random variable $Z_t(\omega) = 1$ if $\omega_t = u$ and $Z_t(\omega) = -1$ if $\omega_t = d$. In other words, $Z_t = 1$ if the market state is an upswing at time t and $Z_t = -1$ if the market state is a downswing at time t. It is easy to check that the random variables $\{Z_t\}$ are pairwise independent. Suppose that a security has price X_0 at time 0, and time t at price $X_t = X_0 + Z_1 + Z_2 + \cdots + Z_t$. Thus, this security's price is mathematically identical to the random walk in the illustration above. At each time t, its price either increases by 1 unit or decreases by 1 unit.

Now, define a second security with price Y_t defined by a given price Y_0 at time 0 and $Y_t = Y_0 + 3Z_1 + 3Z_2 + \cdots + 3Z_t$. This security has greater volatility than the previous security since it increases or decreases by 3 units at a time. Finally, consider a third security with price V_t defined by a given price V_0 at time 0 and $V_t = V_0 - Z_1 - Z_2 - \cdots - Z_t$. This security decreases by 1 unit when there is an upswing in the market state and increases by 1 unit when there is a downswing in the market state. This could serve as a simple model for an inversely correlated security.

The stochastic processes Y_t and V_t are Markov processes. All three of these securities are defined with respect to the same filtration. The expressions $P(X_t|\mathcal{F}_s)$, $P(Y_t|\mathcal{F}_s)$, $P(V_t|\mathcal{F}_s)$, respectively, denote the probabilities that X_t, Y_t, and V_t, respectively, take on particular values given the history of the market states up until time s. But the market states up until time s determine the history of each security's prices up until time s. Thus, $P(X_t|\mathcal{F}_s) = P(X_t|X_0, X_1, \ldots, X_s)$, $P(Y_t|\mathcal{F}_s) = P(Y_t|Y_0, Y_1, \ldots, Y_s)$, and $P(V_t|\mathcal{F}_s) = P(V_t|V_0, V_1, \ldots, V_s)$. This property was noted earlier when Markov processes were defined. The converse is not always true. That is, one might know the history of a security up until time s and this will not be sufficient information to single out a particular history in the filtration. For example, consider the stochastic process W_t defined by $W_0 = 0$, $W_t - W_{t-1} = 1$ whenever t is odd, and $W_t - W_{t-1} = Z_t$ whenever t is even. Suppose we are given W_i up until time 2. For example, assume that $W_1 = 1$ and $W_2 = 2$. This means that there was an upswing at time 2, but at time 1 there could have been either an upswing or a downswing. Thus, even though W_0, W_1, and W_2 are known, the history of the market could have been either (u,u) or (d,u).

Since $X_t = X_{t-1} + Z_t$, then any stochastic process has the following property:

$$E[X_t|X_0, X_1, \ldots, X_{t-1}] = X_{t-1} + E[Z_t|X_0, X_1, \ldots, X_{t-1}]$$

This property motivates the concepts of martingales, submartingales, and supermartingales.

5.1.2.3 Martingales

A *discrete martingale process* is a stochastic process X_t with the properties:

1. $E[X_t|X_0, X_1, \ldots, X_{t-1}] = X_{t-1}$
2. $E[|X_t|] < \infty$

for all $t = 1, 2, \ldots$. In the first property above, regard $X_0, X_1, \ldots, X_{t-1}$ as fixed (history) and X_t as a random variable. Thus, a martingale is a process whose future variations have no specific direction based on the process history $(X_0, X_1, \ldots, X_{t-1})$. A martingale is said to be a "fair game" and will not exhibit consistent trends either up or down. The first property above implies the following:

$$E[X_t|X_0, X_1, \ldots, X_s] = X_s \forall s < t \tag{5.1a}$$

Since this property (5.1a) is more general than property 1 above, we can use it to characterize the martingale instead of in our discrete martingale definition.

In the case where X_t is a *continuous time martingale*, the second property must apply for all positive real numbers t, and the first property is replaced with $E[X_t|X_i, 0 \le i \le s] = X_s$ for all $s < t$ and all positive numbers t.[6] Observe that the definitions of a discrete time martingale and a continuous time martingale are equivalent except that the discrete case indices i, s, and t assume integer values, while the continuous case takes on all real number values. Analogous to our scenario involving Markov processes, the first property of a martingale can be replaced with the more general condition:

$$E[X_t|\mathscr{F}_s] = X_s \ \forall s < t \tag{5.5}$$

In each example involving a martingale in this book, the two expressions $E[X_t|X_i, 0 \le i \le s]$ and $[E[X_t|\mathscr{F}_s]$ will be equivalent. Thus, we will use these notations interchangeably. Since $Z_t = X_t - X_{t-1}$, then for the discrete case, we can also express the first martingale property above as follows:

$$E[X_t|X_0, Z_1, Z_2, \ldots, Z_{t-1}] = X_{t-1}$$

Consider the random walk described above with jumps equal to $+1$ with probability p and -1 with probability $(1-p)$. Since $E[Z_t] = 1 \times p + (-1)(1-p) = 2p - 1$, $E[X_t|X_0, X_1, X_2, \ldots, X_{t-1}] = E[X_t|X_{t-1}] = X_{t-1} + 2p - 1$. Thus, the random walk is a martingale when $p = 1/2$. Since $E[X_t|X_0, Z_1, Z_2, \ldots, Z_{t-1}] = X_{t-1}$ when $p = 1/2$, a martingale's future has no specific direction in its trend from its present state. We also need to verify that the second condition $E[|X_t|] < \infty$ is satisfied. In this case, this is obvious since in time t, the farthest the random walk could have taken us would be t steps from the starting position at S_0. Thus, $E[|X_t|] \le \text{Max}[|S_0 + t|, |S_0 - t|]$. Whenever we have a discrete process in which the change in the value of the process at each time increment is finite, the second condition to be a martingale (submartingale or supermartingale to be covered next) is trivially satisfied.

5.1.2.4 Submartingales

A *submartingale* with respect to probability measure $\mathbb{P}$ is a stochastic process X_t in which the first property to be a martingale is replaced with:

$$E_{\mathbb{P}}[X_t|X_0, X_1, X_2, \ldots, X_{t-1}] \ge X_{t-1}$$

A submartingale will tend either to trend upward over time or is a martingale. The definition of a *supermartingale* replaces the greater than or equal inequality above with a less than or equal inequality. A supermartingale will tend to trend downward over time or is a martingale.[7] In our random walk example above, X_t is a submartingale when $p \geq 1/2$ and is a supermartingale when $p \leq 1/2$. Stock prices are often modeled as submartingales because they trend upward due to time value of money and investor risk aversion.

5.1.3 Equivalent Probabilities and Equivalent Martingale Measures

In Chapter 3, we used the concept of Arrow–Debreu (pure) securities and risk-neutral probabilities to begin to introduce and illustrate the concept of arbitrage-free pricing. Alternatively, physical probabilities are measures that we assign to outcomes that reflect the likelihoods of these outcomes actually occurring. Physical probabilities range between zero and one, they sum to one, and their levels increase as presumed likelihoods of events increase. However, prices of assets need not be functions of these physical probabilities or the expected values based on physical probabilities. First, it might be perfectly reasonable to expect that asset prices will reflect investor preferences, heterogeneous expectations, risk aversion, lexicographic preference orderings, portfolios, and other factors that will be unrelated to physical probabilities. For example, with heterogeneous investor expectations, prices could easily reflect one investor's probability estimates, but not another's. In fact, prices might not fully reflect any individual investor's expectations. In addition, as attractive as models based on physical probabilities might be, there is no market mechanism that forces prices to equal expected values based on physical probabilities.

On the other hand, risk-neutral probabilities arise from the strongest of financial forces—arbitrage. Risk-neutral probabilities are constructed from prices such that any violation of these prices or the risk-neutral probabilities that they imply will create arbitrage opportunities. Thus, pricing relationships implied by appropriate functions of risk-neutral probabilities must hold in the absence of arbitrage opportunities. Strictly speaking, from a mathematical perspective, risk-neutral probabilities are probabilities in that they range from zero to one and sum to one just as do physical probabilities; it is not necessary that they bear strong relationships to physical probabilities. Thus, risk-neutral probabilities do not strictly measure our opinions of likelihood in the way that physical probabilities do. However, they still resemble probabilities, they are essential to arbitrage-free pricing, and they can still be treated as probabilities for relative pricing purposes.

5.1.3.1 Numeraires

Thus far, we have expressed all asset values in terms of dollars and returns relative to some currency or monetary unit such as dollars. Thus, the monetary unit (e.g., dollar) served as the *numeraire*, which is simply the unit in which values are expressed. However, we can just as easily express values in terms of other currencies, other securities such as pure securities as we did in Chapter 3, or riskless bonds as we will do shortly. We can also express returns in terms of units of other securities such as forward contracts or even stocks. Flexibility in selecting the numeraire and an associated equivalent martingale measure affords us the ability of being able to use valuation techniques that otherwise would not be available or would be more complicated.

5.1.3.2 *Equivalent Probability Measures*

Recall that a probability measure $\mathbb{P}$ has the property that it is a mapping from an event space such that all events φ have probabilities $p[\varphi] \in [0, 1]$. An *equivalent probability measure* $\mathbb{Q}$ has the same null space as $\mathbb{P}$. That is, two probability measures are said to be equivalent ($\mathbb{P} \sim \mathbb{Q}$) if the set of events that have probability 0 under measure $\mathbb{P}$ (say, the physical probability measure) is the same as that set under the second measure $\mathbb{Q}$ (say, the risk-neutral probability measure). This also implies that $\mathbb{Q} \leq \mathbb{P}$ ($\mathbb{Q}$ is *absolutely continuous* with respect to $\mathbb{P}$, which means that $p(\varphi) = 0 \Rightarrow q(\varphi) = 0$) and $\mathbb{P} \leq \mathbb{Q}$ ($\mathbb{P}$ is absolutely continuous with respect to $\mathbb{Q}$; $q(\varphi) = 0 \Rightarrow p(\varphi) = 0$).[8] This implies that an equivalent probability measure is consistent with respect to which outcomes are possible. However, the actual nonzero probabilities assigned to events might differ.

5.1.3.3 *Equivalent Martingale Measures*

A probability measure $\mathbb{Q}$ is an *equivalent martingale measure* (also called a risk-neutral measure) to $\mathbb{P}$ in a complete market if $\mathbb{Q}$ and $\mathbb{P}$ are equivalent probability measures, and the price of every security in the market (using the riskless bond as the numeraire) is a martingale with respect to the probability measure $\mathbb{Q}$. We will examine an illustration of this later in the chapter. In a complete market in which there are no arbitrage opportunities, there will always exist a unique equivalent martingale measure. This measure can be used to obtain risk-neutral pricing for every security in that market. There are a number of observations that relate to equivalent martingale measures:

1. For the finite outcome case, we will define a complete market to be one in which every security in the market can be expressed as a portfolio of a finite number of pure securities. The set of all possible securities in the complete market forms a vector space, and the set of pure securities of say n linearly independent vectors forms a basis for this vector space. Once we have determined the price for each pure security with respect to the equivalent martingale measure $\mathbb{Q}$, it is then a simple matter to find the price of any other security as a linear combination of the pure security prices (using the riskless bond as the numeraire).
2. Since the equivalent martingale measure $\mathbb{Q}$ is unique, one is free to choose any set of n linearly independent securities that forms a basis for the market and that has already been priced in the market in order to construct the measure $\mathbb{Q}$. It may seem quite surprising that the measure $\mathbb{Q}$ turns out to be the same regardless of which basis of securities we choose to construct the measure $\mathbb{Q}$. This follows because of the linear relationship between the various securities' payoff and price vectors in a no-arbitrage market.
3. A market has a unique equivalent martingale measure if and only if it is both complete and arbitrage-free.
4. To use the riskless bond as the numeraire is equivalent to discounting the security by the riskless rate to obtain the present value of the security. The discounted security will be a martingale with respect to the measure $\mathbb{Q}$, and the risk-neutral price of the security is the expected value of the discounted security.
5. Since the discounted security price is a martingale, its expected value has the same return as the return on the riskless bond. If this were not the case, there would be an opportunity for arbitrage.

5.1.3.4 *Pricing with Submartingales*

We generally expect that financial securities requiring initial cash outlays will price as submartingales with respect to money because of the time value of money; investors demand compensation for giving up alternative uses of their initial investments:

$$E_{\mathbb{P}}[S_{t+\Delta t}] > S_t$$

However, if there exists an equivalent probability measure $\mathbb{Q}$ such that discounted security prices can be converted into martingales, this will ease the process of valuing derivative instruments relative to their underlying securities:

$$E_{\mathbb{Q}}\left[\frac{S_{t+\Delta t}}{(1+r)^{\Delta t}}\right] = S_t$$

We will discuss such conversions to martingales in this and the following chapters.

5.2 BINOMIAL PROCESSES: CHARACTERISTICS AND MODELING

A *Bernoulli trial* is a single random experiment with two possible outcomes (e.g., 0, 1, or u or d) whose outcomes depend on a probability. We will define a *binomial process* to be any stochastic process based on a series of n statistically independent Bernoulli (0,1) trials, all with the same outcome probability.

5.2.1 Binomial Processes

Consider the following Markov process: $X_t = X_0 + aZ_1 + aZ_2 + \cdots + aZ_t$, where the random variable $Z_t(\omega) = 1$ if $\omega = u$ with probability p and $Z_t(\omega) = -1$ if $\omega = d$ with probability $1 - p$ and a is a positive constant. We also assume that the random variables Z_t are pairwise independent. One can view this process as a random walk starting at X_0 and at each moment of time ($t = 1, 2, \ldots$) taking a step of length a to the right with probability p and taking a step of length a to the left with probability $1 - p$. If it is the price of a security, then X_0 would be its initial value, and at each moment of time there is a probability p that it will increase in value by a and a probability $1 - p$ that it will decrease in value by a. Under the probability measure $\mathbb{P}$, the expected value of X_t at time t is:

$$E_{\mathbb{P}}[X_t] = E_{\mathbb{P}}[X_0 + aZ_1 + aZ_2 + \cdots + aZ_t] = X_0 + aE_{\mathbb{P}}[Z_1] + aE_{\mathbb{P}}[Z_2] + \cdots + aE_{\mathbb{P}}[Z_t]$$
$$= X_0 + a(2p - 1)t$$

since for each i, $E_{\mathbb{P}}[Z_i] = 1(p) + (-1)(1 - p) = 2p - 1$. Thus, the expected value of X_t depends linearly on time. This model is the stochastic version of a deterministic model in which an account pays a fixed and simple interest over time (when $p > 1/2$). If one regards X_s as given and calculates the expected value $E_{\mathbb{P}}[X_t|X_0, X_1, \ldots, X_s]$ for $s < t$, then similar to the calculation above:

$$E_{\mathbb{P}}[X_t|X_0, X_1, \ldots, X_s] = X_s + a(2p - 1)(t - s)$$

Observe that X_t is a martingale when $p = \frac{1}{2}$, is a submartingale when $p > \frac{1}{2}$, and is a supermartingale when $p < \frac{1}{2}$. The process described here does not involve multiplicative factors or compounded returns. We will discuss multiplicative returns and compounding in the next section.

5.2.1.1 Binomial Returns Process

The binomial process described above can be applied to security prices, with prices increasing or decreasing by a specified monetary amount. However, this model does not provide for compounding of returns over time. For example, over a period of time, one might expect that a security with a high price to be subject to greater monetary fluctuation than a security with a low price; a \$500 stock will probably experience greater price fluctuation than a \$2 stock. Thus, it may be more realistic instead to construct a binomial process to security returns, which satisfies the following proportional model of returns:

$$\frac{S_t}{S_{t-1}} = (1 + aZ_t)$$

for $t = 1, 2, \ldots$. Thus, the security increases (upjump) by a factor $1 + a$ with probability p and decreases (downjump) by a factor $1 - a$ with probability $1 - p$ at each moment of time. This implies that:

$$S_t = (1 + aZ_1)(1 + aZ_2)\cdots(1 + aZ_t)S_0$$

From time 0 to time t, the security could have experienced a total of k upjumps for $k = 0, 1, 2, \ldots, t$. The number of upjumps is a binomial process and as we reviewed in Section 2.5.1. The probability that exactly k upjumps occurred from time 0 to time t equals:

$$\binom{t}{k} p^k (1-p)^{t-k}$$

Since k upjumps means the Z_i's equaled one k times and equaled minus one $t - k$ times, this would result in the value of S_t equaling $(1 + a)^k (1 - a)^{t-k} S_0$. Thus, the expected value of S_t given S_0 is:

$$E_{\mathbb{P}}[S_t] = \sum_{k=0}^{t} (1+a)^k (1-a)^{t-k} S_0 \binom{t}{k} p^k (1-p)^{t-k}$$

$$= S_0 \sum_{k=0}^{t} \binom{t}{k} [p(1+a)]^k [(1-p)(1-a)]^{t-k}$$

The binomial theorem states that:

$$\sum_{k=0}^{t} \binom{t}{k} x^k y^{t-k} = (x+y)^t$$

Choosing $x = p(1 + a)$ and $y = (1 - p)(1 - a)$ in the equation above gives the following result for the expected value:

$$E_{\mathbb{P}}[S_t] = S_0 \left(p(1+a)+(1-p)(1-a)\right)^t = S_0[1+a(2p-1)]^t$$

In this case, the expected value has an exponential dependence on time. This model is the stochastic version of a deterministic model of an account paying compound interest or the return of a security on a compound basis. By the same argument we used to obtain $E_\mathbb{P}[S_t]$, we can show:

$$E_\mathbb{P}[S_t | S_0, S_1, \ldots, S_s] = S_s[1 + a(2p - 1)]^{t-s}$$

for $s < t$. Clearly, $1 + a(2p - 1)$ equals 1 when $p = \frac{1}{2}$, and S_t is a martingale. S_t is a submartingale when $p \geq \frac{1}{2}$, and is a supermartingale when $p \leq \frac{1}{2}$.

In the model above, we assumed that the price of the security increased by the multiplicative factor $1 + a$ in the event of an upjump and decreased by the multiplicative factor $1 - a$ in the event of a downjump with $0 < a < 1$. This model can be generalized by allowing the multiplicative upward factor to be u for any $u > 1$ and the multiplicative downward factor to be d for any $0 < d < 1$. The symbols u and d can simply refer to an upjump or downjump; u and d can also refer to particular values for multiplicative jumps. As in the previous model, we assume that the probability of an upjump or downjump at a particular time t is independent of the probability of an upjump or downjump at any other time s. Similar to the calculation earlier, the expected value of the security is:

$$E_\mathbb{P}[S_t] = S_0[pu + (1-p)d]^t$$

Since the security price follows the same probability and pricing law in this model if we shift the time by any value s, then:

$$E_\mathbb{P}[S_t | S_s] = S_s[pu + (1-p)d]^{t-s}$$

for $s < t$. The process S_t is a martingale as long as $pu + (1 - p)d = 1$, or $p = (1 - d)/(u - d)$.

ILLUSTRATION: BINOMIAL OUTCOME AND EVENT SPACES

In this section, we will consider a relatively simple two-time period time binomial process in order to illustrate the construction of risk-neutral pricing for a security. Table 5.1 depicts a sampling of $n = 2$ successive independent and identically distributed jumps, in which each Bernoulli trial can result in one of two potential outcomes. One such sampling might be based on a stock whose price can either rise or fall in each of two sequential transactions. As before, the letter u (upjump) will mean the stock increased in value at time t, and the letter d (downjump) will mean the stock price decreased. For example, the letters ud mean that the price of the stock went from 10 to 15 at time 1 and then from 15 to 7.5 at time 2. Suppose that the *physical probability* (as opposed to risk-neutral probability) associated with a stock price increase in each transaction is given to be p, implying a probability of $(1 - p)$ for a price decline in each transaction. Figure 5.1 and the listing below depict first the sample space Ω and then the filtration:

$$\Omega = \{uu, ud, du, dd\}$$

$$\mathscr{F}_0 = \{\varnothing, \Omega\}$$

$$\mathscr{F}_1 = \{\varnothing, \{uu, ud\}, \{du, dd\}, \Omega\}$$

$$\mathscr{F}_2 = \{\varnothing, \{uu\}, \{ud\}, \{du\}, \{dd\}, \{uu, ud\}, \{uu, du\}, \{uu, dd\}, \{ud, du\}, \{uu, dd\}, \{du, dd\},$$

$$\{uu, ud, du\}, \{uu, ud, dd\}, \{uu, du, dd\}, \{ud, du, dd\}, \Omega\}$$

TABLE 5.1   Pure Security Prices

Pure security price	At time	Maturity	Time 1 outcome	Time 2 outcome	Numerical value
SPOT PRICES					
$\psi_{0,1;u}$	0	1	*upjump*	N/A	0.6
$\psi_{0,1;d}$	0	1	*downjump*	N/A	0.2
$\psi_{0,2;u,u}$	0	2	*upjump*	*upjump*	0.36
$\psi_{0,2;u,d}$	0	2	*upjump*	*downjump*	0.12
$\psi_{0,2;d,u}$	0	2	*downjump*	*upjump*	0.12
$\psi_{0,2;d,d}$	0	2	*downjump*	*downjump*	0.04
FORWARD PRICES					
$\psi_{1,2;u,u}$	1	2	*upjump*	*upjump*	0.6
$\psi_{1,2;u,d}$	1	2	*upjump*	*downjump*	0.2
$\psi_{1,2;d,u}$	1	2	*downjump*	*upjump*	0.6
$\psi_{1,2;d,d}$	1	2	*downjump*	*downjump*	0.2

This table gives pure security prices at time t (second column) for instruments that pay off (mature) in T years (third column) contingent on the outcome in the fourth column and following the jump listed in the fifth column. The final column lists the pure security prices. The subscripting for one-period spot prices (first subscript is zero) identifies the contract as follows: time 0 (first subscript) pure security price paying 1 at time 1 (second subscript) contingent on up/downjump (third subscript). The subscripting for two-period spot prices (first subscript is zero) identifies the contract as follows: pure security price at time 0 (first subscript) paying 1 at time 2 (second subscript) contingent on up/downjump at time 1(third subscript) and on up/downjump at time 2 (fourth subscript). The subscripting for forward prices (first subscript exceeds zero) identifies the contract as follows: pure security price at time 1 (first subscript) paying 1 at time 2 (second subscript) contingent on up/downjump at time 1(third subscript) and on up/downjump at time 2 (fourth subscript).

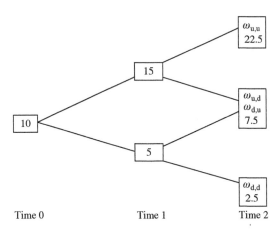

FIGURE 5.1   Two time period binomial model—current and potential stock prices.

Note in Figure 5.1 that the process is *recombining* (multiple paths can lead to the same time 2 price).

Suppose that a stock (nondividend paying) trades in this economy, such that its current and expected prices are shown in Figure 5.1. In addition, three riskless bonds trade in this economy, all of which have face values equal to 1. The first two bonds originate in time 0, mature in years 1 and 2, respectively, and currently trade for 0.8 and 0.64. The third bond will originate in time 1, and in the absence of arbitrage opportunities, its forward contract must trade with a settlement price equal to 0.8 based on known prices of the first two bonds.

5.2.1.2 Pure Security Prices

Next, we will discuss Arrow–Debreu (pure) security prices, one associated with each starting and stopping node combination in our two-time period model (see Table 5.1). First, we will list spot (time 0) prices for investments extending from time 0 to time 1. Let $\psi_{0,1;u}$ be the time 0 price of a pure security that is worth 1 at time 1 if and only if an upjump occurs at time 1. The first subscript 0 refers to the initial time 0, the second subscript 1 refers to the next time 1, and the third subscript u refers to the fact that there is an upjump (at time 1). Let $\psi_{0,1;d}$ be the time 0 price of a pure security that is worth 1 at time 1 if and only if a downjump occurs at time 1. The third subscript d in this case refers to the fact that there is a downjump (at time 1). Subscripts before the semicolon refer to time periods, and subscripts after the semicolon refer to the occurrence of upjumps or downjumps. The vector $[1\ 0]^T$ will denote the payoff of the pure security identified with an upjump occurring at time 1. The vector $[0\ 1]^T$ will denote the payoff of the pure security identified with a downjump occurring at time 1. To purchase the stock at time 0 at a price of 10 and hold to it until time 1 means that at time 1 the portfolio of the owner takes the form of the vector $[15\ 5]^T$ because it will pay 15 if an upjump occurs and will pay 5 if a downjump occurs. Since an investor is willing to pay $\psi_{0,1;u}$ at time 0 for a payoff vector $[1\ 0]^T$ and pay $\psi_{0,1;d}$ at time 0 for a payoff vector $[0\ 1]^T$ and since the value of the stock portfolio $[15\ 5]^T$ is equal to 10 at time 0, then:

$$15\psi_{0,1;u} + 5\psi_{0,1;d} = \begin{bmatrix} 15 & 5 \end{bmatrix} \begin{bmatrix} \psi_{0,1;u} \\ \psi_{0,1;d} \end{bmatrix} = 10 \tag{5.6}$$

Also, if an investor has the riskless portfolio $[1\ 1]^T$, then she is guaranteed to be paid 1 at time 1. As this is equivalent to holding the bond that pays 1 at time 1, and since the time 0 price of the portfolio $[1\ 1]^T$ is 0.8, then:

$$\psi_{0,1;u} + \psi_{0,1;d} = \begin{bmatrix} 1 & 1 \end{bmatrix} \begin{bmatrix} \psi_{0,1;u} \\ \psi_{0,1;d} \end{bmatrix} = 0.8 \tag{5.7}$$

Combining these two vector equations into one matrix equation yields:

$$\begin{bmatrix} 15 & 5 \\ 1 & 1 \end{bmatrix} \begin{bmatrix} \psi_{0,1;u} \\ \psi_{0,1;d} \end{bmatrix} = \begin{bmatrix} 10 \\ 0.8 \end{bmatrix} \tag{5.8}$$

In Chapter 1 we reviewed how to solve this system using inverse matrices. The solution is $\psi_{0,1;u} = 0.6$ and $\psi_{0,1;d} = 0.2$, and the results are listed in Table 5.1.

Before we discuss the two-period spot prices, we also have time 1 forward prices for pure securities that pay off in time 2. As described in Table 5.1, let $\psi_{1,2;u,u}$ be the time 1 price of a pure security that pays 1 at time 2 if and only if upjumps occur at both times 1 and 2 (outcome $\omega_{u,u}$).

The remaining forward prices $\psi_{1,2;u,d}$, $\psi_{1,2;d,u}$, and $\psi_{1,2;d,d}$ are defined similarly. Note that two paths lead to a time 2 price of 7.5. The same procedure can used to calculate pure security prices as we did going from time 0 to 1. These forward pure security prices are calculated from the following and listed in Table 5.1:

$$\begin{bmatrix} 22.5 & 7.5 \\ 1 & 1 \end{bmatrix} \begin{bmatrix} \psi_{1,2;u,u} \\ \psi_{1,2;u,d} \end{bmatrix} = \begin{bmatrix} 15 \\ 0.8 \end{bmatrix}; \quad \begin{matrix} \psi_{1,2;u,u} = 0.6 \\ \psi_{1,2;u,d} = 0.2 \end{matrix} \tag{5.9}$$

$$\begin{bmatrix} 7.5 & 2.5 \\ 1 & 1 \end{bmatrix} \begin{bmatrix} \psi_{1,2;d,u} \\ \psi_{1,2;d,d} \end{bmatrix} = \begin{bmatrix} 5 \\ 0.8 \end{bmatrix}; \quad \begin{matrix} \psi_{1,2;d,u} = 0.6 \\ \psi_{1,2;d,d} = 0.2 \end{matrix} \tag{5.10}$$

We can calculate the two-period pure security spot prices from the one-period spot and forward prices. Again as described in Table 5.1, let $\psi_{0,2;u,u}$ be the time 0 price of a pure security that pays 1 at time 2 if and only if there are consecutive upjumps (denoted by the outcome $\omega_{u,u}$). Thus, the time 0 pure security price associated with time 2 outcome $\omega_{u,u}$ is $\psi_{0,2;u,u}$. The remaining two-period spot prices $\psi_{0,2;u,d}$, $\psi_{0,2;d,u}$, and $\psi_{0,2;d,d}$ are defined similarly. Internal pricing consistency (no-arbitrage pricing) requires that $\psi_{0,2;u,u} = \psi_{0,1;u} \times \psi_{1,2;u,u} = 0.6 \times 0.6 = 0.36$ and $\psi_{0,2;d,d} = \psi_{0,1;d} \times \psi_{1,2;d,d} = 0.2 \times 0.2 = 0.04$. Also, no-arbitrage pricing that $\psi_{0,2;u,d} = \psi_{0,1;u} \times \psi_{1,2;u,d} = 0.6 \times 0.2 = 0.12$ and $\psi_{0,2;d,u} = \psi_{0,1;d} \times \psi_{1,2;d,u} = 0.2 \times 0.6 = 0.12$. For the recombining part of our lattice, the pure security price for the set of outcomes $\{\omega_{u,d}, \omega_{d,u}\}$ is $\psi_{0,2;u,d} + \psi_{0,2;d,u} = 0.12 + 0.12 = 0.24$. These no-arbitrage relationships are captured in the following pure security prices matrix equation and listed in Table 5.1:

$$\begin{bmatrix} \psi_{0,2;u,u} \\ (\psi_{0,2;u,d} + \psi_{0,2;d,u}) \\ \psi_{0,2;d,d} \end{bmatrix} = \begin{bmatrix} \psi_{1,2;u,u} & 0 \\ \psi_{1,2;u,d} & \psi_{1,2;d,u} \\ 0 & \psi_{1,2;d,d} \end{bmatrix} \begin{bmatrix} \psi_{0,1;u} \\ \psi_{0,1;d} \end{bmatrix} \tag{5.11}$$

$$\begin{bmatrix} \psi_{0,2;u,u} \\ (\psi_{0,2;u,d} + \psi_{0,2;d,u}) \\ \psi_{0,2;d,d} \end{bmatrix} = \begin{bmatrix} 0.6 & 0 \\ 0.2 & 0.6 \\ 0 & 0.2 \end{bmatrix} \begin{bmatrix} 0.6 \\ 0.2 \end{bmatrix} = \begin{bmatrix} 0.36 \\ 0.24 \\ 0.04 \end{bmatrix}$$

5.2.1.3 Bond Prices

From these nine pure security prices, we can obtain spot prices for 1- and 2-year zero-coupon bonds with face value 1 along with discount functions and their associated rates as follows:[9]

$$B_{0,1} = d_{0,1} = \psi_{0,1;u} + \psi_{0,1;d} = 0.8; \quad r_{0,1} = \frac{1}{\psi_{0,1;u} + \psi_{0,1;d}} - 1 = 0.25$$

$$B_{0,2} = d_{0,2} = \psi_{0,2;u,u} + \psi_{0,2;u,d} + \psi_{0,2;d,u} + \psi_{0,2;d,d} = 0.64$$

$$r_{0,2} = \sqrt{\frac{1}{\psi_{0,2;u,u} + \psi_{0,2;u,d} + \psi_{0,2;d,u} + \psi_{0,2;d,d}}} - 1 = 0.25$$

We can also obtain forward bond prices, discount functions, and rates contingent on realization of outcomes u and d, respectively, at time 1:

$$B_{1,2;u} = d_{1,2;u} = \psi_{1,2;u,u} + \psi_{1,2;u,d} = 0.8; \quad r_{1,2;u} = \frac{1}{\psi_{1,2;u,u} + \psi_{1,2;u,d}} - 1 = 0.25$$

$$B_{1,2;d} = d_{1,2;d} = \psi_{1,2;d,u} + \psi_{1,2;d,d} = 0.8; \quad r_{1,2;d} = \frac{1}{\psi_{1,2;d,u} + \psi_{1,2;d,d}} - 1 = 0.25$$

5.2.1.4 Physical Probabilities

Suppose that in our process, we define the physical probability of each upjump to be $p_u = 0.8$; the physical probability of each downjump is $p_d = 0.2$. For our process, we will define $p_0(\omega_t)$ to be the time 0 physical probability of a time t outcome ω_t in Ω. For example, at time 0, $p_0(u) = p_u = 0.8$ and $p_0(d) = p_d = 0.2$. Because of independence, at time 0, $p_0(u, u) = p_u^2 = 0.64$, $p_0(u, d) + p_0(d, u) = 2p_u p_d = 0.32$ and $p_0(d, d) = p_d^2 = 0.04$. These are all nonzero (relevant) physical probabilities in probability measure $\mathbb{P}$ and are listed in Table 5.2. However, now that we have set forth these physical probabilities, we will not actually have much use for them for valuation purposes. In fact, these physical probability estimates might vary among individuals, might be unrelated to actual market prices, and really are not enforced by any particular market discipline. By converting to risk-neutral probability measure $\mathbb{Q}$, we will be able to conduct risk-neutral pricing of the stock. We will compute risk-neutral probabilities from pure security prices and will represent the stock as a martingale with respect to the risk-neutral probability measure.

TABLE 5.2   Risk-Neutral Probabilities

Risk-neutral probability	At time	Maturity	Outcome	Physical probability
$q_{0,1;u} = 0.75$	0	1	u	0.8
$q_{0,1;d} = 0.25$	0	1	d	0.2
$q_{0,2;u,u} = 0.5625$	0	2	u,u	0.64
$q_{0,2;u,d} = 0.1875$	0	2	u,d	0.16
$q_{0,2;d,u} = 0.1875$	0	2	d,u	0.16
$q_{0,2;d,d} = 0.0625$	0	2	d,d	0.04
$q_{1,2;u,u} = 0.75$	1	2	u,u	0.8
$q_{1,2;u,d} = 0.25$	1	2	u,d	0.2
$q_{1,2;d,u} = 0.75$	1	2	d,u	0.8
$q_{1,2;d,d} = 0.25$	1	2	d,d	0.2

This table gives risk-neutral (hedging) probabilities in the first column at time t (second column) for instruments that pay off (mature) in T years (third column) contingent on the outcome in the fourth column and following the event listed in the fifth column. The final column lists the numerical values for physical probabilities.

5.2.1.5 *The Equivalent Martingale Measure*

Next, we will characterize the set of risk-neutral probabilities or the *equivalent martingale measure* $\mathbb{Q}$ for this space. We obtain the time t risk-neutral probability for outcome i at time T by dividing the pure security price $\psi_{t,T;i}$ by the time (t, T) riskless discount function $d_{t,T}$:

$$q_{t,T;i} = \psi_{t,T;i}/d_{t,T} = \psi_{t,T;i}/B_{t,T} \tag{5.12}$$

In Chapter 3, we learned that:

$$q_{t,T;i} = \frac{\psi_{t,T;i}}{\sum_{j=1}^{n} \psi_{t,T;j}}$$

but these results are equivalent since $B_{t,T} = \sum_{j=1}^{n} \psi_{t,T;j}$. As you can see from Eq. (5.12), the risk-neutral probability is the price of the pure security associated with that time period and outcome in terms of the riskless bond serving as the numeraire. Thus, the numeraire is the price of the one dollar face value bond maturing in the time period from t to T. These risk-neutral probabilities are all listed in Table 5.2 and they can be obtained from Table 5.1. For example, $q_{0,1;u} = \psi_{0,1;u}/B_{0,1} = 0.6/0.8 = 0.75$, $q_{1,2;u,d} = \psi_{1,2;u,d}/B_{1,2} = 0.2/0.8 = 0.25$, and $q_{0,2;d,d} = \psi_{0,2;d,d}/B_{0,2} = 0.04/0.64 = 0.0625$.

Even though the risk-neutral probabilities in Table 5.2 differ substantially from the physical probabilities listed above, the two sets of four probability measures are equivalent. This is simply because at each node on our lattice, we see that each nonzero probability in probability measure $\mathbb{P}$ corresponds to a nonzero probability in probability measure $\mathbb{Q}$. Thus, our risk-neutral probability measure $\mathbb{Q}$ is equivalent to our original physical probability measure $\mathbb{P}$. Next, we will demonstrate that the stock price is a martingale with respect to our risk-neutral probability measure $\mathbb{Q}$ using our alternative numeraire, the riskless bond.

5.2.1.6 *Change of Numeraire and Martingales*

To express the stock price as a martingale, we will change the numeraire for valuing our securities. Rather than express security values in terms of monetary units (e.g., dollars), we will express values in terms of some security, namely, the riskless bond. Note that at time 0, the value of a single share of stock S_0, 10, is $S_0/B_0 = 10/0.8 = 12.5$ times the value of the bond, $B_0 = 0.8$. We also point out that the bond's value at time 0 in bond units is 1 since $B_0/B_0 = 1$. Thus, the value of one share of stock at time 0 equals that of 12.5 bonds. In addition, note that with our equivalent probability measure $\mathbb{Q}$, the time 0 expected future value of the share at time 1 given the time 0 value of the share of stock is also 12.5 times that of the bond:

$$\begin{bmatrix} 15 & 5 \\ 1 & 1 \end{bmatrix} \begin{bmatrix} q_{0,1;u} \\ q_{0,1;d} \end{bmatrix} = \begin{bmatrix} 12.5 \\ 1 \end{bmatrix}; \qquad \begin{matrix} q_{0,1;u} = 0.75 \\ q_{0,1;d} = 0.25 \end{matrix}$$

$$E_{\mathbb{Q}}[S_{0,1}|S_0/B_0] = 15q_{0,1;u} + 5q_{0,1;d} = 12.5 = S_0/B_0$$

Since the time 0 expected value of the stock at time 1 (given the time 0 stock value in bond numeraire) under the probability measure $\mathbb{Q}$ equals the time 0 stock value (in bond numeraire), the stock price has the martingale property at time 1. We can demonstrate similar martingale properties going from time 1 to time 2, and from time 0 to time 2. This will be explained in more detail in the next section. Thus, in this two time period process, the probability measure $\mathbb{Q}$ is an equivalent martingale measure to $\mathbb{P}$.

Also note that if we convert from the bond numeraire back to the original currency (such as dollars), we get the right values at time 0. The time 0 value of the stock is 12.5 times that of the bond. The bond is worth 0.8 dollars at time 0; therefore, the stock has a time 0 value equal to 10 dollars:

$$S_0 = \frac{E_{\mathbb{Q}}[S_{0,1}]}{(1+r)^{\Delta t}} = \frac{15q_{0,1;u} + 5q_{0,1;d}}{1 + 0.25} = 10$$

$$B_0 = \frac{E_{\mathbb{Q}}[B_{0,1}]}{(1+r)^{\Delta t}} = \frac{1q_{0,1;u} + 1q_{0,1;d}}{1 + 0.25} = 0.8$$

5.2.2 Binomial Pricing, Change of Numeraire, and Martingales

Here, we will prove that for an arbitrary risk-free binomial pricing model, the stock price S_t is a martingale with respect to the price of a bond (the riskless bond will be the numeraire). We will price the stock and bond at each time period given their potential subsequent cash flows. After each time period, we will verify that the stock price, when priced with the bond as the numeraire, satisfies the martingale property.

5.2.2.1 Pricing the Stock and Bond from Time 0 to Time 1

Let $S_{0,1;u}$ and $S_{0,1;d}$ denote the time 1 prices of the stock after an upjump and downjump, respectively, based on the price path from time 0. $\psi_{0,1;u}$ and $\psi_{0,1;d}$ will denote the prices that an investor is willing to pay at time 0 for pure securities that pay off 1 at time 1 if and only if the stock has an upjump (respectively downjump) to the value $S_{0,1;u}$ (respectively $S_{0,1;d}$) from time 0 to time 1. Let $B_{0,1}$ be the time 0 price of a bond that pays 1 at time 1. To purchase the stock at time 0 at a price of S_0 and hold to it until time 1 means that at time 1 the portfolio of the owner takes the form of the vector $\begin{bmatrix} S_{0,1;u} & S_{0,1;d} \end{bmatrix}^T$ because it will pay $S_{0,1;u}$ if an upjump occurs and will pay $S_{0,1;d}$ if a downjump occurs. Since an investor is willing to pay $\psi_{0,1;u}$ at time 0 for a payoff vector $\begin{bmatrix} 1 & 0 \end{bmatrix}^T$ and pay $\psi_{0,1;d}$ at time 0 for a payoff vector $\begin{bmatrix} 0 & 1 \end{bmatrix}^T$, then:

$$S_{0,1;u}\psi_{0,1;u} + S_{0,1;d}\psi_{0,1;d} = \begin{bmatrix} S_{0,1;u} & S_{0,1;d} \end{bmatrix} \begin{bmatrix} \psi_{0,1;u} \\ \psi_{0,1;d} \end{bmatrix} = S_0 \tag{5.13}$$

Also, if an investor has the riskless portfolio $\begin{bmatrix} 1 & 1 \end{bmatrix}^T$, then she is guaranteed to be paid 1 at time 1. As this is equivalent to holding the bond that pays 1 at time 1, then the time 0 price of the portfolio $\begin{bmatrix} 1 & 1 \end{bmatrix}^T$ is:

$$\psi_{0,1;u} + \psi_{0,1;d} = \begin{bmatrix} 1 & 1 \end{bmatrix} \begin{bmatrix} \psi_{0,1;u} \\ \psi_{0,1;d} \end{bmatrix} = B_{0,1} \tag{5.14}$$

Combining these two vector equations into one matrix equation yields:

$$\begin{bmatrix} S_{0,1;u} & S_{0,1;d} \\ 1 & 1 \end{bmatrix} \begin{bmatrix} \psi_{0,1;u} \\ \psi_{0,1;d} \end{bmatrix} = \begin{bmatrix} S_0 \\ B_{0,1} \end{bmatrix} \tag{5.15}$$

Using inverse matrices, this system can be solved for $\psi_{0,1;u}$ and $\psi_{0,1;d}$.

5.2.2.2 *Verifying the Martingale Property*

Now we will verify that the stock price relative to the bond (the bond will be the numeraire) satisfies the martingale property from time 0 to 1. Since $B_{0,1}$ is the time 0 value of a bond that pays 1 at time 1, the value of the stock at time 0 expressed in units of the bond as a numeraire equals $S_0/B_{0,1}$. Dividing both sides of Eq. (5.15) by $B_{0,1}$, we have:

$$\begin{bmatrix} S_{0,1;u} & S_{0,1;d} \\ 1 & 1 \end{bmatrix} \begin{bmatrix} \psi_{0,1;u}/B_{0,1} \\ \psi_{0,1;d}/B_{0,1} \end{bmatrix} = \begin{bmatrix} S_0/B_{0,1} \\ B_{0,1}/B_{0,1} \end{bmatrix} \tag{5.16}$$

However, by Eq. (5.12), $q_{0,1;u} = \psi_{0,1;u}/B_{0,1}$ is the risk-neutral probability that the stock will go from $S_0/B_{0,1}$ (the price in bond numeraire) at time 0 to the price $S_{0,1;u}$ at time 1. The quotient $q_{0,1;d} = \psi_{0,1;d}/B_{0,1}$ is the risk-neutral probability that the stock will go from $S_0/B_{0,1}$ at time 0 to the price $S_{0,1;d}$ at time 1. Note that $S_{0,1;u}$ and $S_{0,1;d}$ are already expressed in bond numeraire since the value of the bond equals 1 at time 1. The first row of matrix Eq. (5.16) can be written in the form:

$$S_{0,1;u} q_{0,1;u} + S_{0,1;d} q_{0,1;d} = \frac{S_0}{B_{0,1}} \tag{5.17}$$

The left side of Eq. (5.17) equals the expected value of the stock at time 1 (given that its value was $S_0/B_{0,1}$ at time 0 in bond numeraire). Recalling that S_t is the Markov binomial process that characterizes the price of the stock, we can express Eq. (5.17) succinctly in the form:

$$E_{\mathbb{Q}} \left[S_1 \,\middle|\, \frac{S_0}{B_{0,1}} \right] = \frac{S_0}{B_{0,1}} \tag{5.18}$$

Equation (5.18) states that the expected value with respect to the probability measure $\mathbb{Q}$ of the stock price at time 1 given its value at time 0 (in bond numeraire) is equal to the stock's price at time 0 (in bond numeraire). This demonstrates the requirement for the stock price to satisfy the martingale property under measure $\mathbb{Q}$. Thus, $q_{0,1;u}$ and $q_{0,1;d}$ will define the risk-neutral probabilities going from time 0 to 1. These risk-neutral probabilities must be used in order to guarantee that there are no arbitrage opportunities because they are the only probabilities that are consistent with current market prices (see the companion website for a verification of the martingale property into subsequent periods).

Now, we will "verify by contradiction" our no-arbitrage risk-neutral probabilities with an illustration. Suppose that there were a group of investors that used the physical probability measure $\mathbb{P}$ with probability $p_{0,1;u} = 0.8$ and $p_{0,1;d} = 0.2$ to calculate security prices. Also suppose that we could create and sell a contingent claim security on our stock that would allow these investors to take ownership of the stock at time 1 if and only if its value rises to 15. In probability space $\mathbb{P}$, the present value of this contingent claim would equal $PV_{15} = (15 \times 0.8)/1.25 = 9.60$ at a 25% discount rate. Further suppose that we could sell a second contingent claim security on our stock that would allow these investors to take ownership at time 1 if and only if its value declines to 5. In probability space $\mathbb{P}$, the present value of this second contingent claim equals $PV_5 = (5 \times 0.2)/1.25 = 0.80$. The pair of contingent claims securities has a sum value equal to $9.60 + 0.80 = 10.40$. Obviously, this creates an arbitrage opportunity since the stock sells for 10.00 while its contingent claims equivalent sells for 10.40. Only the risk-neutral probabilities $q_{0,1;u} = 0.75$ and $q_{0,1;d} = 0.25$ implied by current market prices eliminate this arbitrage opportunity: $PV_Q = (15 \times 0.75)/1.25 + (5 \times 0.2)/1.25 = 10.00$. Use of physical or any other probabilities will contradict our characterization of no arbitrage.

5.2.3 Binomial Option Pricing

The *binomial option pricing model* is based on the assumption that the underlying stock price is a Bernoulli trial in each period, such that it follows a binomial multiplicative return generating process. This means that for any period following a particular outcome, the stock's value will be only one of two potential constant values. For example, the stock's value at time $t+1$ will be either u (multiplicative upward movement) times its prior value S_t or d (multiplicative downward movement) times its prior value S_t.

Note that we have not specified probabilities of a stock price increase or decrease during the period prior to option expiration. Nor have we specified a discount rate for the option or made inferences regarding investor risk preferences. We will value this call based on the fact that during this single time period, we can construct a riskless hedge portfolio consisting of a position in a single call and offsetting positions in α shares of stock. This means that by purchasing a single call and by selling α shares of stock, we can create a portfolio whose value is the same regardless of whether the underlying stock price increases or decreases. Let us first define the following terms:

$X =$ Exercise price of the stock
$S_0 =$ Initial stock value
$u =$ Multiplicative upward stock price movement
$d =$ Multiplicative downward stock price movement
$c_u = \text{Max}[0, uS_0 - X]$; Value of call if stock price increases
$c_d = \text{Max}[0, dS_0 - X]$; Value of call if stock price decreases
$\alpha =$ Hedge ratio
$r =$ Riskless return rate

5.2.3.1 One-Time Period Case

In Chapter 3, we valued a call as payoff structure identical to a portfolio comprised of underlying shares and bonds. We use a similar method here, where the equivalent martingale measure is calculated from bond and stock payoffs, and the call is valued based on those risk-neutral probabilities. Valuing the one-time period call in order to obtain its risk-neutral pricing in a single time period binomial framework is straightforward given the risk-neutral probability measure $\mathbb{Q}$, where q represents the probability of an upjump:

$$E_\mathbb{Q}[c_1] = [c_u q + c_d(1-q)] = c_0(1+r)$$

Solving for c_0 gives the price:

$$c_0 = \frac{c_u q + c_d(1-q)}{1+r}$$

We value shares of the stock relative to the bond as follows:

$$E_Q[S_1] = uS_0q + dS_0(1-q) = S_0(1+r)$$

which implies that:

$$q = \frac{1+r-d}{u-d}$$

5.2.3.2 Multitime Period Case

Suppose that we express our outcomes in terms of u and d, such that the numbers of upjumps and downjumps over time determine stock prices. If we are willing to assume that the probability measure is the same in each of these T time periods, by invoking the binomial theorem, we see that valuing the call in the multiperiod binomial setting is similarly straightforward:[10]

$$c_0 = \frac{\sum_{j=0}^{T} \frac{T!}{j!(T-j)!} q^j (1-q)^{T-j} \text{Max}[u^j d^{T-j} S_0 - X, 0]}{(1+r)^T}$$

The number of computational steps required to solve this equation is reduced if we eliminate from consideration all of those outcomes where the option's expiration date price is zero. Thus, a, the smallest nonnegative integer for j where $S_T > X$, is given as follows:[11]

$$a = \text{Int}\left[\text{Max}\left[\frac{\ln\left(\frac{X}{S_0 d^T}\right)}{\ln\left(\frac{u}{d}\right)}, 0\right] + 1\right] \tag{5.19}$$

We can simplify the binomial model further by substituting a and rewriting as follows:

$$c_0 = \frac{\sum_{j=a}^{T} \frac{T!}{j!(T-j)!} q^j (1-q)^{T-j} \left[u^j d^{T-j} S_0 - X\right]}{(1+r)^T} \tag{5.20}$$

or:

$$c_0 = S_0\left[\sum_{j=a}^{T} \frac{T!}{j!(T-j)!} q^j (1-q)^{T-j} \frac{u^j d^{T-j}}{(1+r)^T}\right] - \frac{X}{(1+r)^T}\left[\sum_{j=a}^{T} \frac{T!}{j!(T-j)!} q^j (1-q)^{T-j}\right]$$

or, in shorthand form:[12]

$$c_0 = S_0 B[T, q'] - X(1+r)^{-T} B[T, q]$$

where $q' = qu/(1+r)$ and $1 - q' = d(1-q)/(1+r)$. The values q', q, and T are the parameters for the two binomial distributions. Three points are worth further discussion regarding this simplified binomial model:

1. First, as T approaches infinity, the binomial distribution will approach the normal distribution, and the binomial model will approach the Black–Scholes model, which we will discuss later in this chapter and in Chapter 6.
2. The current value of the option is:

$$c_0 = \frac{E[c_T]}{(1+r)^T} = P[S_T > X]\frac{E[S_T|S_T > X]}{(1+r)^T} - \frac{X}{(1+r)^T}P[S_T > X]$$

First, this implies that the binomial distribution $B[T,q] = P[S_T > X]$ provides the probability that the stock price will be sufficiently high at the expiration date of the option to warrant its exercise. Second, $S_0 B[T, q']/B[T, q]$ can be interpreted as the discounted expected future value of the stock conditional on its value exceeding X.

3. The call is replicated by a portfolio comprised of a long position in $B[T, q'] < 1$ shares of stock and borrowings. Investment in stock totals $S_0 B[T, q']$ and borrowings total $X(1+r)^{-T}B[T, q]$. The replication amounts must be updated at each time period.

5.2.3.3 *The Dynamic Hedge*

In a one-time period binomial framework where there exists a riskless asset, we can hedge the call against the stock such the resultant portfolio with one call and α shares of stock produces the same cash flow whether the stock increases or decreases:

$$c_u + \alpha u S_0 = c_d + \alpha d S_0 = (c_0 + \alpha S_0)(1 + r)$$

This one-time period hedge implies a hedge ratio α, which provides for the number of shares per call option position to maintain the perfect hedge (portfolio that replicates the bond):

$$\alpha = \frac{c_u - c_d}{S_0(d - u)}$$

Multiperiod models lead to 2^T potential outcomes without recombining, or, in many instances, $T + 1$ potential outcomes with recombining. Thus, complete capital markets require a set of 2^T or T priced securities (stocks or options) with payoff vectors in the same payoff space such that the set of payoff vectors is independent. In multiple period frameworks, hedging with many securities guarantees hedged portfolios at the portfolio termination or liquidation dates. In a binomial framework, there is an important exception. If the hedge α_t can be updated each period t, we write the hedge ratio for period t as follows:

$$\alpha_t = \frac{c_{u,t} - c_{d,t}}{S_t(d - u)}$$

The dynamic hedge is updated each period. Any portfolio employing and updating this dynamic hedge in each period in a binomial framework will be riskless at each period.

ILLUSTRATION: BINOMIAL OPTION PRICING

5.2.3.4 One-Time Period Case

Consider a stock currently selling for 10 and assume for this stock that u equals 1.5 and d equals 0.5 (this is a continuation of our numerical example from Section 5.2.2). The stock's value in the forthcoming period will be either 15 (if outcome u is realized) or 5 (if outcome d is realized). Consider a one-period European call trading on this particular stock with an exercise price of 9. If the stock price were to increase to 15, the call would be worth 6 ($c_u = 6$); if the stock price were to decrease to 5, the value of the call would be zero ($c_d = 0$). In addition, recall that the current riskless 1-year return rate is 0.25. Based on this information, we should be able to determine the value of the call.

In our numerical example offered above, we use the following to determine the value of the call in the binomial framework:

$$S_0 = 10 \qquad u = 1.5 \qquad d = 0.5$$
$$c_u = 6 \qquad c_d = 0 \qquad r = 0.25$$
$$X = 9$$

The risk-neutral probability of an upjump, the hedge ratio, and the time 0 value of the call in the one-period framework are calculated as follows:

$$q = \frac{1 + r - d}{u - d} = \frac{1 + 0.25 - 0.5}{1.5 - 0.5} = 0.75$$

$$\alpha = \frac{c_u - c_d}{S_0(d - u)} = \frac{6 - 0}{10(0.5 - 1.5)} = -0.6$$

$$c_0 = \frac{c_u q + c_d(1 - q)}{1 + r} = \frac{6 \times 0.75 + 0 \times (1 - 0.75)}{1 + 0.25} = 3.6$$

Recall that the time 0 value of the bond was 0.8. The time 0 value of the call is $3.6/0.8 = 4.5$ times that of the bond; the time 0 expected value of the call at time 1 is also 4.5 times the value of the time 1 bond value:

$$E_\mathbb{Q}[c_1] = [c_u q + c_d(1 - q)] = 6 \times 0.75 + 0 \times (1 - 0.75) = 4.5$$

Thus, under probability measure $\mathbb{Q}$ with the bond as the numeraire, the call price process is a martingale, just as the stock price process is.

5.2.3.5 Extending the Binomial Model to Two Periods

Now, we will extend our illustration above from a single period to two periods, each with a riskless return rate equal to 0.25. As before, the stock currently sells for 10 and will change to either 15 or 5 in one time period ($u = 1.5$, $d = 0.5$). However, in the second period, the stock will change a second time by a factor of either 1.5 or 0.5, leading to potential values of either 22.5 (up then up again), 10 (up once and down once), or 2.5 (down twice). Recall the lattice associated

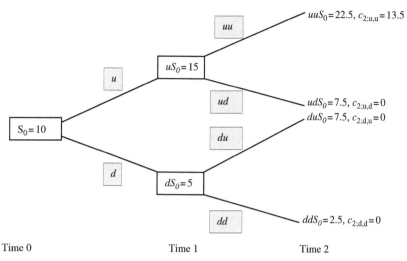

FIGURE 5.2   Two-period binomial model with option values.

with this stock price process is depicted in Figure 5.1. These stock prices are listed with call option values in Figure 5.2.

Since u, d, and r are the same for each period, probability measure $\mathbb{Q}$ will be the same for each period. Thus, $q = 0.75$ and $(1 - q) = 0.25$ in each period. However, the hedge ratio α_t will change for each period, depending on the share price movement in the prior period:

$$\alpha_0 = \frac{c_{1;u} - c_{1;d}}{S_0(d - u)} = \frac{6 - 0}{10(0.5 - 1.5)} = -0.6$$

$$\alpha_{1,u} = \frac{c_{2;u,u} - c_{2;u,d}}{uS_0(d - u)} = \frac{13.5 - 0}{15(0.5 - 1.5)} = -0.9$$

$$\alpha_{1,d} = \frac{c_{2;d,u} - c_{2,d,d}}{dS_0(d - u)} = \frac{0 - 0}{7.5(0.5 - 1.5)} = 0 \quad \text{(no hedge)}$$

Thus, the hedge ratio must be adjusted after each price change. Actually, there is no hedge or hedge ratio after the stock price decreases after time 0. The call option has no value in this event and cannot be used to hedge the stock risk. The hedge ratio for time 1, assuming that the stock price increased after time 0 will be -0.9, meaning that 0.9 shares of stock must be short sold for each purchased call to maintain the hedged portfolio.

The time 0 two-period binomial call price is calculated as follows:

$$c_0 = \frac{\sum_{j=0}^{2} \frac{2!}{j!(2-j)!} 0.75^j (1-0.75)^{2-j} \text{Max}\left[1.5^j(0.75^{2-j}) \times 10 - 9, 0\right]}{(1+0.25)^2} = \frac{c_{2;uu}q^2}{(1+r)^2} = \frac{13.5 \times 0.5625}{1.5625} = 4.86$$

Note that, in the two-period framework, the call has the same value at time 0 (4.86) as 7.59375 bonds at time 0. In addition, its time 0 expected value in time 2 is also the same as 7.59375 bonds:

$$E_Q[c_2] = [c_{2;u,u}\, q^2 + 2c_{2;u,d}q(1-q) + c_{2;d,d}(1-q)^2] = (22.5 - 9) \times 0.5625 = 7.59375$$

Thus, under the two-period probability measure Q with the bond as the numeraire, the call price process is a martingale, just as the stock price process is:

$$E_Q[S_{0,2}] = [u^2 S_0 q_u^2 + 2ud S_0 q(1-q) + d^2 S_0 (1-q)^2] = 15.625 = S_0/b_0$$

where the current value of the stock in this two-time period framework is 15.625 times the time 0 value (0.64) of the bond.

5.3 BROWNIAN MOTION AND ITÔ PROCESSES

In Chapter 2, we discussed the central limit theorem and the approach of a binomial distribution to a normal distribution as the number of Bernoulli trials approach infinity. Here, we discuss the approach of a binomial stochastic process to a continuous normal process as the sizes of intervals become smaller and their number approaches infinity. Brownian motion is derived as a limit of a discrete random walk in the companion website for this chapter. We will focus on continuous time–space models, beginning with Brownian motion, which is a Markov and martingale process. A continuous time–space Markov process is also known as a *diffusion process*. Brownian motion processes are probably the most simple of diffusion processes. We will introduce the processes and certain features of them here and perform various operations on them in the next chapter.

5.3.1 Brownian Motion Processes

One particular version of a continuous time–space random walk is a standard *Brownian motion process Z_t*, also called a Wiener process. A process Z_t is a standard Brownian motion process if:

1. Changes in Z_t over time are independent over disjoint intervals of time; that is, $\text{Cov}(Z_s - Z_\tau, Z_u - Z_v) = 0$ when $s > \tau > u > v$.
2. Changes in Z_t are normally distributed with $E[Z_s - Z_\tau] = 0$ and $E[(Z_s - Z_\tau)^2] = s - \tau$ for $s > \tau$. Thus, $Z_s - Z_\tau \sim N(0, s - \tau)$ with $s > \tau$.
3. Z_t is a continuous function of t.
4. The process begins at zero, $Z_0 = 0$.

Standard Brownian motion is a martingale, since for $s < t$, the process satisfies the two conditions for our martingale definition set forth in Section 5.1.2. The first is:

$$E[Z_t|\mathscr{F}_s] = E[(Z_t - Z_s) + Z_s|\mathscr{F}_s] = E[(Z_t - Z_s)|\mathscr{F}_s] + E[Z_s|\mathscr{F}_s] = Z_s$$

Since $Z_t \sim N(0, t)$, the following verifies our second martingale condition:

$$E[|Z_t|] = \frac{1}{\sqrt{2\pi t}} \int_{-\infty}^{\infty} |z|\, e^{-\frac{1}{2t}z^2}\, dz = \frac{2}{\sqrt{2\pi t}} \int_{0}^{\infty} z\, e^{-\frac{1}{2t}z^2}\, dz = \frac{-\sqrt{2t}}{\sqrt{2\pi}} e^{-\frac{1}{2t}z^2}\Big|_{0}^{\infty} = \frac{\sqrt{2t}}{\sqrt{2\pi}} < \infty$$

A random walk consists of taking random discrete unit steps at discrete times $t = 0, 1, 2, \ldots$ along a one-dimensional number line (axis). Brownian motion consists of continuous random movement along a one-dimensional number line over continuous time $t \geq 0$. If we let z denote the possible values on the z-axis for Brownian motion Z_t, then as time progresses (increases) the values of Z_t will move up and down the z-axis in a random but continuous way. Just as a random walk results in a particular path for each specific history in its sample space, a specific history ω in the sample space Ω for Brownian motion results in a particular Brownian motion path $Z_t(\omega)$. One can then graph Brownian motion $Z_t(\omega)$ versus time t for a specific history ω. Figure 5.3 depicts a plot of $Z_t(\omega)$ versus time t for a particular specific history ω in its sample space along with a close-up of a short segment of the process (note the changes in axis scaling).

Brownian motion has a number of interesting traits. First, it is continuous everywhere and differentiable nowhere (it never smooths) under Newtonian calculus; the Brownian motion process is not smooth and does not become smooth as time intervals decrease. This is because along any interval of time Δt, the change in height of the Brownian motion, $Z_{t+\Delta t} - Z_t$, is on the order of its standard deviation, which equals $\sqrt{\Delta t}$. Thus, $(Z_{t+\Delta t} - Z_t)/\Delta t$ is on the order of $1/\sqrt{\Delta t}$, which approaches infinity as $\Delta t \to 0$. This means that the average rate of change of Brownian motion over any time interval approaches infinity as the width of the time interval approaches zero. We see in Figure 5.3 that Brownian motion is a *fractal*, meaning that regardless of the length of the observation time period, the process can be defined equivalently by a simple change in the timescale. Graphically, this results in Brownian motion appearing similar no matter what scale of time one studies the graph if one is careful to scale the horizontal timescale quadratically faster

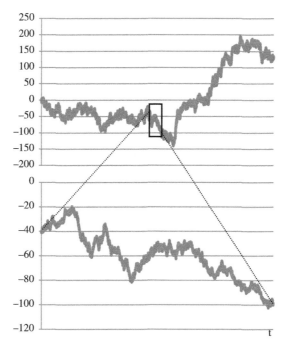

FIGURE 5.3 Brownian motion: a fractal.

than the vertical scale. This is because the horizontal scale is $\Delta t = (\sqrt{\Delta t})^2$, which is the square of the vertical scale. Consider the Brownian motion process represented by the top graph in Figure 5.3. If a short segment is cut out and magnified as in the bottom graph in Figure 5.3, the graphs look very similar to one another in that they do not smooth and the variance is proportional to time. Further magnifications of cutouts would continue to result in the same phenomenon. In addition, once a Brownian motion hits a given value, it will return to that value infinitely often. More generally, we will scale the variance by a multiple σ^2 so that the change in the process S_t from time t to time $t + \Delta t$ takes the form:

$$\Delta S_t = \sigma \Delta Z_t = \sigma(Z_{t+\Delta t} - Z_t) \sim N(0, \sigma^2 \Delta t)$$

Consider the (arithmetic) Brownian motion process S_t of the form $S_t = S_0 + \sigma Z_t$ where S_0 and $\sigma > 0$ are constants, and Z_t is standard Brownian motion. The graph of a particular possible path of S_t versus time t is similar to the graph of standard Brownian motion. It is continuous but nondifferentiable (zigzag shape). The difference is that it can start (time 0) at $S_0 \neq 0$ rather than at the origin 0. Furthermore its variance may be different (when $\sigma^2 \neq 1$) from the variance for standard Brownian motion. S_t has a normal distribution with mean S_0 and variance $\sigma^2 t$ ($S_t \sim N(S_0, \sigma^2 t)$) since:

$$\frac{S_t - S_0}{\sigma \sqrt{t}} = \frac{Z_t}{\sqrt{t}} = Z$$

where $Z \sim N(0,1)$. Note that $\Delta S_t = S_{t+\Delta t} - S_t \sim N(0, \sigma^2 \Delta t)$. This Brownian motion process is said to have a unit variance of σ^2 since when $\Delta t = 1$ the variance of $S_{t+1} - S_t$ equals σ^2.

ILLUSTRATION: BROWNIAN MOTION

Suppose a stock price S_t follows a Brownian motion process with an initial stock value of $50 and a unit variance of $\sigma^2 = 4$. Suppose that we want to find the probability that the stock price is less than $56 at time 3. Since $S_t \sim N(50, 4t)$, $S_3 \sim N(50, 12)$. Using a standard z-table, we find that this probability equals:

$$P(S_3 < 56) = P\left(\frac{S_2 - 50}{\sqrt{12}} < \frac{56 - 50}{\sqrt{12}}\right) = P(Z < 1.73) = 0.9582$$

If one were to graph this stock price as a function of time, the shape might look similar to Figure 5.3 until the present time. After the present time, we would not yet know the shape of the graph since the values of the stock are unknown. What we can say is that the graph will be one of an infinite number of possible choices of Brownian motion paths of the type in Figure 5.3. At any given moment of time t in either the present or future, the stock price S_t will change by unknown random amount $\Delta S_t = S_{t+\Delta t} - S_t$ either up or down over the time interval $[t, t + \Delta t]$ so that this change is normally distributed with variance Δt.

5.3.2 Stopping Times

A *stopping time* is the length of time required for a stochastic process to first satisfy some condition. A stopping time for a stochastic process X_t with filtration $\{\mathscr{F}_t\}$ is a random variable τ with the property that the event $\{\tau = t\}$ is measurable with respect to the σ-algebra $\mathscr{F}_t$.

This definition implies that the event $\{\tau = t\}$ depends only on the history of X_i, up until time $i = t$. For example, let τ be the first time that the discrete random walk X_t with increments ± 1 starting at integer X_0 reaches (hits) the value a where a is any integer. Certainly, the event $\{\tau = t\}$ depends only on $X_0, X_1, \ldots, X_t$. Since every distinct outcome $\{(X_0, X_1, \ldots, X_t)\}$ is in the σ-algebra $\mathscr{F}_t$, then the event $\{\tau = t\}$ must be measurable with respect to $\mathscr{F}_t$. Thus, τ is a stopping time.

Next, suppose that we let τ be the first time that the random walk hits either a or b with $a \leq X_0 \leq b$. Once again, τ is a stopping time. For a continuous version of this example, let τ be the first time that a Brownian motion $S_t = S_0 + \sigma Z_t$ hits either S_{low} or S_{high} such that $S_{\text{low}} \leq S_0 \leq S_{\text{high}}$. Analogously, τ is a stopping time.

Now, let τ_{S^*} be the first time that Brownian motion S_t hits fixed value S^* with $S^* > 0$. The event $\{\tau_{S^*} = t\}$ depends only on S_i for $i \leq t$. We will derive the probability that the hitting time (stopping time) τ_{S^*} is less than or equal to t (the distribution function for τ_{S^*}) for any $t \geq 0$. First, consider the probability that $S_t \geq S^*$:

$$P(S_t \geq S^*) = P(\tau_{S^*} \leq t)P(S_t \geq S^* | \tau_{S^*} \leq t)$$
$$+ P(\tau_{S^*} > t)P(S_t \geq S^* | \tau_{S^*} > t)$$

In other words, we account for all possible hitting times τ_{S^*} for S^*. That is, the probability that $S_t \geq S^*$ equals:

1. The probability that the hitting time for S^* is not more than t, multiplied by the probability that $S_t \geq S^*$, conditional on the hitting time already having been reached
2. plus the probability that the hitting time for S^* exceeds t, multiplied by the probability that $S_t \geq S^*$, conditional on the hitting time not already having been reached.

Let us first discuss part 1. Once the hitting time τ_{S^*} for S^* has already been reached ($\tau_{S^*} \leq t$) the probability that $S_t \geq S^*$ is 0.5; that is, $P(S_t \geq S^* | \tau_{S^*} \leq t) = 0.5$ because Brownian motion S_t is equally likely to go up as to go down from position S^*. Thus, once S^* has been reached, there is a 50% probability that any given subsequent value will exceed S^*. Moving on to part 2, if hitting time τ_{S^*} has not yet been reached ($\tau_{S^*} > t$), then $P(S_t \geq S^* | \tau_{S^*} > t) = 0$. So, the probability above simplifies to:

$$P(S_t \geq S^*) = 0.5 P(\tau_{S^*} \leq t)$$

We will rewrite this probability to obtain a distribution function for our hitting time as follows:

$$P(\tau_{S^*} \leq t) = 2P(S_t \geq S^*)$$

Since this Brownian motion S_t is distributed normally $S_t \sim N(S_0, \sigma^2 t)$, $(S_t - S_0)/\sigma\sqrt{t} \sim Z \sim N(0, 1)$. Thus, when $S_t \geq S^*$, the random variable $Z \geq (S^* - S_0)/(\sigma\sqrt{t})$, which implies:

$$P(\tau_{S^*} \leq t) = 2P(S_t \geq S^*) = 2P\left(Z > \frac{S^* - S_0}{\sigma\sqrt{t}} \right)$$
$$= 2\left[1 - P\left(Z \leq \frac{S^* - S_0}{\sigma\sqrt{t}} \right) \right] = 2\left(1 - N\left(\frac{S^* - S_0}{\sigma\sqrt{t}} \right) \right) \tag{5.21}$$

ILLUSTRATION: STOCK PRICE HITTING TIMES WITH ARITHMETIC BROWNIAN MOTION

An investor has purchased a stock whose price follows a Brownian motion process, with $S_0 = 3$ and unit variance $\sigma^2 = 4$, and time is measured in days. The investor has placed a limit order at 5 with his broker for the stock, meaning that the broker will sell the stock as soon she can obtain a selling price of 5. What is the probability that the hitting time $\tau_{S^*} = \tau_5$ for $S^* = 5$ is less than 8? That is, what is the probability that the limit sell order at 5 will be executed prior to 8 days?

We solve this problem for hitting time as follows:

$$P(\tau_5 \le 8) = 2P\left(Z > \frac{5-3}{\sqrt{4\sqrt{8}}}\right) = 2\left[1 - P\left(Z \le \frac{1}{\sqrt{8}}\right)\right] = 0.72367$$

Thus, the probability that the limit order executes within 8 days is 0.72367.

5.3.3 The Optional Stopping Theorem

When S_t is a martingale, $E[S_t] = S_0$ for any fixed t. However, the stopping time τ is a random variable, so that we cannot necessarily conclude that $E[S_\tau] = S_0$. However, with appropriate assumptions for bounds on the stopping time τ, it will be true that $E[S_\tau] = S_0$. These bounds lead to the optional stopping theorem, which allows us to conclude that $E[S_\tau] = S_0$. In addition, the optional stopping theorem allows us to calculate various probabilities and expectations associated with the stopping time. For example, suppose that an investor plans to invest until he either bankrupts himself ($S_{\tau_1} = 0$) at a stopping time τ_1, or has a million dollars ($S_{\tau_2} = 1,000,000$) at a stopping time τ_2. Let τ be the time when the first of these outcomes that occurs ($\tau = \text{Min}[\tau_1, \tau_2]$). Note that τ is also a stopping time. What is the probability that the investor makes a million dollars before he goes bankrupt? What is the expected value of the investor's terminal wealth S_τ? Since we do not know whether the investor will bankrupt himself or make a million first, his stopping time τ is a random variable. In order to compute these and other stopping time expectations and probabilities, the optional stopping theorem can be a powerful tool. We will state the theorem for discrete martingales. There is an analogous statement for continuous martingales.

The optional stopping theorem: Suppose that a stopping time τ for a martingale X_t satisfies any one the following assumptions:

1. $|X_t| \le c$ for all $t \le \tau$
2. $\tau \le c$
3. $\lim_{a \to \infty} \sup_{s \ge 0} E\left[|X_{min(\tau,s)}| \ge a\right] = 0$,

where c is some positive finite constant. Then, the expected value of the random variable X_τ is $E[X_\tau] = X_0$.

This theorem roughly holds that any gambling rule based on some stopping time τ has an expected payoff equal to zero (in other words equal to the starting capital X_0), as long as the gambler either has limited funds with which to gamble (assumption 1) or limited time to gamble (assumption 2). Thus, one may apply the optional stopping theorem for any gambling stopping time strategy based on any finite time constraint (I'll stop gambling and go to bed at 2 a.m.), or based on a finite level of winnings (I'll quit the moment I am either ahead by $100 or behind by $50).

Assumption 3 is the more general of the three assumptions, which in some respects imposes limits on bets, with assumptions 1 and 2 being special cases of it. However, the third assumption can be difficult to directly verify in most practical scenarios, so that much effort can be avoided if either of the first two can be verified. Also, if one replaces the condition that X_t is a martingale with X_t being either a submartingale or a supermartingale, the conclusion of the theorem becomes either $E[X_\tau] \geq X_0$ or $E[X_\tau] \leq X_0$, respectively.

The proof of this theorem will not be shown here, but can be found in any advanced text on stochastic processes. However, there is a natural intuition that gives some insight as to why this theorem ought to be true. Consider the special case that the martingale X_t is the following random walk. A gambler plays a game, with repeated independent coin tosses. Before each coin toss, the gambler bets \$1 on the outcome, with a probability of $\frac{1}{2}$ of winning the bet and a probability of $\frac{1}{2}$ of losing the bet. The stochastic variable X_t is the net winnings (the winnings will be negative if the losses more than offset the wins) of the gambler after playing the game t times. Clearly, X_t is a martingale, and after playing t times, the expected winnings of the gambler $E[X_t] = X_0$. Let τ be any type of stopping time strategy used by a player seeking winnings $E[X_\tau] > X_0$. Intuitively, if X_t is a martingale (a fair game), then there should be no winning stopping time strategy if one is restricted to having either a finite bankroll ($|X_t|$ is bounded for $t \leq \tau$) or at most a fixed finite amount of time before stopping (τ is bounded). If a gambler had an infinite bankroll, she could keep playing until she was guaranteed to be ahead by, say, \$1. This indicates that a stopping time strategy with these types of constraints cannot be a winning strategy for games based on a martingale process.

5.3.3.1 High and Low Hitting Times

Next, suppose that we are interested in the probability that our Brownian motion process S_t hits a lower barrier S_{Low}^* before hitting an upper barrier S_{High}^* where $S_{Low}^* < S_0 < S_{High}^*$:

$$P[\tau_{Low}^* < \tau_{High}^*]$$

To find this probability, denote $\tau = \text{Min}[\tau_{Low}^*, \tau_{High}^*]$. First, we find the expected value of S_τ. We have already shown that Brownian motion S_t is a martingale in Section 5.3.1. Since $|S_t| \leq \text{Max}[S_{Low}, S_{High}]$ for all $t \leq \tau$, then assumption 1 of the optional stopping theorem is satisfied. Thus, the optional stopping theorem applies and we conclude that $E[S_\tau] = E[S_0] = S_0$. If $\tau_{Low}^* < \tau_{High}^*$, then obviously $S_\tau = S_{Low}^*$. If $\tau_{Low}^* > \tau_{High}^*$, then $S_\tau = S_{High}^*$. Therefore:

$$S_0 = E[S_\tau] = S_{Low}^* P[\tau_{Low}^* < \tau_{High}^*] + S_{High}^* P[\tau_{Low}^* > \tau_{High}^*]$$

We also have:

$$P[\tau_{Low}^* > \tau_{High}^*] = 1 - P[\tau_{Low}^* < \tau_{High}^*]$$

Substituting this result into the equation above gives:

$$S_0 = S_{Low}^* P[\tau_{Low}^* < \tau_{High}^*] + S_{High}^* \left(1 - P[\tau_{Low}^* < \tau_{High}^*]\right)$$

Solving algebraically for $P[\tau_{Low}^* < \tau_{High}^*]$ gives:[13]

$$P\left[\tau_{Low}^* < \tau_{High}^*\right] = \frac{S_{High}^* - S_0}{S_{High}^* - S_{Low}^*} \tag{5.22}$$

Similarly, we have:

$$P\left[\tau_{High}^* < \tau_{Low}^*\right] = 1 - \frac{S_{High}^* - S_0}{S_{High}^* - S_{Low}^*} = \frac{S_0 - S_{Low}^*}{S_{High}^* - S_{Low}^*} \qquad (5.23)$$

ILLUSTRATION: STOCK PRICE HITTING TIMES WITH ARITHMETIC BROWNIAN MOTION

Earlier, we discussed the investor who purchased a stock whose price follows a Brownian motion process, with $S_0 = 3$ and $\sigma^2 = 4$. The investor has placed a limit order at 5. Now, suppose that the investor also placed a stop order at 2, meaning that the stock will sell when its price hits 2. What is the probability that the stock price hits 2 before it hits 5? We compute as follows:

$$P\left[\tau_{Low}^* < \tau_{High}^*\right] = \frac{S_{High}^* - S_0}{S_{High}^* - S_{Low}^*} = \frac{5-3}{5-2} = 0.667$$

5.3.3.2 *Expected Stopping Time*

Another important result that draws from the optional stopping theorem is the expected stopping time for a process with stopping time $\tau = \text{Min}[\tau_{Low}^*, \tau_{High}^*]$:

$$E[\tau] = \frac{(S_{High}^* - S_0)(S_0 - S_{Low}^*)}{\sigma^2}$$

We will prove this result for the case where S_t is standard Brownian motion; that is, when $S_0 = 0$ and $\sigma = 1$. First, consider a related stochastic process $S_t^2 - t$, which we will now verify to be a martingale. Using the definition of a martingale, we will show that $E[X_t | X_i, 0 \le i \le s] = X_s$ for all $s < t$ for the stochastic process $X_t = S_t^2 - t$. We begin our verification by calculating $E\left[S_t^2 - t | S_i^2 - i, 0 \le i \le s\right] \forall s < t$. Since S_t is a Markov process, it is obvious that $S_t^2 - t$ is also a Markov process. Thus:

$$E\left[S_t^2 - t | S_i^2 - i, 0 \le i \le s\right] = E\left[S_t^2 - t | S_s^2 - s\right] \forall s < t$$

Furthermore, if we are given $S_s^2 - s$ then we are given S_s and vice versa. So, we have:

$$E\left[S_t^2 - t | S_s^2 - s\right] = E\left[S_t^2 - t | S_s\right] = E\left[S_t^2 | S_s\right] - t$$

We calculate $E\left[S_t^2 | S_s\right]$ with a little algebra:

$$E\left[S_t{}^2 | S_s\right] = E\left[(S_t - S_s + S_s)^2 | S_s\right] = E\left[(S_t - S_s)^2 | S_s\right] + 2S_s E[S_t - S_s | S_s] + S_s^2$$

Note that $E[S_t - S_s | S_s] = 0$ because S_t is a martingale. We also have the obvious fact that $E\left[(S_t - S_s)^2 | S_s\right] = \text{Var}[S_t - S_s | S_s]$ by the definition of variance. Because $\sigma = 1$ for each unit period in the process, $\text{Var}[S_t - S_s | S_s] = t - s$. Putting all of this together gives:

$$E[S_t^2 - t | S_i^2 - i, 0 \le i \le s] = E\left[S_t^2 - t | S_s^2 - s\right] = E\left[S_t^2 - t | S_s\right] = E\left[S_t^2 | S_s\right] - t = t - s + S_s^2 - t = S_s^2 - s$$

Further, observe that:

$$E\left[|S_t^2 - t|\right] \le E\left[S_t^2 + t\right] = E[S_t^2] + t = 2t$$

From the two martingale conditions above, we conclude that $S_t^2 - t$ is a martingale. It can also be shown that $S_t^2 - t$ satisfies assumption 2 of the optional stopping theorem, but we will not show the details. By the optional stopping theorem, we have $E[S_\tau^2 - \tau] = 0$, or equivalently $E[S_\tau^2] = E[\tau]$. Next, consider a random variable defined by $X_\tau = (S_\tau - S_{Low}^*)(S_\tau - S_{High}^*)$. Since S_τ can only take on the values S_{Low}^* or S_{High}^*, then $E[X_\tau] = 0$. This can be expressed as:

$$0 = E[X_\tau] = E[(S_\tau - S_{Low}^*)(S_\tau - S_{High}^*)]$$

$$= E[S_\tau^2] - (S_{Low}^* + S_{High}^*)E[S_\tau] + S_{Low}^* S_{High}^*$$

$$= E[\tau] + S_{Low}^* S_{High}^*$$

In the last equation we used the fact that $E[S_\tau] = 0$ since we showed earlier that S_τ satisfies the requirements of the optional stopping theorem. We conclude that $E[\tau] = -S_{Low}^* S_{High}^*$ for our standard Brownian motion process. When we generalize the Brownian motion process for a nonzero starting point S_0 and a nonunit variance σ^2:

$$E[\tau] = -(S_{Low}^* - S_0)(S_{High}^* - S_0)/\sigma^2 = (S_{High}^* - S_0)(S_0 - S_{Low}^*)/\sigma^2 \tag{5.24}$$

ILLUSTRATION: EXPECTED MINIMUM HITTING TIME

We return once again to our illustration involving the investor who purchased a stock whose price follows a Brownian motion process, with $S_0 = 3$ and $\sigma^2 = 4$. The investor has placed a limit order at 5 and a stop order at 2. What is the expected length of time that will elapse before the investor sells his stock? That is, what is the expected minimum hitting time for either of our two barriers? We calculate this expected minimum hitting time as follows:

$$E\left[Min[\tau_{Low}^*, \tau_{High}^*]\right] = (S_{High}^* - S_0)(S_0 - S_{Low}^*)/\sigma^2$$

$$E\left[Min[\tau_{Low}^*, \tau_{High}^*]\right] = \frac{(5-3)(3-2)}{4} = 0.5$$

5.3.4 Brownian Motion Processes with Drift

In Sections 4.2.1 and 4.2.2, we studied and solved a number of financial models based on deterministic differential equations and obtained their solutions. However, since the future is unknown, more realistic differential models will require terms that are probabilistic and random in nature. Just as $dx(t) = x(t + dt) - x(t)$ denotes the infinitesimal change of a real valued function $x(t)$ resulting from an infinitesimal change dt in time, $dZ_t = Z_{t+dt} - Z_t$ will denote an infinitesimal change in Brownian motion Z_t resulting from the time change dt. For each fixed t and dt, by property 2 of Brownian motion, $dZ_t = Z_{t+dt} - Z_t \sim N(0, dt)$ is a random variable having a normal distribution with mean 0 and variance dt.

In order to create probabilistic models for a security S_t, we will generalize further on the Brownian motion process to allow for drift. For now, we provide a basic introduction here where we will now allow for drift a in the process as follows:

$$dS_t = a\, dt + b\, dZ_t$$

where a represents the drift tendency in the value of S_t, dZ_t is the infinitesimal change in the standard Brownian motion process, and b is a scaling factor for standard deviation in this process. In a sense, b can represent the instantaneous standard deviation of returns for a stock whose returns follow this Wiener process. If a and b are constants, then this process is called arithmetic Brownian motion with drift. Because prices of many securities such as stocks tend to have a predictable drift component in addition to randomness, generalized Wiener processes might be more practical for modeling purposes than standard Brownian motion, which only includes a random element.[14] The generalized Wiener process expression can be applied to stock returns as follows:

$$dS_t/S_t = \mu\,dt + \sigma\,dZ_t \tag{5.25}$$

The drift term, μ, represents the instantaneous expected rate of return for the stock per unit of time and σ is the instantaneous stock return standard deviation. A process S_t that follows Eq. (5.25) is called a geometric Wiener process (geometric Brownian motion), and it is the primary process that is used in finance to model stock prices S_t. We will derive the Black–Scholes option pricing model from this Wiener process shortly.

This geometric Wiener process can be interpreted in a stock environment as a return generating process. Again, the Brownian path $\sigma\,dZ_t$ of this process is not Newtonian differentiable. This means that the path does not smooth, so that we cannot draw tangent lines that we would otherwise associate with first derivatives.

5.3.5 Itô Processes

An *Itô process*, defined as a function of one or more stochastic variables such as S_t and one or more deterministic variables such as t, can be characterized similarly to the following function of S_t and t:

$$dS_t = a(S_t, t)dt + b(S_t, t)dZ_t \tag{5.26}$$

where a and b represent drift and variability terms that may change over time. Note that both the drift and variance terms, a and b, are functions of both S_t and t, and may change over time. We will not attempt to solve stochastic differential equations that take the form of Eq. (5.25) or (5.26) in this chapter. We merely seek to motivate the concept to more smoothly transition to in-depth coverage in Chapters 6–8.

5.4 OPTION PRICING: A HEURISTIC DERIVATION OF BLACK–SCHOLES[15]

Here, we will make our first effort at deriving the Black–Scholes option pricing model, reasoning through a rather heuristic derivation, and follow up with more rigorous derivations in Chapter 7. The derivation of the price for the call option provided in this section is not based on the powerful techniques that will be developed in later chapters, yet it produces the same results. The advantage of this heuristic approach is that the derivation is easier. However, it does not develop the rationale underlying risk-neutral pricing aspects for options. We will make all standard Black–Scholes assumptions, including that investors' price options as though they are risk neutral. In this section, we will derive the value of a call. The expected future value of the call is:

$$E[c_T] = E\left[\text{Max}(S_T - X), 0\right] = \int_X^\infty (S_T - X)p(S_T)\,dS_T$$

$$= \int_X^\infty S_T p(S_T)\,dS_T - X\int_X^\infty p(S_T)\,dS_T = \int_X^\infty S_T p(S_T)\,dS_T - XP(S_T > X)$$

where $p(S_T)$ is the density function for the random variable S_T.[16]

5.4.1 Estimating Exercise Probability in a Black–Scholes Environment

The value of a stock option is directly related to the probability that it will be exercised. That is, the option value is related to the probability that expiry date stock price S_T exceeds X, the exercise price of the option. The assumption that the stock price follows a geometric Brownian motion process means that the logarithmic return of a stock follows an arithmetic Brownian motion and is normally distributed with an upward drift $\hat{\mu}T$ taking the following form:

$$\ln\left(\frac{S_T}{S_0}\right) = \hat{\mu}T + \sigma Z_T \tag{5.27}$$

with S_T having the following probability distribution:

$$S_T = S_0\, e^{\hat{\mu}T + \sigma\sqrt{T}Z}$$

where $Z \sim N(0,1)$.[17] The drift constant $\hat{\mu}$ is called the mean logarithmic stock return rate. In order to obtain the risk-neutral price of a stock option, we will show in Chapter 7 that $\hat{\mu} = r - (\sigma^2/2)$, with r being the riskless return rate. We will assume this result for now.

To price the option, we begin by finding the probability that $S_T > X$:

$$P(S_T > X) = P\left(S_0\, e^{\hat{\mu}T + \sigma\sqrt{T}Z} > X\right) = P\left(\hat{\mu}T + \sigma\sqrt{T}Z > \ln\left(\frac{X}{S_0}\right)\right)$$

$$= P\left(Z > \frac{\ln\left(\frac{X}{S_0}\right) - \hat{\mu}T}{\sigma\sqrt{T}}\right) = P\left(Z > -\frac{\ln\left(\frac{S_0}{X}\right) + \hat{\mu}T}{\sigma\sqrt{T}}\right) = P\left(Z < \frac{\ln\left(\frac{S_0}{X}\right) + \left(r - \frac{1}{2}\sigma^2\right)T}{\sigma\sqrt{T}}\right) \tag{5.28}$$

$$= N\left(\frac{\ln\left(\frac{S_0}{X}\right) + \left(r - \frac{1}{2}\sigma^2\right)T}{\sigma\sqrt{T}}\right) = N(d_2)$$

if we define d_2 to equal $\frac{\ln\left(\frac{S_0}{X}\right) + (r - \frac{1}{2}\sigma^2)T}{\sigma\sqrt{T}}$. Thus, $N(d_2)$ is the probability that the option will be exercised.

5.4.2 The Expected Expiry Date Call Value

Next, we will focus on the term:

$$\int_X^\infty S_T p(S_T)\, dS_T$$

Since $S_T = S_0\, e^{\hat{\mu}T + \sigma\sqrt{T}Z}$ and $S_T > X$ is equivalent to $Z > -d_2$, it follows that:

$$\int_X^\infty S_T p(S_T)\, dS_T = S_0 \int_{-d_2}^\infty e^{\hat{\mu}T + \sigma\sqrt{T}z}\, e^{-\frac{z^2}{2}} \frac{dz}{\sqrt{2\pi}}$$

Next, we rewrite based on an algebraic manipulation involving "completing the square":[18]

$$\int_X^\infty S_T p(S_T)\, dS_T = S_0\, e^{rT} \int_{-d_2}^\infty e^{-\frac{1}{2}(z - \sigma\sqrt{T})^2} \frac{dz}{\sqrt{2\pi}}$$

Make the change of variables $y = z - \sigma\sqrt{T}$, which yields:

$$\int_X^\infty S_T p(S_T)\, dS_T = S_0 e^{rT} \int_{-d_2 - \sigma\sqrt{T}}^\infty e^{-\frac{1}{2}y^2} \frac{dy}{\sqrt{2\pi}}$$

$$= S_0 e^{rT} \int_{-\infty}^{d_2 + \sigma\sqrt{T}} e^{-\frac{1}{2}y^2} \frac{dy}{\sqrt{2\pi}} = S_0 e^{rT} N(d_1)$$

where we define $d_1 = d_2 + \sigma\sqrt{T}$. Thus, the expected expiry date call value equals:

$$E[c_T] = \int_X^\infty S_T p(S_T)\, dS_T - XP(S_T > X) = S_0 e^{rT} N(d_1) - XN(d_2)$$

Now discount this expected future value to obtain the Black–Scholes option pricing model:

$$c_0 = S_0 N(d_1) - \frac{X}{e^{rT}} N(d_2) \tag{5.29}$$

5.4.3 Observations Concerning $N(d_1)$, $N(d_2)$, and c_0

With some minor manipulation, the Black–Scholes option pricing model provides several useful interpretations concerning call value:

1. The probability that the stock price at time T will exceed the exercise price X of the call is $P[S_T > X] = N(d_2)$.
2. The expected value of the stock conditional on the stock's price S_T exceeding the exercise price of the call is:

$$E[S_T|S_T > X] = \frac{\int_X^\infty S_T p(S_T)\, dS_T}{\int_X^\infty p(S_T)\, dS_T} = \frac{S_0 e^{rT} N(d_1)}{N(d_2)}$$

3. The expected expiry date call value is simply the product of the probability of call exercise and the expected value of the stock conditional on the stock's price exceeding the exercise price of the call, minus the expected exercise value paid at time T:

$$E[c_T] = E[S_T|S_T > X]\, P[S_T > X] - XP(S_T > X)$$

$$= \frac{S_0 e^{rT} N(d_1)}{N(d_2)} N(d_2) - XN(d_2) = S_0 e^{rT} N(d_1) - XN(d_2)$$

4. The present value of the call is simply the discounted value of its expected future value:

$$c_0 = E[c_T]e^{-rT} = e^{-rT}\left[S_0 e^{rT} N(d_1) - XN(d_2)\right] = S_0 N(d_1) - Xe^{-rT} N(d_2)$$

5.5 THE TOWER PROPERTY

The *tower property* concerns expectations that are conditioned at different points in time. This property is very useful for valuing securities and estimating term structures based on information that is revealed over time. Suppose X_t is a stochastic process with an adapted filtration $\{\mathscr{F}_t\}$. If $s \leq t$, then for any $T \geq t$, we have:

$$E[E[X_T|\mathscr{F}_t]|\mathscr{F}_s] = E[X_T|\mathscr{F}_s]$$

Remember that expected value is just a weighted average. Essentially, this property can be interpreted to suggest that our average now (time s) of tomorrow's (time t) averages of the following day's (time T) prices is the same as our average today of the prices the day following (time T). Similarly, but still rough, whatever we might gain by analyzing the content of the anticipated but still unknown information set that we will access tomorrow is already accounted for in our information set for today in order to calculate the following day's average price. We state and verify the related tower theorem on the companion website for the special case when the filtrations are finite. We also demonstrate in Chapter 6 that the tower property can be used to show that the expectation of a stochastic process that is itself an integral of a stochastic process with respect to Brownian motion equals zero.

5.6 EXERCISES

5.1. Suppose that a particular swap contract is currently valued at 0 and that the probability equals p that the price of the contract will increase by 1 at times 1 and 2. The probability that the price of the contract will decrease by 1 equals $1 - p$ at times 1 and 2. Assume that the event that the price increases or decreases at time 1 is independent of the event that the price increases or decreases at time 2. No numerical calculations are needed for this example.

    a. What are all of the possible distinct price two-period change outcomes for the contract through time 2? For example, let the notation $(1, -1)$ mean that the contract's price will have increased by 1 at time 1 and decreased by 1 at time 2.

    b. What is the sample space for this process?

    c. What are the potential time 0 events? What are the probabilities associated with each of the potential time 0 events?

    d. What is $P(X_1 = 1)$?

    e. What is $P(\{(1, 1), (1, -1)\})$?

    f. Is the following statement true: $P(X_1 = -1) = P(\{(-1, 1), (-1, -1)\}) = 1 - p$?

    g. Is the following statement true: $P(X_1 = 1 \text{ or } X_1 = -1) = P(\varnothing) = 0$?

    h. What is $P(X_2 = 0 \text{ or } X_2 = 2) = P(\{(1, 1), (1, -1), (-1, 1)\}) = 2p(1 - p) + p^2 = p(2 - p)$.

    i. What is $P(X_1 = 1$ and $X_2 = 0)$?

5.2. Suppose that a brokerage firm uses an algorithm so that the number of portfolios that it assigns to any given broker satisfies the following model. Let X_t denote the number of portfolios that it assigns a broker on day t. Assume that the firm will assign either 1, 2, or 3 portfolios per day to a broker. At time $t = 1$ the random variable X_1 has probabilities equal to $1/3$ of taking on the values of 1, 2, or 3. For all subsequent times $t = 2, 3, 4, \ldots$, the variable X_t satisfies the following conditions. If $X_{t-1} = 1$, then $P(X_t = 2) = \frac{1}{2}$ and $P(X_t = 3) = \frac{1}{2}$. If $X_{t-1} = 2$, then $P(X_t = 1) = \frac{1}{2}$ and $P(X_t = 3) = \frac{1}{2}$. If $X_{t-1} = 3$, then $P(X_t = 1) = \frac{1}{2}$ and $P(X_t = 2) = \frac{1}{2}$.

    a. Is the process X_t stochastic?

    b. Is the process X_t a Markov process?

    c. Are the increments $Z_t = X_t - X_{t-1}$ independent over time starting with $t = 2$?

5.3. Cards are dealt one at a time from a standard 52-card randomly shuffled deck and points are awarded to the lone recipient based on the number on the card (2–10) or 11 if the dealt card is a "face card" or ace. Let S_t represents the number of points to be held by the recipient after t cards have been dealt by the dealer. For parts a through c, suppose that the cards have been dealt without replacement. For parts d, e, f, and g, assume that the cards have been dealt with replacement and that 1 point is awarded if the number on the card is a 2 through a 6, 0 points are awarded if the number on the card is 7, 8, or 9, and −1 point is awarded if the card is a 10, a face card, or an ace. We point out that this is the most common point system used by card counters playing blackjack. Note that we are assuming a finite process since the "time" remains finite, running from 0 to 52.

    a. Is this process stochastic?

    b. Is this process Markov?

c. Is this process a submartingale?

d. Is this process Markov?

e. Is this process a submartingale?

f. Is this process Markov?

g. Is this process a martingale?

5.4. a. Consider two probabilities $\mathbb{P}$ and $\mathbb{Q}$ defined on the sample space $\Omega = \{\omega_1, \omega_2, \omega_3\}$ in the following way: $p(\omega_1) = 0$, $p(\omega_2) = 0.4$, $p(\omega_3) = 0.6$, and $q(\omega_1) = 0.4$, $q(\omega_2) = 0$, $q(\omega_3) = 0.6$. Are $\mathbb{P}$ and $\mathbb{Q}$ equivalent probability measures?

  b. Now, we change our exercise values. Suppose that, instead, $p(\omega_1) = 0$, $p(\omega_2) = 0.4$, $p(\omega_3) = 0.6$, and $q(\omega_1) = 0$, $q(\omega_2) = 0.3$, $q(\omega_3) = 0.7$. Are $\mathbb{P}$ and $\mathbb{Q}$ equivalent probability measures now?

5.5. Suppose that a particular stock will experience $n = 4$ consecutive transactions over a given period and that we wish to know the probability that the stock price will have increased in exactly y^* of those four transactions. Assume that the transactions follow a binomial process.

  a. What is the number of potential orderings of price increases (+) and decreases (−) in these stock prices over the four transactions?

  b. List all potential orderings of the directional changes of the stock price (+) or (−).

  c. Suppose that each transaction is equally likely to result in a price increase or decrease, then each trial has an equal probability of an upjump or downjump ($p = 1/2$). What is the probability that any one of the potential orderings listed in part c will be realized?

  d. Suppose that the probability of a price increase in any given transaction equals $p = 0.6$. What is the probability of realizing three price increases followed by a single decrease?

  e. What is the probability that exactly $y^* = 3$ price increases will result from $n = 4$ transactions where $p = 0.6$?

  f. What is the probability that more than three price increases will result from $n = 4$ transactions where $p = 0.6$?

5.6. Examination of trade-by-trade data for a given stock reveals that the stock has a 51% probability of increasing by $0.0625 on any given transaction and a 49% probability of decreasing by $0.0625 on any given transaction. The stock has a current market value equal to $100 and is expected to trade 10 times per day starting today.

  a. What is the probability that the stock's price will exceed $99.99 at the end of today?

  b. What is the probability that the stock's price will exceed $100.49 at the end of today?

  c. What is the probability that the stock's price will exceed $110 at the end of 10 days?

5.7. Consider a Markov process S_t that produces one of two potential outcomes at each time $t = n \, \Delta t$ for $n = 0, 1, 2, \ldots$. For example, suppose a stock price can increase (uptick) by $a\sqrt{\Delta t}(a > 0)$ with physical probability p or decrease (downtick) by $a\sqrt{\Delta t}$ with physical probability $(1 - p)$. This process applies to each time period $t \geq \Delta t$. Assume that each uptick or downtick at a particular time is independent of the uptick or downtick at any other time, and so S_t follows a binomial process.

  a. Write a function that provides the expected value of $S_{t+\Delta t}$ at time $t + \Delta t$ given S_t.

  b. Under what circumstances is this Markov process also a martingale?

  c. Derive a formula to obtain the variance of the process $S_{t+\Delta t} - S_t$.

  d. Simplify your formula in parts a and c for $p = 0.5$.

  e. What is the standard deviation of this process?

5.8. Examination of price data for a given stock reveals that the stock has a 41% probability of increasing by 42% in any given 6-month period and a 59% probability of decreasing by 30% during any 6-month period. The stock has a current market value equal to $60; a 1-year call option exists on the stock with an exercise price equal to $50. Assume an annual riskless return rate equal to 5% and that it is compounded continuously.

  a. What is the probability that the stock's price will exceed $100 at the end of 6 months?

  b. What is the probability that the stock's price will exceed $100 at the end of 1 year? (One year represents two 6-month intervals.)

  c. What are the two potential stock prices at the end of the first 6-month period? What are the probabilities associated with each of these prices?

  d. What are the three potential stock prices at the end of the second 6-month period? What are the probabilities associated with each of these prices?

  e. What are the three potential call option values at the end of 1 year? That is, what would be the value of the call option conditional on each of the three potential stock prices being realized?

  f. If the stock price increases during the first 6-month period, two potential stock prices are possible at the end of 1 year. What are these two prices? What are the potential call values? Based on these two potential prices and associated probabilities, what would be the value of the call option if the stock price increases during the first 6-month period? Note that we are pricing the call based upon the physical probabilities, so that we will not obtain risk-neutral pricing.

  g. If the stock price decreases during the first 6-month period, two potential stock prices are possible at the end of 1 year. What are these two prices? What are the potential call values? Based on these two potential prices and associated probabilities, what would be the value of the call option if the stock price decreases during the first 6-month period?

  h. Based on the two potential call option values estimated in parts f and g and their associated probabilities, what is the current value of the call?

5.9. A stock currently selling for $100 has a 75% probability of increasing by 20% in each time period and a 25% chance of decreasing by 20% in a given period. Assume a binomial process. What is the expected value of the stock after four periods?

5.10. In Section 5.2.4 (see also endnote 9), we claimed that in a multiperiod binomial framework affecting stock prices, $c_0 = \dfrac{\sum_{j=0}^{T} \frac{T!}{j!(T-j)!} q^j (1-q)^{T-j} \mathrm{Max}[u^j d^{T-j} S_0 - X, 0]}{(1+r_f)^T}$. Derive this formula for call valuation. You may assume that the riskless hedge can be created and that valuation occurs in a riskless framework over T periods of time.

5.11. Consider the binomial returns process with upjump u and downjump d covered at the end of Section 5.2.1. In that section, we claimed that if $pu + (1-p)d = 1$, then S_t is a martingale.

  a. Demonstrate that S_t is indeed a martingale under the circumstances given above.

  b. Suppose that the riskless bond prices are $B_0 = 1$ and $B_t = (1+r)^t$. Further assume that S_t follows the binomial returns process described above with probability q of an upjump. Demonstrate that if $qu + (1-q)d = 1+r$, then S_t is a martingale with respect to the bond (change of numeraire).

  c. Under the same circumstances as in part b above, demonstrate that if $q = (1+r-d)/(u-d)$, then S_t is a martingale with respect to the bond.

5.12. Assume that a binomial pricing model holds for all securities with a probability p of an upjump $1 + u$ and a probability $1 - p$ of a downjump $1 - d$. Also suppose that a risk-free rate of return r holds at each time interval.

    a. Derive an expression that solves for the discounted expected value of the price S_n at time n given the price S_m at time m.

    b. Find the value of p so that the discounted expected value satisfies $E[S_n \mid S_m]/(1+r)^{n-m} = S_m$ for $m < n$; that is, find the value of p so that S_n is a martingale (in bond numeraire).

5.13. Consider a one-time period, two-potential-outcome framework where there exists Company Q stock currently selling for $50 per share and a riskless $100 face value T-bill currently selling for $90. Suppose Company Q faces uncertainty, such that it will pay its owner either $30 or $70 in 1 year. Further assume that a call with an exercise price of $55 exists on one share of Q stock.

    a. Define the equivalent probability measure for this payoff space.

    b. Is this measure an equivalent martingale (risk-neutral probability measure)?

    c. Is this market complete?

    d. What are the two potential values the call might have at its expiration?

    e. What is the riskless rate of return for this example? Remember, the Treasury bill pays $100 and currently sells for $90.

    f. What is the hedge ratio for this call option?

    g. What is the current value of this option?

    h. What is the value of a put with the same exercise terms as the call?

5.14. Rollins Company stock currently sells for $12 per share and is expected to be worth either $10 or $16 in 1 year. The current riskless return rate is 0.125.

    a. Define the equivalent probability measure for this payoff space.

    b. Is this market complete?

    c. What would be the value of a 1-year call with an exercise price of $8?

5.15. The riskless return rate equals 0.08 per year in a given economy. A stock, which currently sells for $50, has an expected per-jump multiplicative upward movement of 1.444, and $d = 1/u$. Under the binomial framework, what would be the value of 9-month (0.75 year) European calls and European puts with striking prices equal to $80 if the number of tree steps (n) over the 9-month period were:

    a. 2?

    b. 3?

    c. 8?

    Assume the riskless return rate is compounded over each relevant time increment for parts a, b, and c.

5.16. Ibis Company stock is currently selling for $50 per share and has a multiplicative upward movement equal to 1.2776 and a multiplicative downward movement equal to 0.7828. What is the value of a 9-month (0.75 years) European call and a European put with striking prices equal to $60 if the number of tree steps were 2? Assume a riskless return rate equal to 0.081.

5.17. Show that in the binomial pricing model, the expectation:

$$E[S_t \mid S_s] = S_s[pu + (1-p)d]^{t-s}$$

for $s < t$ can be derived from the equation:

$$E[S_t \mid S_0] = S_0[pu + (1-p)d]^t$$

5.18. An investor has taken a long position in a futures contract in oil. The spot price of oil is expected to follow a Brownian motion process, with $S_0 = 90$ and a monthly variance equal to 25, and time is measured in months.

   a. What is the probability that the price of oil will exceed $100 in 25 months?

   b. What is the probability that the oil price will reach 100 in less than 25 months?

   c. What is the median hitting time for $S^* = 100$? That is, what hitting time t corresponds to the probability $P[\tau_{100} \leq t] = 0.5$?

   d. What is the probability that the oil price hits 50 before it hits 100?

   e. What is the expected hitting time for either of the two barriers?

5.19. Suppose we have a Brownian motion process S_t with an initial value of $S_0 = 0$ and a unit variance of $\sigma^2 = 4$. What is the probability that S_t hits $S^*_{\text{Low}} = -2$ before hitting $S^*_{\text{High}} = 5$? How does this probability change if the variance changes?

5.20. A gambler plays a game, with repeated independent coin tosses. Before each coin toss, the gambler bets $1 on the outcome, with a probability of $1/2$ of winning the bet and a probability of $1/2$ of losing the bet. Let X_t be the winnings of the gambler betting on this martingale process. Define a stopping time τ to be the first time that the winnings X_t reach $1.

   a. Let τ be the stopping time for this game. What is $E[X_\tau]$?

   b. What is $E[X_t]$ for this game?

5.21. Kestrel Company stock is currently selling for $40 per share. Its historical standard deviation of returns is 0.5; this historical standard deviation will be used as a forecast for its standard deviation of returns. Assume that the logarithmic return on the stock follows a Brownian motion with drift process given by Eq. (5.27). The 1-year Treasury bill rate is currently 5%. What is the probability that the value of the stock will be less than $30 in 1 year?

5.22. Verify that the following two expressions to define d_2 are identical. That is, verify that $d_2 = d_1 - \sigma\sqrt{T}$:

$$\frac{\ln\left(\frac{S_0}{X}\right) + (r - \frac{1}{2}\sigma^2)T}{\sigma\sqrt{T}} = \frac{\ln\left(\frac{S_0}{X}\right) + (r + \frac{1}{2}\sigma^2)T}{\sigma\sqrt{T}} - \sigma\sqrt{T}$$

5.23. Verify endnote 17 for Section 5.4.2, which states the following:

$$\hat{\mu}T + \sigma\sqrt{T}z - \frac{z^2}{2} = \left(\hat{\mu} + \frac{\sigma^2}{2}\right)T - \frac{(z - \sigma\sqrt{T})^2}{2}$$

5.24. Let $P[S_T > X] = \int_X^\infty p(S_T)\,dS_T$ equal the probability that a given call will be exercised at expiry, time T, where S_T is the expiry date random stock price and X is the exercise price of the option. Do not assume that the Black–Scholes assumptions are necessarily true (avoid use of $N(d_1)$ and $N(d_2)$).

   a. Write a function that gives the expected expiry date value of the stock conditional on its value exceeding X.

   b. Write a function that gives the expected expiry date value of the call conditional on its underlying stock value exceeding X.

   c. Based on your answers to parts a and b, write a function that gives the expected expiry date value of the call.

5.25. A *gap option* has two exercise prices, one (X) that triggers the option exercise and the second (M) that represents the price or cash flow that exchanges hands at exercise. For example, a gap call with exercise prices $X = 50$ and $M = 40$ enables its owner to purchase its underlying stock for 40 when its price rises above 50. The European version of this option can be exercised only when the underlying stock price exceeds 50 at expiration, but at a price of $M = 40$. Assume that the underlying stock price follows the following process:

$$S_T = S_0 \, e^{[\mu T + \sigma \sqrt{T} z]}$$

The current stock price is $S_0 = 45$, the variance of underlying stock returns equals $\sigma^2 = 0.16$, the riskless return rate equals $r = 0.05$.

a. What is the probability that the underlying stock will be worth more than 50 in 1 year?

b. What is the expected value of this stock contingent on its price exceeding 50?

c. The owner of the gap option has the right to pay 40 for the underlying stock if its price rises above 50. What is the expected value of this call contingent on its exercise?

d. What is the current value of this gap call?

NOTES

1. Filtrations can be characterized as the information set required for valuation at any time or state.
2. Pearson (1905) described the optimal search process for finding a drunk left in the middle of a field. Left to stagger one step at a time in an entirely unpredictable fashion, she is more likely to be found where she was left than in any other position on the field.
3. Note that we move from the second line in the equation set to the third line by making use of the following equality from Chapter 2: $P[A|B] = \frac{P[A \cap B]}{P[B]}$. In addition, observe that the independence of the random variables $\{Z_t\}$ was used to move from the third to the fourth line.
4. See end-of-chapter Exercise 5.2 for an illustration of this.
5. Since any member in the σ-algebra $\mathscr{F}_t$ must also be a member of the σ-algebra $\mathscr{F}$, the notation $(\omega_1, \omega_2, \ldots, \omega_t)$ really represents the set of all elements in Ω so that the first t coordinates are $\omega_1, \omega_2, \ldots, \omega_t$. The point is that at time t, we can only know the market state (u or d) for times 1 to t. For times $t + 1, t + 2, \ldots$, the market states are not yet known. More precisely, $(\omega_1, \omega_2, \ldots, \omega_t)$ represents:

$$(\omega_1, \omega_2, \ldots, \omega_t) = \bigcup_{\omega_{t+1}, \omega_{t+2}, \ldots \in \{u,d\}} \{(\omega_1, \omega_2, \ldots, \omega_t, \omega_{t+1}, \omega_{t+2}, \ldots)\}$$

6. The second property is a technical condition that will be satisfied for every stochastic process that we study in this text. Thus, we will focus only on the first condition, excepting for Brownian motion, the most important continuous process in finance, in which case we will focus on both.
7. A submartingale has increments whose expected values equal or exceed zero; expected values of increments from a supermartingale are equal to or less than zero. Martingales are also both submartingales and supermartingales.
8. If $\mathbb{Q} \leq \mathbb{P}$ and $\mathbb{P} \leq \mathbb{Q}$, then $\mathbb{Q} \sim \mathbb{P}$.
9. Alternatively, we might assume continuous compounding of interest such that $r_{t,T} = -\ln \sum \psi_{0,t,i}$.
10. See end-of-chapter Exercise 5.10 for a derivation. See also Cox and Rubenstein (1985).
11. We obtain a by first determining the minimum number of price increases j needed for S_T to exceed X:

$$S_T = u^j d^{T-j} S_0 > X$$

We then solve this inequality for the minimum positive integer value for j with $u^j d^{T-j} S_0 > X$ (note: if $j \leq 0$, a will equal 0). First, divide both sides of the inequality by $S_0 d^T$ so that $(u/d)^j > X/(S_0 d^T)$. Next, take logs on both sides to obtain $j \ln(u/d) > \ln(X/(S_0 d^T))$. Finally, divide both sides by $\ln(u/d)$ to get the desired result. Thus, a is the smallest positive integer for j such that $S_T > X$.

12. As the lengths of time periods approach zero, d must approach $1/u$.

13. The intermediate steps are:

$$S_0 = S^*_{Low} P\left[\tau^*_{Low} < \tau^*_{High}\right] + S^*_{High} - S^*_{High} P\left[\tau^*_{Low} < \tau^*_{High}\right]$$

$$S_0 = \left[S^*_{Low} - S^*_{High}\right] P\left[\tau^*_{Low} < \tau^*_{High}\right] + S^*_{High}$$

14. A Wiener process is a continuous time−space Markov process with normally distributed increments. A positive drift would be consistent with time value of money and investor risk aversion.

15. This section assumes that the reader has been introduced to the Black−Scholes (Black and Scholes, 1972, 1973; Merton, 1973) option pricing model elsewhere. If this is not the case, the model will be thoroughly discussed in Chapter 7 and the essential mathematics are introduced here. This heuristic derivation is in the spirit of Boness (1964) and as described by Jarrow and Rudd (1983).

16. The derivation improves if we merely substitute risk-neutral densities $q(S_T)$ for physical densities $p(S_T)$. This will be a focus of the next chapter.

17. In the text website for this chapter, we show how this model can be derived as a limit of a certain discrete process with drift and its volatility based on a random walk.

18. If we focus on the exponents of the equations above and below, with a little algebra, we can verify that $\hat{\mu} T + \sigma \sqrt{T} z - \frac{z^2}{2} = \left(\hat{\mu} + \frac{\sigma^2}{2}\right) T - \frac{(z - \sigma \sqrt{T})^2}{2} = rT - \frac{(z - \sigma \sqrt{T})^2}{2}$. See Exercise 5.23 at the end of the chapter.

References

Black, F., Scholes, M., 1972. The valuation of options contracts and a test of market efficiency. J. Finance. 27, 399−417.

Black, F., Scholes, M., 1973. The pricing of options and corporate liabilities. J. Polit. Econ. 81, 637−654.

Boness, A.J., 1964. Elements of a theory of stock-option value. J. Polit. Econ. 72, 163−175.

Cox, J.C., Rubinstein, M., 1985. Options Markets. Prentice Hall, Englewood Cliffs, NJ.

Jarrow, R., Rudd, A., 1983. Option Pricing. Dow Jones-Irwin, Homewood, IL.

Merton, R.C., 1973. Theory of rational option pricing. Bell J. Econ. Manage. Sci. 4, 141−183.

Pearson, K., 1905. The problem of the random walk. Nature. 72, 342.

6

Fundamentals of Stochastic Calculus

6.1 STOCHASTIC CALCULUS: INTRODUCTION

In Chapters 7 and 8, we will apply stochastic differential equations to modeling the pricing behavior of derivative instruments and bonds. In this chapter, we will introduce the theory and the methods of stochastic calculus. Although stochastic and ordinary calculus share many common properties, there are fundamental differences. The probabilistic nature of stochastic processes distinguishes them from the deterministic functions associated with ordinary calculus. Since stochastic differential equations so frequently involve Brownian motion, second-order terms in the Taylor series expansion of functions become important, in contrast to ordinary calculus where they can be ignored. This attribute is reflected in Itô's lemma, which is a powerful tool used to solve stochastic differential equations.

Arbitrage pricing (pricing securities relative to securities or portfolios producing identical pay-off structures) can be accomplished in numerous ways. A most useful feature of arbitrage pricing is that it does not require that we forecast security prices or even calculate risk premiums or expected returns. Such forecasts are, at best, unreliable and difficult to make. We use option pricing models such as the Black–Scholes or binomial models that require no inputs for expected future security prices, risk premiums, or expected returns (aside from the riskless security). However, such models do require knowledge of underlying security volatility, so it will be convenient to be able to delete references to expected security returns while maintaining information concerning volatility. This will entail creation of martingales.

Our analysis in this chapter will generally focus on continuous-time stochastic processes, though for development purposes, we will make references in certain sections to discrete-time processes for the sake of simplicity. In this section, we will introduce stochastic differentiation and stochastic integration. In the following section, we will discuss the decomposition of sub-martingales into drift and martingale components. We continue by discussing shifting probability measures of submartingales to produce martingales, in many respects similar to what we did in previous chapters. We will discuss Itô's lemma along with a number of applications and further discussions concerning stochastic integration. Much of the mathematics in this chapter will be applied to financial problems in Chapters 7 and 8.

6.1.1 Differentials of Stochastic Processes

In many respects, differentials of stochastic processes mirror differentials of real-valued functions from ordinary calculus, sharing many of their properties. However, there are also important differences. In ordinary calculus, if X_t is a real-valued function of the real variable t, then the derivative X_t' exists for a large class of functions. If X_t is a stochastic process, then it is usually not possible to well define the derivative of X_t with respect to t, at least for the class of stochastic processes that are relevant in finance. This is because normally X_t involves Brownian motion, and Brownian motion is not differentiable. Nevertheless, we can still study the differential of a stochastic process, which is the change of a stochastic process X_t resulting from a small change in t. Define the differential dX_t of a stochastic process X_t to be a quantity that satisfies the property:

$$\lim_{dt \to 0} \frac{X_{t+dt} - X_t - dX_t}{dt} = 0$$

The differential dX_t is used to approximate $X_{t+dt} - X_t$, and any terms that approach zero after dividing by dt as $dt \to 0$ can be ignored. This is the same requirement that we have seen in ordinary calculus. In the following sections, we will compare an example of a real-valued (ordinary) differential with that of a stochastic differential.

6.1.1.1 An Example of a Real-Valued Differential from Ordinary Calculus

Consider the (ordinary) real-valued function $X(t) = t^3$, with t and y being real variables. Then:

$$X(t + dt) - X(t) = (t+dt)^3 - t^3 = 3t^2 \, dt + 3t(dt)^2 + (dt)^3$$

Note that $[3t(dt)^2 + (dt)^3]/dt = 3t \, dt + (dt)^2 \to 0$ as $dt \to 0$. This means that for small values of dt, $3t(dt)^2 + (dt)^3$ is much smaller than $3t^2 \, dt$ and we can approximate $X(t + dt) - X(t)$ by $3t^2 \, dt$. So, the differential of $X(t) = t^3$ is $dX = 3t^2 \, dt$. As we saw in the review of the differential for real-valued continuously differential functions in Chapter 1, the differential of a function $X(t)$ equals the derivative of the function times dt. For both real-valued and stochastic differentials, one can always choose $X_{t+dt} - X_t$ itself as the differential dX_t, but it is often better to use another choice that is either more useful or leads to a simpler expression. We saw this illustrated earlier in this book, and we will see this situation again in the example below.

6.1.1.2 An Example of a Stochastic Differential

Next, consider the following stochastic process $X_t = tZ_t$, with Z_t being standard Brownian motion. We calculate that:

$$X_{t+dt} - X_t = (t + dt)Z_{t+dt} - tZ_t = t(Z_{t+dt} - Z_t) + Z_{t+dt} \, dt$$
$$= t(Z_{t+dt} - Z_t) + Z_{t+dt} \, dt - Z_t \, dt + Z_t \, dt = t(Z_{t+dt} - Z_t) + (Z_{t+dt} - Z_t) \, dt + Z_t \, dt$$
$$= t \, dZ_t + dZ_t \, dt + Z_t \, dt$$

Now, consider choosing $dX_t = t \, dZ_t + Z_t \, dt$ as a differential of X_t. With this choice, note that:

$$\frac{X_{t+dt} - X_t - dX_t}{dt} = \frac{t \, dZ_t + dZ_t \, dt + Z_t \, dt - t \, dZ_t - Z_t \, dt}{dt} = \frac{dZ_t \, dt}{dt} = dZ_t$$

Since $dZ_t = Z_{t+dt} - Z_t$, dZ_t is distributed normally with mean 0 and variance dt. Thus, with probability approaching 1, the values of dZ_t must be on the order of the standard deviation $\sqrt{dt}$. This means that the quotient above approaches 0 with probability 1 as $dt \to 0$. This confirms that $dX_t = t\,dZ_t + Z_t\,dt$ is a differential of X_t.

One of the most important applications of the differential is to evaluate integrals. Suppose a term in the integrand of an integral has the property that if one divides it by dt, the result will approach 0 as dt approaches 0. Then, the term itself will not contribute at all to the integral. This means that such a term can be ignored.

Next, we state three elementary but important properties of the differential:

1. *Linearity:* If X_t and Y_t are stochastic processes, and a and b are constants, then $d(aX_t + bY_t) = a\,dX_t + b\,dY_t$.
2. *General product rule:* If X_t and Y_t are stochastic processes, then $d(X_tY_t) = X_t\,dY_t + Y_t\,dX_t + dX_t\,dY_t$.
3. *Special product rule:* Suppose that $dX_t = \mu_t\,dt + \sigma_t\,dZ_t$ and $dY_t = \rho_t\,dt$ where σ_t, μ_t, and ρ_t are stochastic processes, then $d(X_tY_t) = X_t\,dY_t + Y_t\,dX_t$.

First, the linearity property is stated in terms of a sum of just two processes, but it can be extended to any finite sum of constant multiples of stochastic processes. The special product rule is expressed in differential form. It has the same form as in ordinary calculus. We note that the result would be different if dY_t had a Brownian motion component, as we will demonstrate later in our discussion of Itô's lemma and in the solution to end-of-chapter Exercise 6.24.

We prove our three properties here. As to property 1 above, note that

$$d(aX_t + bY_t) = (aX_{t+dt} + bY_{t+dt}) - (aX_t + bY_t)$$

$$= a(X_{t+dt} - X_t) + b(Y_{t+dt} - Y_t) = a\,dX_t + b\,dY_t$$

As to property 2, since $X_{t+dt} - X_t = dX_t$ and $Y_{t+dt} - Y_t = dY_t$, then $X_{t+dt} = X_t + dX_t$ and $Y_{t+dt} = Y_t + dY_t$. Thus, the differential of X_tY_t is:

$$d(X_tY_t) = X_{t+dt}Y_{t+dt} - X_tY_t = (X_t + dX_t)(Y_t + dY_t) - X_tY_t$$

$$= X_t\,dY_t + Y_t\,dX_t + dX_t\,dY_t$$

To derive property 3, we make use of property 2. The term $dX_t\,dY_t$ can be ignored in the differential as we now demonstrate:

$$dX_t\,dY_t = (\mu_t\,dt + \sigma_t\,dZ_t)(\rho_t\,dt) = \mu_t\rho_t(dt)^2 + \sigma_t\rho_t\,dZ_t\,dt$$

We saw earlier that the values of dZ_t are on the order of $\sqrt{dt}$, which implies that the values for $\sigma_t\rho_t\,dZ_t\,dt/dt = \sigma_t\rho_t\,dZ_t$ are on the order of $\sqrt{dt}$. This implies that the term $\sigma_t\rho_t\,dZ_t\,dt/dt$ approaches 0 as dt approaches 0. Obviously, $\mu_t\rho_t(dt)^2/dt = \mu_t\rho_t\,dt$ approaches 0 as dt approaches 0. This proves that the term $dX_t\,dY_t$ can be ignored in the differential, which implies that $d(X_tY_t) = X_t\,dY_t + Y_t\,dX_t$.

6.1.2 Stochastic Integration

In many respects, the theory of integration of stochastic processes mirrors integration of real-valued functions. If $f(x)$ is a real-valued continuous function defined on the interval $a \leq x \leq b$, recall from the review of the definite integral in Chapter 1 that:

$$\int_a^b f(x)\,dx = \lim_{\Delta x \to 0} \sum_{i=0}^{n-1} f(x_i)\Delta x$$

Integration of a stochastic process X_t with respect to the real variable t can be defined in a way analogous to that of a real-valued continuous function in ordinary calculus. In order to take the integral X_t over the interval $a \leq t \leq b$, divide the interval into n equal parts so that $\Delta t = (b-a)/n$ and $t_i = a + i\Delta t$ for $i = 0, 1, \ldots, n$. Define:

$$\int_a^b X_t\,dt = \lim_{\Delta t \to 0} \sum_{i=0}^{n-1} X_{t_i}\Delta t$$

Observe that whenever this integral exists the result is a random variable, since it is a limiting sum of random variables.

Next, we define the integral of a stochastic process X_t with respect to a stochastic process Y_t. Divide the interval $a \leq t \leq b$ in the same way as above. Then:

$$\int_a^b X_t\,dY_t = \lim_{\Delta t \to 0} \sum_{i=0}^{n-1} X_{t_i}\left(Y_{t_{i+1}} - Y_{t_i}\right)$$

Once again, whenever this integral exists, the result is a random variable, since it is a limiting sum of random variables. Note that our first definition of an integral is a special case of the second by choosing $Y_t = t$ so that $dY_t = dt$. Although $Y_t = t$ is a real-valued function, it is also a special case of a stochastic process. This follows from the observation that for each fixed value of t, $X_t = t$ with probability 1. It is a random variable that can only take on one value; namely, t at time t.

Consider the following particular case of stochastic integration that will arise frequently in this book. Suppose that σ_t and μ_t are stochastic processes (note that σ_t and μ_t can be chosen so that each takes on single constant values) and Z_t is standard Brownian motion. Then:

$$\int_a^b (\sigma_t\,dZ_t + \mu_t\,dt) = \lim_{\Delta t \to 0} \sum_{i=0}^{n-1} \left[\sigma_{t_i}\left(Z_{t_{i+1}} - Z_{t_i}\right) + \mu_{t_i}\,\Delta t\right]$$

where $\Delta t = (b-a)/n$, $t_0 = a$, $t_n = b$, and $t_{i+1} - t_i = \Delta t$ for $i = 1, 2, \ldots, n$. Observe that the expression on the right-hand side above is a sum of random variables, and if the limit exists, then the result will be a random variable. In fact, for a fairly general class of stochastic processes, including processes discussed in this text, the limit above does exist and the result is a well-defined random variable.

6.1.2.1 Contrasting Integration of Real-Valued Functions with Integration of Stochastic Processes

Consider the following two illustrations involving integration, the first a real-valued function and the second a stochastic process where Z_t is standard Brownian motion:

1. $\displaystyle\int_0^2 5\,dX_t$ with $X_t = t^2$

$$\int_0^2 5\,dX_t = 5X_t\big|_0^2 = 5X_2 - 5X_0 = 5(2)^2 - 5(0)^2 = 20$$

2. $\displaystyle\int_0^2 5\,dZ_t$

$$\int_0^2 5\,dZ_t = 5Z_t\big|_0^2 = 5Z_2 - 5Z_0 = 5Z_2$$

The solution in the first illustration is a number. The solution in the second illustration is a random variable since Z_2 is a normally distributed random variable, with mean 0 and variance 2. Thus, $5Z_2$ is a normally distributed random variable with mean 0 and variance $(5)^2(2) = 50$.

Suppose one keeps the lower limit of integration a constant and replaces the upper limit of integration with a variable. In the case of integration of real-valued functions, the result is a real-valued function. Analogously, in the case of integration of stochastic processes, the result is a stochastic process (see end-of-chapter Exercise 6.3).

6.1.3 Elementary Properties of Stochastic Integrals

6.1.3.1 Integral of a Stochastic Differential

If X_t is a stochastic process, the integral of a stochastic differential dX_t is simply:

$$\int_a^b dX_t = X_b - X_a$$

We prove this by noting that, by definition, this integral is defined as:

$$\int_a^b dX_t = \lim_{\Delta t \to 0} \sum_{i=0}^{n-1}(X_{t_{i+1}} - X_{t_i})$$

$$= \lim_{\Delta t \to 0} \left[(X_{t_1} - X_{t_0}) + (X_{t_2} - X_{t_1}) + \ldots + (X_{t_{n-1}} - X_{t_{n-2}}) + (X_{t_n} - X_{t_{n-1}})\right]$$

$$= \lim_{\Delta t \to 0} \left[X_{t_n} + \left(-X_{t_{n-1}} + X_{t_{n-1}}\right) + \left(-X_{t_{n-2}} + X_{t_{n-2}}\right) + \ldots + \left(-X_{t_1} + X_{t_1}\right) - X_{t_0}\right]$$

$$= \lim_{\Delta t \to 0} (X_{t_n} - X_{t_0}) = X_b - X_a$$

Alternatively, if we define the notation $X_t|_a^b = X_b - X_a$ as is used in ordinary calculus, then we can write this property as:

$$\int_a^b dX_t = X_t|_a^b = X_b - X_a$$

6.1.3.2 Linearity

If X_t, Y_t, U_t, and V_t are stochastic processes, and C and D are real numbers, then:

$$\int_a^b (CU_t \, dX_t + DV_t \, dY_t) = C \int_a^b U_t \, dX_t + D \int_a^b V_t \, dY_t$$

as long as each of the integrals is defined. This rule constitutes the familiar "sum" and "constant multiple" rules for integrals of real-valued functions in calculus carrying over to stochastic processes.

We prove this linearity property as follows:

$$\int_a^b (CU_t \, dX_t + DV_t \, dY_t) = \lim_{\Delta t \to 0} \left[\sum_{i=0}^{n-1} [CU_{t_i}(X_{t_{i+1}} - X_{t_i}) + DV_{t_i}(Y_{t_{i+1}} - Y_{t_i})] \right]$$

$$= \lim_{\Delta t \to 0} C \sum_{i=0}^{n-1} U_{t_i}(X_{t_{i+1}} - X_{t_i}) + \lim_{\Delta t \to 0} D \sum_{i=0}^{n-1} V_{t_i}(Y_{t_{i+1}} - Y_{t_i})$$

$$= C \int_a^b U_t \, dX_t + D \int_a^b V_t \, dY_t$$

In the last line, we used the fact that the limit of a sum is the sum of the limits, as long as each of the limits exists. We also assumed that we used the usual partition of the time interval $[a,b]$ so that $\Delta t = (b - a)/n$, $t_{i+1} - t_i = \Delta t$, $t_0 = a$, and $t_n = b$.

Consider the following illustration. Suppose that the price of a security follows the following arithmetic Brownian motion process with drift: $dX_t = 0.06 \, dZ_t + 0.02 \, dt$. Further suppose that the price of the security at time 0 equals 20: $X_0 = 20$. Now we seek to find X_t, the price of the security at time t: First observe that:

$$\int_0^t dX_s = \int_0^t (0.06 \, dZ_s + 0.02 \, ds)$$

By the first property, the left side equals:

$$\int_0^t dX_s = X_s|_0^t = X_t - 20$$

By linearity and the first property, the right side equals:

$$\int_0^t (0.06\,dZ_s + 0.02\,ds) = (0.06Z_s + 0.02s)|_0^t = 0.06Z_t + 0.02t - 0.06Z_0 = 0.06Z_t + 0.02t$$

Setting these results equal and solving for X_t, we find that $X_t = 20 + 0.02t + 0.06Z_t$. The security price at time t equals its price at time 0 plus 0.02 multiplied by the elapsed time plus 0.06 times the Brownian motion. The price is a normally distributed random variable with mean $20 + 0.02t$ and variance $(0.06)^2 t$.

6.1.4 More on Defining Stochastic Integrals

In the definition of a stochastic integral, we will always require that the change in Y_t, $Y_{t_{i+1}} - Y_{t_i}$, is over the time interval $[t_i, t_{i+1}]$ immediately following time t_i, which is the time in which X_t is evaluated; namely, X_{t_i}. This is a natural requirement for many models in finance since typically X_{t_i} will be the value (or a function of the value) of a security at time t_i, and $Y_{t_{i+1}} - Y_{t_i}$ will be a random proportional factor in the immediate future time interval $[t_i, t_{i+1}]$ that changes the value X_{t_i} of the security. If X_t and Y_t were continuously differentiable real-valued functions, we would not need to be so precise in the definition of the integral. For example, one could define the following integral to be:

$$\int_a^b X_t\,dY_t = \lim_{\Delta t \to 0} \sum_{i=0}^{n-1} X_{t_i^*} \left(Y_{t_{i+1}} - Y_{t_i} \right)$$

with t_i^* to be any value in the interval $[t_i, t_{i+1}]$. The answer would be unaffected of where each t_i^* is chosen in its time interval. This will not be the case for many stochastic integrals. Furthermore, there can be other significant differences between real-valued integration and stochastic integration. For example, consider the following integral:

$$\int_0^T x(t)\,dx(t)$$

with $x(t)$ a continuously differentiable real-valued function of t. This integral can be easily evaluated:

$$\int_0^T x(t)\,dx(t) = \frac{1}{2}[x(t)]^2 \Big|_0^T = \frac{1}{2}[x(T)]^2 - \frac{1}{2}[x(0)]^2$$

As a contrast, consider the following stochastic integral:

$$\int_0^T Z_t\,dZ_t$$

with Z_t being standard Brownian motion. As a first attempt, try the analogous solution:

$$\frac{1}{2}Z_T^2 - \frac{1}{2}Z_0^2 = \frac{1}{2}Z_T^2$$

and see if it works. Since $Z_T \sim N(0,T)$, then:

$$E\left[\frac{1}{2}Z_T^2\right] = \frac{1}{2}\operatorname{Var}[Z_T] = \frac{1}{2}T$$

If $\frac{1}{2}Z_T^2$ is the correct solution, we would need:

$$E\left[\int_0^T Z_t\, dZ_t\right] = \frac{1}{2}T$$

But, in fact:

$$E\left[\int_0^T Z_t\, dZ_t\right] = E\left[\lim_{\Delta t \to 0}\sum_{i=0}^{n-1} Z_{t_i}\left(Z_{t_{i+1}} - Z_{t_i}\right)\right] = \lim_{\Delta t \to 0}\sum_{i=0}^{n-1} E\left[Z_{t_i}(Z_{t_{i+1}} - Z_{t_i})\right]$$

Observe that we switched the order of the limit and expectation operation, which is valid in this case. Since Z_{t_i} is Brownian motion in the time interval $[0,t_i]$ and $Z_{t_{i+1}} - Z_{t_i}$ is Brownian motion in the time interval $[t_i,t_{i+1}]$, and these time intervals are nonoverlapping, then they are independent random variables. Thus:

$$E\left[Z_{t_i}(Z_{t_{i+1}} - Z_{t_i})\right] = E[Z_{t_i}]E[Z_{t_{i+1}} - Z_{t_i}] = 0$$

This shows that:

$$E\left[\int_0^T Z_t\, dZ_t\right] = 0$$

So, it is not possible that:

$$\int_0^T Z_t\, dZ_t = \frac{1}{2}Z_T^2$$

This example also illustrates that we have less freedom in how we define the integral. Suppose we tried the following incorrect definition instead:

$$\int_0^T Z_t\, dZ_t = \lim_{\Delta t \to 0}\sum_{i=0}^{n-1} Z_{t_{i+1}}\left(Z_{t_{i+1}} - Z_{t_i}\right)$$

Note that we have replaced Z_{t_i} with $Z_{t_{i+1}}$. If Z_t were a continuously differentiable real-valued function, the two different definitions would equal each other. This is not the case for Brownian motion Z_t. To see this, first express:

$$Z_{t_{i+1}}(Z_{t_{i+1}} - Z_{t_i}) = (Z_{t_{i+1}} - Z_{t_i})(Z_{t_{i+1}} - Z_{t_i}) + Z_{t_i}(Z_{t_{i+1}} - Z_{t_i})$$

Next, calculate the expectation of this quantity:

$$E[Z_{t_{i+1}}(Z_{t_{i+1}} - Z_{t_i})] = E[(Z_{t_{i+1}} - Z_{t_i})(Z_{t_{i+1}} - Z_{t_i})] + E[Z_{t_i}(Z_{t_{i+1}} - Z_{t_i})]$$

The first term on the right-hand side is:

$$E[(Z_{t_{i+1}} - Z_{t_i})(Z_{t_{i+1}} - Z_{t_i})] = Var[Z_{t_{i+1}} - Z_{t_i}] = \Delta t$$

since $Z_{t_{i+1}} - Z_{t_i}$ has a normal distribution with mean 0 and variance Δt.
We already showed that:

$$E[Z_{t_i}(Z_{t_{i+1}} - Z_{t_i})] = 0$$

Thus:

$$E[Z_{t_{i+1}}(Z_{t_{i+1}} - Z_{t_i})] = \Delta t$$

We use this result to calculate the following expectation:

$$E\left[\int_0^T Z_t \, dZ_t\right] = E\left[\lim_{\Delta t \to 0} \sum_{i=0}^{n-1} Z_{t_{i+1}}(Z_{t_{i+1}} - Z_{t_i})\right]$$

$$\lim_{\Delta t \to 0} \sum_{i=0}^{n-1} E[Z_{t_{i+1}}(Z_{t_{i+1}} - Z_{t_i})] = \lim_{\Delta t \to 0} \sum_{i=0}^{n-1} \Delta t = \lim_{\Delta t \to 0} n\Delta t = T$$

However, had we used our original correct definition for the integral, we saw that the resulting expectation would have been 0.

6.1.5 Significant Results Based on Stochastic Integration

In this section, we will focus on a few important expectation results based on stochastic integration. These results are very useful to finding results related to option pricing and term structure models. We start with a preliminary result.

For any integrable stochastic process f_t that is adapted to the same filtration $\{\mathscr{F}_t\}$ as standard Brownian motion Z_t, we have:

$$E\left[\int_s^T f_t \, dZ_t | \mathscr{F}_s\right] = 0 \tag{6.1}$$

whenever $s < T$. See end-of-chapter Exercise 6.5.a. for a verification of this result.
Now we extend this result a bit, and state an important theorem that is used to construct martingales in order to price derivatives in Chapter 7.

6.1.5.1 *The Martingale Theorem*

If $dX_t = f_t\, dZ_t$, then X_t is a martingale, where f_t is an integrable stochastic process that is adapted to the same filtration $\{\mathscr{F}_t\}$ as a standard Brownian motion Z_t. See end-of-chapter Exercise 6.5.b for a verification of this result.

6.1.5.2 *Itô Isometry*

If f_t is a square integrable stochastic process, then:

$$E\left[\left(\int_0^T f_t\, dZ_t\right)^2\right] = \int_0^T E[f_t^2]\, dt$$

We will find this result very useful for our bond pricing and yield curve models in Chapter 8. This result is proven on the companion website.

6.1.5.3 *Characteristics of Integrals of Real-Valued Functions with Respect to Brownian Motion*

The last theorem in this section will also be used in the derivation of the solutions to pricing bonds in Chapter 8.

If $f(s)$ is a real-valued function of s, then for each t the random variable X_t defined by:

$$X_t = \int_0^t f(s)\, dZ_s$$

has a normal distribution with mean 0 and variance:

$$\int_0^t [f(s)]^2\, ds$$

We prove this theorem by noting that the random variable X_t can be approximated by the following sum:

$$X_t \approx \sum_{i=0}^{n-1} f(s_i)(Z_{s_{i+1}} - Z_{s_i})$$

where $0 = s_0 < s_1 < \cdots < s_n = t$, and $s_{i+1} - s_i = t/n$. Since $Z_{s_{i+1}} - Z_{s_i}$ for i from 0 to $n-1$ are pairwise independent random variables, each with mean 0, then by the theorem at the end of Section 2.6.3 the sum above is a random variable that has a normal distribution with mean 0 and variance:

$$\sum_{i=0}^{n-1} [f(s_i)]^2$$

By the fundamental theorem of calculus, in the limit as n approaches infinity the variance approaches:

$$\int_0^t [f(s)]^2 \, ds$$

A rigorous proof would also need to show that the limit has a normal distribution since each approximating sum has a normal distribution. This step is beyond the level of this book.

6.2 CHANGE OF PROBABILITY AND THE RADON–NIKODYM DERIVATIVE

If a market is arbitrage-free under one probability space, it is arbitrage-free under any equivalent probability space. Thus, if the price of an asset in a market with a given set of physical probabilities does not produce an arbitrage opportunity, it will not produce an arbitrage opportunity under any equivalent probability measure. However, it is often easier to price a security under a particular measure, such as an equivalent measure that produces a martingale, than it is to price a security under a physical probability that is not a martingale. Thus, it is useful to be able to efficiently convert from one measure to another, particularly one that produces a martingale. The Radon–Nikodym process is useful for converting physical probability measures to equivalent martingales. This process is essentially converting from a pricing framework in which investors have risk preferences to a risk-neutral pricing framework.

6.2.1 The Radon–Nikodym Process

In the previous chapter, we converted physical probability measures to risk-neutral measures. Here, we will introduce and use the *Radon–Nikodym process* to convert discrete probability measures and then to convert continuous measures. The *Radon–Nikodym derivative* is a "bridge" that converts one probability measure to an equivalent measure. One of the most important financial applications of Radon–Nikodym derivatives is to compute expected values under different measures. These will prove to be very useful in the analysis of many types of derivative securities, including exotic options (e.g., barrier options) and swap contracts, particularly in environments involving continuous prices and time.

6.2.2 The Radon–Nikodym Derivative

Let $\mathbb{P}$ and $\mathbb{Q}$ be two probability measures on $(\Omega, \mathscr{F})$ such that $\mathbb{Q} \sim \mathbb{P}$ ($\mathbb{Q}$ is equivalent to $\mathbb{P}$) on $\mathscr{F}$. Then there exists a random variable $\frac{d\mathbb{Q}}{d\mathbb{P}}$ known as the Radon–Nikodym derivative of $\mathbb{Q}$ with respect to $\mathbb{P}$ on $\mathscr{F}$ that is defined by the property:

$$\int_{\omega \in \phi} \frac{d\mathbb{Q}}{d\mathbb{P}}(\omega) \, d\mathbb{P}(\omega) = \mathbb{Q}(\phi) \tag{6.2}$$

for every measurable set $\phi \in \mathscr{F}$. $\mathbb{Q}(\phi)$ denotes the probability that the event ϕ occurs with respect to the probability measure $\mathbb{Q}$. So, we see from its definition that the Radon–Nikodym derivative allows us to find probabilities with respect to $\mathbb{Q}$ in terms of integrals with respect to the measure $\mathbb{P}$. If the probability measures are discrete, then the integral is replaced with a sum. It is easy to see that the Radon–Nikodym derivative has the following additional properties:

1. $q(\omega) = \dfrac{d\mathbb{Q}}{d\mathbb{P}} p(\omega)$ for any $\omega \in \Omega$ as long as $p(\omega) > 0$

2. $\dfrac{d\mathbb{Q}}{d\mathbb{P}}(\omega) \geq 0 \; \forall \omega \in \Omega$

3. $\sum_{\omega \in \Omega} p(\omega) \dfrac{d\mathbb{Q}}{d\mathbb{P}}(\omega) = 1$

4. $\dfrac{d\mathbb{Q}}{d\mathbb{P}}$ is a measurable function on the σ-algebra $\mathscr{F}$

The notation $p(\omega)$ refers to the $\mathbb{P}$ probability that the outcome ω occurs, and an analogous interpretation of $q(\omega)$. If the probability measures are continuous, then the sum in property 3 is replaced with an integral and $p(\omega)$ is replaced with $p(x)\,dx$ where $p(x)$ is the density function for the probability space $\mathbb{P}$. If the sample space Ω has at most a countable number of elements and $p(\omega) > 0$ for every $\omega \in \Omega$, then $\frac{d\mathbb{Q}}{d\mathbb{P}}(\omega) = \frac{q(\omega)}{p(\omega)}$ for every $\omega \in \Omega$. We verify this result in end-of-chapter Exercise 6.7 for the case that Ω is finite.

If the probability measures $\mathbb{P}$ and $\mathbb{Q}$ are continuous with continuous density functions, say $p(x)$ and $q(x) > 0$, then the Radon–Nikodym derivative is simply $\frac{d\mathbb{Q}}{d\mathbb{P}}(x) = \frac{q(x)}{p(x)}$. We verify this result as end-of-chapter Exercise 6.8.

The Radon–Nikodym derivative $\xi_t(\omega)$ of $\mathbb{Q}$ with respect to $\mathbb{P}$ on a filtration $\{\mathscr{F}_t\}$ at time t is defined as:

$$\int_{\omega \in \phi} \xi_t(\omega)\, d\mathbb{P}(\omega) = \mathbb{Q}(\phi) \tag{6.3}$$

for every measurable set $\phi \in \mathscr{F}_t$. We can loosely describe $\xi_t(\omega)$ as the Radon–Nikodym derivative of the probability measure $\mathbb{Q}$ restricted to the σ-algebra $\mathscr{F}_t$ with respect to the probability measure $\mathbb{P}$ restricted to the σ-algebra $\mathscr{F}_t$. The point is that when one is "observing" the situation at time t, one can only have information up to time t. It is not possible to distinguish all elements in the entire σ-algebra $\mathscr{F}$, but only those members in the smaller σ-algebra $\mathscr{F}_t$. Effectively, this means that the Radon–Nikodym derivative at time t need only be defined on a smaller sample space than Ω. In the illustrations to follow, these notions will be clarified.

For a given stochastic process X_t that is a price of a security, we will choose $\mathbb{P}$ as the physical probability measure and $\mathbb{Q}$ is the risk-neutral probability measure. Intuitively, the Radon–Nikodym derivative $\xi_t(\omega)$ is the ratio of the risk-neutral probability (or density) to the physical probability (or density) associated with the filtration $\mathscr{F}_t$ at time t. Observe that when dealing with a stochastic process, we have a Radon–Nikodym derivative defined for each distinct time $t = 0, 1, 2, \ldots$ for the discrete case and any real time $t \geq 0$ for continuous time.

We can use the Radon–Nikodym derivative to obtain the risk-neutral probability measure of any event $\phi \in \mathscr{F}_t$ from the physical measure from Eq. (6.3). In the case that the sample space Ω is finite, it is sufficient to define the Radon–Nikodym derivative ξ_t for any outcome $\omega \in \Omega$:

$$q(\omega) = \xi_t(\omega)p(\omega)$$

The Radon–Nikodym derivative for each fixed time t is a random variable itself because it is a measurable function on the probability space $\mathbb{P}$. It tells us how we transform $\mathbb{P}$ to obtain an equivalent probability measure $\mathbb{Q}$. Furthermore, ξ_t is a stochastic process since it is also a function of t. If X_t is a martingale with respect to $\mathbb{Q}$, we call $\mathbb{Q}$ either the *risk-neutral probability measure* or the *equivalent martingale measure*.

The first of the three illustrations that follow portrays a simple calculation of discrete outcome Radon–Nikodym derivatives in a one-period environment. The second illustration calculates discrete outcome Radon–Nikodym derivatives in a one-period environment without changing the variance of the distribution. The third illustration calculates discrete outcome Radon–Nikodym derivatives in a multiperiod environment. In the section that follows, we will work with continuous outcomes, then continuous time. It should be noted that these examples are intended to demonstrate how the Radon–Nikodym derivative is constructed. Its utility and importance will be clear in later sections and chapters, such as when we discuss the Cameron–Martin–Girsanov theorem.

ILLUSTRATION 1: THE RADON–NIKODYM DERIVATIVE AND TWO COIN TOSSES

Suppose that at time 0 we pay 2.2 to toss a coin twice, with time 1 payoffs given as follows: $S_{HH} = 4$, $S_{HT} = 3$, $S_{TH} = 2$, and $S_{TT} = 1$, with probabilities given by the following physical probability measure $\mathbb{P}$ and risk-neutral probability measure $\mathbb{Q}$:

$$p(HH) = 0.25; p(HT) = 0.25; p(TH) = 0.25; p(TT) = 0.25$$

$$q(HH) = 0.16; q(HT) = 0.24; q(TH) = 0.24; q(TT) = 0.36$$

Time 1 values for the Radon–Nikodym derivative are calculated as follows:

$$\xi(HH) = \frac{q(HH)}{p(HH)} = \frac{0.16}{0.25} = 0.64 \qquad \xi(HT) = \frac{q(HT)}{p(HT)} = \frac{0.24}{0.25} = 0.96$$

$$\xi(TH) = \frac{q(TH)}{p(TH)} = \frac{0.24}{0.25} = 0.96 \qquad \xi(TT) = \frac{q(TT)}{p(TT)} = \frac{0.36}{0.25} = 1.44$$

We can use the Radon–Nikodym derivative to compute expected values under the physical probability measure $\mathbb{P}$ and under the new probability measure $\mathbb{Q}$. Expected values for time 1 payoffs under probability measures $\mathbb{P}$ and $\mathbb{Q}$ are calculated as follows:

$$E_{\mathbb{P}}[S] = \sum_i S_i p_i = (4 \times 0.25) + (3 \times 0.25) + (2 \times 0.25) + (1 \times 0.25) = 2.5$$

$$E_{\mathbb{Q}}[S] = \sum_i S_i q_i = \sum_i S_i \xi_i p_i = (4 \times 0.64 \times 0.25) + (3 \times 0.96 \times .25) + (2 \times 0.96 \times 0.25)$$

$$+ (1 \times 1.44 \times 0.25) = 2.2$$

Note that the process S is a martingale with respect to risk-neutral measure $\mathbb{Q}$; that is, we pay at time 0 for the toss exactly its expected time 1 payoff in this one-time period scenario. That is, its value now equals its expected value after the coins are tossed twice; the risk-neutral measure $\mathbb{Q}$ converts S into a martingale.

In this illustration, we changed our probability measure while changing both its expected value and variance. The variances of the two distributions are:

$$E_{\mathbb{P}}\left[S - E_{\mathbb{P}}[S]\right]^2 = \sum_i \left[S_i - E_{\mathbb{P}}[S]\right]^2 p_i$$

$$= (4 - 2.5)^2 \times 0.25 + (3 - 2.5)^2 \times 0.25 + (2 - 2.5)^2 \times 0.25 + (1 - 2.5)^2 \times 0.25$$

$$= 1.25$$

$$E_{\mathbb{Q}}\left[S - E_{\mathbb{Q}}[S]\right]^2 = \sum_i \left[S_i - E_{\mathbb{Q}}[S]\right]^2 q_i$$

$$= (4 - 2.2)^2 \times 0.16 + (3 - 2.2)^2 \times 0.24 + (2 - 2.2)^2 \times 0.24 + (1 - 2.2)^2 \times 0.36$$

$$= 1.2$$

In many applications involving the change of measure, it will be convenient to change the expected value while leaving the variance intact. This will enable us to neutralize risk premiums associated with asset prices without changing their volatilities. For example, we will be able to change the expected return of a stock while retaining its variance when pricing an option written on that stock. We will conduct such a change of measure in the next illustration.

ILLUSTRATION 2: CALCULATING RISK-NEUTRAL PROBABILITIES AND THE RADON–NIKODYM DERIVATIVE

In the previous illustration, we were given both the physical probability space $\mathbb{P}$ and the risk-neutral probability $\mathbb{Q}$ space before calculating Radon–Nikodym derivatives. In previous chapters, we used different securities to complete the security payoff space, then calculated risk-neutral probabilities from pure security prices. With this information, we could have easily calculated Radon–Nikodym derivatives from the ratios of physical and risk-neutral probabilities. Now, we will calculate the risk-neutral probability space and the Radon–Nikodym derivatives directly from the outcome and physical probability spaces. Instead of completing the payoff space with payoffs from an appropriate number of securities as we did earlier, we will stipulate that the change of probability measure does not change the variance of security returns.

Suppose that we pay $S_0 = 6$ to participate in a gamble, with potential payoffs given as follows: $S_1 = 3$, $S_2 = 7$, and $S_3 = 11$, with probabilities given by the following physical probability measure $\mathbb{P}$:

$$p_1 = 1/3; \quad p_2 = 1/3; \quad p_3 = 1/3$$

Obviously, $E_{\mathbb{P}}[S] = 7$; the variance equals $E_{\mathbb{P}}\left[S - E_{\mathbb{P}}[S]\right]^2 = 10^2/3$. How would we obtain the risk-neutral probability measure $\mathbb{Q}$ from physical probability measure $\mathbb{P}$ without knowing the risk-neutral probabilities in advance? Consider the following three statements, which state that the risk-neutral probabilities must sum to one, the expected value of the $\mathbb{Q}$ distribution must equal $S_0 = 6$ (it is an equivalent martingale to $\mathbb{P}$), and the variance of both distributions will equal $10^2/_3$:

$$\sum_i q_i = q_1 + q_2 + q_3 = 1$$

$$\sum_i S_i q_i = 3q_1 + 7q_2 + 11q_3 = E_{\mathbb{Q}}[S] = 6$$

$$E_{\mathbb{Q}}\left[S - E_{\mathbb{Q}}[S]\right]^2 = \sum_i \left[S_i - E_{\mathbb{Q}}[S]\right]^2 q_i = (3-6)^2 \times q_1 + (7-6)^2 \times q_2 + (11-6)^2 \times q_3 = 10^2/_3$$

We solve this system simultaneously as follows:

$$\begin{bmatrix} 1 & 1 & 1 \\ 3 & 7 & 11 \\ 9 & 1 & 25 \end{bmatrix} \begin{bmatrix} q_1 \\ q_2 \\ q_3 \end{bmatrix} = \begin{bmatrix} 1 \\ 6 \\ 10^2/_3 \end{bmatrix}$$

$$\begin{bmatrix} 1.28125 & -0.1875 & 0.03125 \\ 0.1875 & 0.125 & -0.0625 \\ -0.46875 & 0.0625 & 0.03125 \end{bmatrix} \begin{bmatrix} 1 \\ 6 \\ 10^2/_3 \end{bmatrix} = \begin{bmatrix} q_1 \\ q_2 \\ q_3 \end{bmatrix} = \begin{bmatrix} 0.48958333 \\ 0.27083333 \\ 0.23958333 \end{bmatrix}$$

Thus, values of the Radon–Nikodym derivative ξ_1, ξ_2, and ξ_3 are:

$$\xi_1 = \frac{q_1}{p_1} = \frac{0.489583}{0.333333} = 1.46875 \quad \xi_2 = \frac{q_2}{p_2} = \frac{0.270833}{0.333333} = 0.8125 \quad \xi_3 = \frac{q_3}{p_3} = \frac{0.239583}{0.333333} = 0.71875$$

Note that in this example, we changed our notation slightly, so that ξ_i for $i = 1$, 2, and 3 refer to the different possible values of the Radon–Nikodym derivative at time 1, rather than looking at the Radon–Nikodym derivative at different times. Here there should be no confusion, since we only deal with time 1 in this example. In this illustration, we have calculated the Radon–Nikodym derivative while changing the drift (expected value) of our initial measure and maintaining its variance. However, with more than three potential outcomes, we would require more than a single priced security to accomplish this. In the next section, we will discuss how we can change our measure and its drift without affecting the measure's variance. We will assume a stable variance over time and a limited number of potential outcomes each time period relative to the number of priced securities.

6.2.3 The Radon–Nikodym Derivative and Binomial Pricing

Recall that in Chapters 3 and 5, we discussed equivalent probabilities q_i and applied them to arbitrage-free pricing in binomial environments. Typically, we do not know physical probabilities p_i, nor would we have much use for them in pricing even if we did. To avoid arbitrage opportunities, we price based on arbitrage-free probabilities q_i. Consider a binomial model over one-time period with the price of a security potentially increasing from price S_0 to uS_0 with probability p_u

and decreasing to price dS_0 with probability $p_d = 1 - p_u$. The expected value of the price of the stock at time 1 is simply:

$$E_\mathbb{P}[S_1] = uS_0 p_u + dS_0 p_d$$

However, it might well be that $E_\mathbb{P}[S_1] \neq S_0$, which means that the security price is not a martingale. However, as in Chapters 3 and 5, we can use the equivalent probabilities for pricing:

$$q_u = \frac{1 - d}{u - d} \text{ and } q_d = 1 - q_u$$

such that $E_\mathbb{Q}[S_1] = uS_0 q_u + dS_0 q_d$. After a change in the numeraire, these arbitrage-free probabilities will ensure consistent pricing in which $E_\mathbb{Q}[S_1] = S_0$ in terms of the riskless bond. This methodology results in arbitrage-free pricing and the martingale property is satisfied for the time period from time 0 to time 1. As we discussed earlier, it can be very useful to represent the change from the physical probabilities $\mathbb{P} = \{p_1, p_2\}$ to the equivalent probabilities $\mathbb{Q} = \{q_1, q_2\}$ by means of the Radon–Nikodym derivative $\frac{d\mathbb{Q}}{d\mathbb{P}}$, and use these derivatives to represent the expected value of the security S_1 with respect to the probabilities from space $\mathbb{Q}$ in terms of the appropriate expectation with respect to the probabilities from space $\mathbb{P}$. In our illustration above, we would have:

$$\frac{d\mathbb{Q}}{d\mathbb{P}} = \begin{cases} \dfrac{q_u}{p_u} \text{ if } S_1 = uS_0 \\ \dfrac{q_d}{p_d} \text{ if } S_1 = dS_0 \end{cases}$$

Observe that:

$$E_\mathbb{P}\left[\frac{d\mathbb{Q}}{d\mathbb{P}} S_1\right] = \frac{q_u}{p_u} uS_0 p_u + \frac{q_d}{p_d} dS_0 p_d = q_u uS_0 + q_d dS_0 = E_\mathbb{Q}[S_1]$$

6.2.3.1 Radon–Nikodym Derivatives in a Multiple Period Binomial Environment

Recall from the Chapter 5 our two-time period binomial model illustration where we converted from one-period physical probabilities (all upjumps had physical probabilities of 0.6) to one-period risk-neutral probabilities (all upjumps had risk-neutral probabilities of 0.75). This two-time period model had the sample space $\Omega = \{uu, ud, du, dd\}$. The filtration $\mathcal{F}$ consists of all possible subsets of Ω:

$$\mathcal{F} = \{\emptyset, \{uu\}, \{ud\}, \{du\}, \{dd\}, \{uu, ud\}, \{uu, du\}, \{uu, dd\}, \{ud, du\}, \{ud, dd\}, \{du, dd\},$$
$$\{uu, ud, du\}, \{uu, ud, dd\}, \{uu, du, dd\}, \{ud, du, dd\}, \Omega\}$$

At time 2, since each possible history of the stock can be known, then $\mathcal{F}_2 = \mathcal{F}$. We will work backward in time shortly to obtain our time 1 σ-algebra $\mathcal{F}_1$. The time 2 physical probability measure $\mathbb{P}$ was defined as $p(uu) = (0.6)^2 = 0.36$, $p(ud) = (0.6)(0.4) = 0.24$, $p(du) = (0.4)(0.6) = 0.24$, and $p(dd) = (0.4)^2 = 0.16$. The risk-neutral probability measure $\mathbb{Q}$ was defined as $q(uu) = (0.75)^2 = 0.5625$, $q(ud) = (0.75)(0.25) = 0.1875$, $q(du) = (0.25)(0.75) = 0.1875$, and $q(dd) = (0.25)^2 = 0.0625$.[1] At time 2, we have complete knowledge of the possible outcomes in the sample space Ω, so that $p_2(\omega) = p(\omega)$

and $q_2(\omega) = q(\omega)$ for every $\omega \in \Omega$. The Radon–Nikodym derivative ξ_2 with respect to the σ-algebra $\mathscr{F}_2$ is:

$$\xi_2(uu) = \frac{q_2(uu)}{p_2(uu)} = \frac{0.5625}{0.36} = 1.5625, \; \xi_2(ud) = \frac{q_2(ud)}{p_2(ud)} = \frac{0.1875}{0.24} = 0.78125$$

$$\xi_2(du) = \frac{q_2(du)}{p_2(du)} = \frac{0.1876}{0.24} = 0.78125, \; \xi_2(dd) = \frac{q_2(dd)}{p_2(dd)} = \frac{0.0625}{0.16} = 0.390625$$

One can use the values of $\xi_2(\omega)$ to compute the probability of any event $\phi \in \mathscr{F}_2$ with respect to probability measure $\mathbb{Q}$ in terms of the probability measure $\mathbb{P}$. For example, suppose that a given event $\phi = \{uu, ud, dd\}$. Then, $\mathbb{Q}(\phi)$ for $\phi = \{uu, ud, dd\}$ is:

$$\mathbb{Q}(\phi) = \sum_{\omega \in \phi} \xi_2(\omega) p_2(\omega) = 1.5625(0.36) + 0.78125(0.24) + 0.390625(0.16)$$

$$= 0.5625 + 0.1875 + 0.0625 = 0.8125$$

We obtain the same result by directly computing this probability with respect to the probability measure $\mathbb{Q}$:

$$\mathbb{Q}(\phi) = \sum_{\omega \in A} q_2(\omega) = 0.5625 + 0.1875 + 0.0625 = 0.8125$$

Next, we find the Radon–Nikodym derivative $\xi_1(\omega)$ with respect to the σ-algebra $\mathscr{F}_1$. Here, we only have knowledge up to time 1. We can only know whether there has been an upjump or downjump at time 1. We do not know what will happen at time 2. So, we cannot distinguish uu from ud, or du from dd. So the σ-algebra for possible events at time 1 is:

$$\mathscr{F}_1 = \{\varnothing, \{uu, ud\}, \{du, dd\}, \Omega\}$$

The event $\{uu, ud\}$ means that an upjump (u) occurred at time 1, but we do not know what the jump will be at time 2. The event $\{du, dd\}$ means that a downjump (d) occurred at time 1, and either a u or a d can occur at time 2. So, the only possible distinguishable outcomes that can occur at time 1 are either u or d. The probabilities for these two outcomes with respect to probability measures $\mathbb{P}$ and $\mathbb{Q}$ are: $p_1(u) = 0.6$, $p_1(d) = 0.4$, $q_1(u) = 0.75$, and $q_1(d) = 0.25$. The Radon–Nikodym derivative $\xi_1(\omega)$ with respect to $\mathscr{F}_1$ is:

$$\xi_1(u) = \frac{q_1(u)}{p_1(u)} = \frac{0.75}{0.6} = 1.25, \; \xi_1(d) = \frac{q_1(d)}{p_1(d)} = \frac{0.25}{0.4} = 0.625$$

We can use the Radon–Nikodym derivative $\xi_1(\omega)$ to calculate $\mathbb{Q}$ measure probabilities at time 1 in terms of the probability measure $\mathbb{P}$. For example, choose the event $\phi = \{uu, ud\}$ in the σ-algebra $\mathscr{F}_1$. The event ϕ can be identified with just u occurring at time 1. The probability that ϕ occurs under measure $\mathbb{Q}$ is:[2]

$$\mathbb{Q}(\phi) = \xi_1(u) p_1(u) = 1.25(0.6) = 0.75$$

We could also have calculated $\mathbb{Q}(\phi)$ directly:[2]

$$\mathbb{Q}(\phi) = q_1(u) = 0.75$$

Since capital markets were complete in this illustration, there will be only a single set of one-period Radon–Nikodym derivatives, just as there will be only a single equivalent martingale. In other scenarios, incomplete markets may produce multiple sets of Radon–Nikodym derivatives. We can use the Radon–Nikodym derivative to compute expected values under the new probability measure $\mathbb{Q}$. Expected values for time 1 stock prices (recall that $S_0 = 10$, $u = 1.5$, and $d = 0.5$) under probability measures $\mathbb{P}$ and $\mathbb{Q}$ are calculated as follows:

$$E_\mathbb{P}[S_1] = \sum_{w \in \{u,d\}} S_1(w)p_1(w) = (1.5 \times 10)(0.6) + (0.5 \times 10)(0.4) = 11$$

$$E_\mathbb{Q}[S_1] = \sum_{w \in \{u,d\}} S_1(w)q_1(w) = \sum_{w \in \{u,d\}} S_1(w)\xi_1(w)p_1(w)$$

$$= (1.5 \times 10)(1.25)(0.6) + (0.5 \times 10)(0.625)(0.4) = 12.50$$

Note that while probability measure $\mathbb{P}$ prices the stock expected value in terms of currency, probability measure $\mathbb{Q}$ prices using riskless bonds (with face value $1) as the numeraire. The price of the stock under risk-neutral measure $\mathbb{Q}$ is 12.5 bonds, each worth $0.8 at time 0. Discounting these bonds reveals that the monetary value of the stock is $0.8 \times 12.50 = \$10.00$. Thus, the discounted value of the stock exhibits the martingale property at time 1 with respect to the probability measure $\mathbb{Q}$. Similarly, we compute expected values for time 2 stock prices under probability measures $\mathbb{P}$ and $\mathbb{Q}$ as follows (note that $p(uu) = 0.36$, $p(ud) = p(du) = 0.24$, $p(dd) = 0.16$, $u^2 = 2.25$, $ud = du = 0.75$, and $d^2 = 0.25$):

$$E_\mathbb{P}[S_2] = \sum_{w \in \Omega} S_2(w)p_2(w) = (1.5 \times 1.5 \times 10)(0.36) + (1.5 \times 0.5 \times 10)(0.24)$$

$$+ (0.5 \times 1.5 \times 10)(0.24) + (0.5 \times 0.5 \times 10)(0.16) = 1.21$$

$$E_\mathbb{Q}[S_2] = \sum_{w \in \Omega} S_2(w)q_2(w) = \sum_{w \in \Omega} S_2(w)\xi_2(w)p_2(w)$$

$$= (1.5 \times 1.5 \times 10)(1.25)^2(0.36) + (1.5 \times 0.5 \times 10)(0.78125)(0.24)$$

$$+ (0.5 \times 1.5 \times 10)(0.625)^2(0.24) + (0.5 \times 0.5 \times 10)(0.390625)(0.16) = 15.625$$

where $\xi_2(u,u) = 1.25^2$, $\xi_2(u,d) = \xi_2(d,u) = 1.25 \times 0.625$ and $\xi_2(d,d) = 0.625^2$. These results are depicted in Figure 6.1. The price of the stock under risk-neutral measure $\mathbb{Q}$ is 15.625 bonds, each worth $0.64 (there are two time periods before payoff) at time 0. Discounting these bonds reveals that the monetary value of the stock is $0.64 \times 15.625 = \$10.00$. Thus, at time 2, the discounted value of the stock once again exhibits the martingale property with respect to $\mathbb{Q}$.

These concepts can be extended to multiple time periods in the binomial situation as well as to cases when there are more than two possible choices at each stage. Suppose $p_{s,t}(w)$ is the physical probability that, given the stock price S_s at time s, it will have the value $S_t(w)$ at time t for any outcome w. Let $q_{s,t}(w)$ be the equivalent probability that, given the stock price S_s at time s, it will have the value $S_t(w)$ at time t for any outcome w, so that the

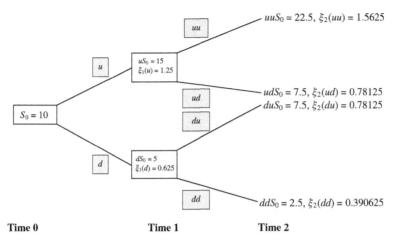

Time 0   **Time 1**   **Time 2**

FIGURE 6.1   Two-period binomial model with Radon–Nikodym derivatives.

martingale property is satisfied. To make this precise, define the Radon–Nikodym derivative $\frac{dQ}{dP}(\omega) = \frac{q_{s,t}(\omega)}{p_{s,t}(\omega)}$. This gives:

$$E_\mathbb{P}\left[\frac{dQ_{s,t}}{dP_{s,t}}S_t\Big|S_s\right] = \sum_\omega \frac{q_{s,t}(\omega)}{p_{s,t}(\omega)}S_{s,t}p_{s,t} = \sum_\omega S_{s,t}q_{s,t}(\omega) = E_Q[S_t|S_s] = S_s$$

Thus, in this case, the Radon–Nikodym derivative $\frac{dQ_{s,t}}{dP_{s,t}}$ is the multiplicative factor of the stochastic process S_t that converts the expectation on a probability space $\mathbb{P}$ to a probability space Q, so that the stochastic process S_t becomes a martingale in the probability space Q. We point out that the Radon–Nikodym derivative is a function of the times s and t.

6.2.4 Radon–Nikodym Derivatives and Multiple Time Periods

Suppose that there is a physical probability $p_{0,1}$ to go from a given price S_0 at time 0 to a particular price S_1 at time 1. Of course, $\Delta S_1 = S_1 - S_0$ is a random variable and there can be a countable number of possible values of ΔS_1 with each particular choice of its value associated with a particular probability $p_{0,1}$. Assume that the change in the price of the security at time t, ΔS_t, is independent of the change of the price ΔS_s at any other time s. Thus, the price of the stock S_n at time n is a sum of n independent random variables $\Delta S_1, \Delta S_2, \ldots, \Delta S_n$. The probability $p_{0,n}$ that the price takes on the value S_n by going through a particular choice of price changes ΔS_1, $\Delta S_2, \ldots, \Delta S_n$ equals:

$$p_{0,n} = p_{0,1}p_{1,2}\ldots p_{n-1,n} = \prod_{i=1}^{n}p_{i-1,i}$$

Now, suppose that $q_{i-1,i}$ is the equivalent probability measure to replace $p_{i-1,i}$ such that the price S_i has the martingale property going from time $i-1$ to time i: $E_Q[S_i | S_{i-1}] = S_{i-1}$. The equivalent probability measure that the price goes from S_0 to S_n equals:

$$q_{0,n} = q_{0,1}q_{1,2} \cdots q_{n-1,n} = \prod_{i=1}^{n} q_{i-1,i} = \prod_{i=1}^{n} \frac{q_{i-1,i}}{p_{i-1,i}} p_{i-1,i} = \prod_{i=1}^{n} \frac{q_{i-1,i}}{p_{i-1,i}} \prod_{i=1}^{n} p_{i-1,i}$$

$$= \left[\prod_{i=1}^{n} \frac{q_{i-1,i}}{p_{i-1,i}} \right] p_{0,n} = \xi_n p_{0,n}$$

(6.4)

This shows that the Radon−Nikodym derivative to use to change the measure from time 0 to time n is:

$$\xi_n = \prod_{i=1}^{n} \frac{q_{i-1,i}}{p_{i-1,i}}$$

(6.5)

Choosing the measure $\mathbb{Q}$ appropriately, we can arrange that S_n has the martingale property: $E_{\mathbb{Q}}[S_n | S_0] = S_0$. After working through an illustration, discussing random variables with continuous distributions, we will return to Radon−Nikodym derivatives in multiple time period settings in Sections 6.2.6 and 6.3.1.

ILLUSTRATION: THE BINOMIAL PRICING MODEL

Suppose the physical probability measure $\mathbb{P}$ is associated with a given value p and $p \neq (1-d)/(u-d)$. Then, as we showed earlier, the risk-neutral probability measure $\mathbb{Q}$ is associated with the value $q = (1-d)/(u-d)$, where q replaces p in the equations for the binomial pricing model. This will give $E_{\mathbb{Q}}[S_n | S_m] = S_m$ for any $m < n$. Next, we find the Radon−Nikodym derivative associated with this change of measure from $\mathbb{P}$ to $\mathbb{Q}$ at time n. The original probability measure $\mathbb{P}_{i-1,i}$ for the price change from time $i-1$ to i associated with the value p is:

$$\mathbb{P}_{i-1,i} = \begin{cases} p & \text{if the price increases by the factor } u \\ 1-p & \text{if the price decreases by the factor } d \end{cases}$$

The risk-neutral probability measure $\mathbb{Q}_{i-1,i}$ for the price change from time $i-1$ to time i associated with the value q is:

$$\mathbb{Q}_{i-1,i} = \begin{cases} q & \text{if the price increases by the factor } u \\ 1-q & \text{if the price decreases by the factor } d \end{cases}$$

where $q = d/(u+d)$. The Radon−Nikodym derivative for the change of measure is:

$$\frac{d\mathbb{Q}_{i-1,i}}{d\mathbb{P}_{i-1,i}} = \begin{cases} \dfrac{q}{p} & \text{if the price increases by the factor u from time } i-1 \text{ to } i \\ \dfrac{1-q}{1-p} & \text{if the price decreases by the factor d from time } i-1 \text{ to } i \end{cases}$$

Since the changes in the price over the different time intervals are independent of one another, then the Radon–Nikodym derivative (ξ_n) of the price changes from time 0 to n equals:

$$\xi_n = \prod_{i=1}^{n} \frac{d\mathbb{Q}_{i-1,i}}{d\mathbb{P}_{i-1,i}}$$

If there were exactly k increases and $n-k$ decreases in the price from time 0 to n then:

$$\xi_n = \left(\frac{q}{p}\right)^k \left(\frac{1-q}{1-p}\right)^{n-k} \tag{6.6}$$

This is the Radon–Nikodym derivative for change of measure for the binomial pricing model. The probability of obtaining any one particular outcome of k upjumps and $n-k$ downjumps equals $p^k(1-p)^{n-k}$ in probability measure $\mathbb{P}$ and equals $q^k(1-q)^{n-k}$ in probability measure $\mathbb{Q}$. Note that ξ_n provides the correct multiplicative factor to the change in measure from $\mathbb{P}$ to $\mathbb{Q}$ since:

$$\xi_n p^k(1-p)^{n-k} = \left(\frac{q}{p}\right)^k \left(\frac{1-q}{1-p}\right)^{n-k} p^k(1-p)^{n-k} = q^k(1-q)^{n-k}$$

Recall that there are $\binom{n}{k}$ possible permutations so that we obtain exactly k upjumps and $n-k$ downjumps in the price from time 0 to n. So, the probability of obtaining exactly k upjumps and $n-k$ downjumps in the price from time 0 to n equals:

$$\binom{n}{k} p^k(1-p)^{n-k}$$

with respect to probability measure $\mathbb{P}$, and equals:

$$\binom{n}{k} q^k(1-q)^{n-k}$$

with respect to probability measure $\mathbb{Q}$. Obviously, we also have:

$$\xi_n \binom{n}{k} p^k(1-p)^{n-k} = \binom{n}{k} q^k(1-q)^{n-k}$$

6.2.5 Change of Normal Density

Thus far, we have focused on changes of measure for discrete distributions with only a finite number of potential outcomes per period. Now, we will extend our analysis, focusing on random variables with continuous distributions, in particular, the normal distribution. First, consider a random variable X with the normal distribution $X \sim N(\mu, 1)$. Denote its density function under probability measure $\mathbb{P}$ as:

$$f_\mathbb{P}(x) = \frac{1}{\sqrt{2\pi}} e^{-(x-\mu)^2/2}$$

For any continuous function F, the expectation of the random variable $F(X)$ is given by:

$$E_\mathbb{P}[F(X)] = \frac{1}{\sqrt{2\pi}} \int_{-\infty}^{\infty} F(x) e^{-(x-\mu)^2/2} \, dx$$

If $F(X) = X$, we recognize this as the mean $E_\mathbb{P}[X] = \mu$. If $F(X) = (X - \mu)^2$, we recognize this as the variance $E_\mathbb{P}\left[(X - \mu)^2\right] = 1$.

Next, we will create a random variable that is equivalent to the random variable X except that we will shift its mean to equal 0 rather than μ. Thus, the new random variable will still have a normal distribution and will have a variance equal to 1. We will accomplish this by changing its probability measure, which is reflected in the appropriate change in its density function. To accomplish this, suppose that we were to multiply both sides of the density function $f_\mathbb{P}(x)$ by $e^{-\mu x + \mu^2/2}$:

$$e^{-\mu x + \frac{\mu^2}{2}} f_\mathbb{P}(x) = \frac{1}{\sqrt{2\pi}} e^{-\frac{(x-\mu)^2}{2}} \cdot e^{-\mu x + \frac{\mu^2}{2}}$$

$$= \frac{1}{\sqrt{2\pi}} e^{-\frac{x^2}{2} - \frac{\mu^2}{2} + \frac{2\mu x}{2} - \mu x + \frac{\mu^2}{2}} = \frac{1}{\sqrt{2\pi}} e^{-x^2/2}$$

Note from our exponents that $-x^2/2$ has been obtained from $-(x - \mu)^2/2 - \mu x + \mu^2/2$. By multiplying both sides of our density function in $\mathbb{P}$-space by $e^{-\mu x + \mu^2/2}$, we produced a new probability measure for our random variable X such that the new random variable still has a normal distribution with variance 1, but with a mean of 0 as we will see in a moment. We will call this new measure $\mathbb{Q}$, and its density function is:

$$f_\mathbb{Q}(x) = \frac{1}{\sqrt{2\pi}} e^{-\frac{x^2}{2}}$$

The new random variable assumes the same values as the original random variable X and over the same σ-algebra. However, it has a new probability measure $\mathbb{Q}$. We will also label this new random variable as X. For any continuous function F, the expectation of $F(X)$ with respect to $\mathbb{Q}$ is given by:

$$E_\mathbb{Q}[F(X)] = \frac{1}{\sqrt{2\pi}} \int_{-\infty}^{\infty} F(x) e^{-x^2/2} \, dx$$

If $F(X) = X$, we recognize this as the mean $E_\mathbb{Q}[X] = 0$. If $F(X) = (X - \mu)^2$, we recognize this as the variance $E_\mathbb{Q}\left[(X - \mu)^2\right] = 1$.

In a manner analogous our change of measure for discrete probability spaces, we can define a change of measure for a continuous probability space as follows:

$$f_\mathbb{Q}(x) = \frac{1}{\sqrt{2\pi}} e^{-\frac{x^2}{2}} = e^{-\mu x + \mu^2/2} f_\mathbb{P}(x) = \xi(x) f_\mathbb{P}(x) = \frac{d\mathbb{Q}}{d\mathbb{P}} f_\mathbb{P}(x)$$

Here, we have defined the Radon–Nikodym derivative for our change of measure:

$$\frac{d\mathbb{Q}}{d\mathbb{P}}(x) = \xi(x) = e^{-\mu x + \frac{\mu^2}{2}} \tag{6.7}$$

This change of measure allows us to take a random variable that had a nonzero mean with respect to one probability space and convert the probability space so that the random variable has a zero mean with respect to the new probability space, while leaving its variance unchanged. After illustrating and generalizing this change, we will perform a similar change of measure on a Brownian motion process.

ILLUSTRATION: CHANGE OF NORMAL DENSITY

Suppose that we were to begin with a random variable X with mean μ under probability measure $\mathbb{P}$, and we seek to transform this measure to an equivalent probability measure $\mathbb{Q}$ with a mean of zero by employing the change of measure process that we just worked through. More specifically, suppose that we wish to change a measure with distribution $N(0.5, 1)$ to another with distribution $N(0, 1)$. Figure 6.2 depicts the change that we wish to make and Table 6.1 provides a somewhat skeletal version of related computations.

In Table 6.1, the normal density $p(x_i)$ associated with value x_i, along with Radon–Nikodym derivatives and equivalent densities $q(x_i)$, are computed as follows:

$$f_{\mathbb{P}}(x_i) = \frac{1}{\sqrt{2\pi}} e^{-(x_i - 0.5)^2/2}$$

$$\xi(x_i) = e^{-0.5x_i + 0.5^2/2}$$

$$q(x_i) = f_{\mathbb{Q}}(x_i) = f_{\mathbb{P}}(x_i)\xi(x_i)$$

In Table 6.1, the $\mathbb{P}$ densities are computed with a mean equal to $\mu = 0.5$ (the darker densities in Figure 6.2) and $\mathbb{Q}$ densities with mean $\mu = 0$ (the lighter densities in Figure 6.2). The Radon–Nikodym derivative is computed as $\xi(x_i) = e^{-\mu x_i + \mu^2/2}$. Note that each density under $\mathbb{Q}$ is

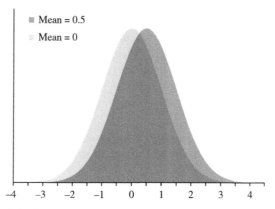

FIGURE 6.2 Change of normal density.

TABLE 6.1 Normal Densities and Distributions Under $\mathbb{P}$ and $\mathbb{Q}$

x_i	$p(x_i)$	$\xi(x_i)$	$q(x_i)$	Cumulative $\mathbb{P}$	Cumulative $\mathbb{Q}$
−3.0	0.000872683	5.078419037	0.004431848	0.000261805	0.001329555
−2.7	0.002384088	4.371035773	0.010420935	0.000750320	0.003557472
−2.4	0.005952532	3.762185355	0.022394530	0.002000814	0.008479792
−2.1	0.013582969	3.238142944	0.043983596	0.004931139	0.018436511
−1.8	0.028327038	2.787095461	0.078950158	0.011217640	0.036876574
−1.5	0.053990967	2.398875294	0.129517596	0.023565340	0.068146737
−1.2	0.094049077	2.064731100	0.194186055	0.045771347	0.116702285
−0.9	0.149727466	1.777130527	0.266085250	0.082337829	0.185742980
−0.6	0.217852177	1.529590420	0.333224603	0.137474775	0.275639458
−0.3	0.289691553	1.316530675	0.381387815	0.213606334	0.382831321
0	0.352065327	1.133148453	0.398942280	0.309869866	0.499880835
0.3	0.391042694	0.975309912	0.381387815	0.421336069	0.616930350
0.6	0.396952547	0.839457021	0.333224603	0.539535356	0.724122212
0.9	0.368270140	0.722527354	0.266085250	0.654318759	0.81401869
1.2	0.312253933	0.621885056	0.194186055	0.756397370	0.883059386
1.5	0.241970725	0.535261429	0.129517596	0.839531069	0.931614934
1.8	0.171368592	0.460703781	0.078950158	0.901531966	0.962885097
2.1	0.110920835	0.396531419	0.043983596	0.943875380	0.981325160
2.4	0.065615815	0.341297755	0.022394530	0.970355877	0.991281879
2.7	0.035474593	0.293757700	0.010420935	0.985519439	0.996204199
3.0	0.017528300	0.252839596	0.004431848	0.993469873	0.998432116

simply the product of the corresponding values for the density under $\mathbb{P}$ and the Radon–Nikodym derivative $\xi(x_i)$. Reimann integrals are computed to obtain both sets of densities.[3]

6.2.6 More General Shifts

Now, we further extend our analysis to allow for nonunit variances and more general shifts in the mean. Consider a normal density function for a random variable $X \sim N(\mu, \sigma^2)$ under probability measure $\mathbb{P}$:

$$f_{\mathbb{P}}(x) = \frac{1}{\sigma\sqrt{2\pi}} e^{-\frac{1}{2}\left(\frac{x-\mu}{\sigma}\right)^2} \tag{6.8}$$

Suppose that we wish to shift our mean left ($\lambda > 0$) or right ($\lambda < 0$) by some arbitrary constant λ. By multiplying both sides of our density function in $\mathbb{P}$-space by $e^{\left(-\frac{\lambda x}{\sigma^2} + \frac{\mu\lambda}{\sigma^2} - \frac{\lambda^2}{2\sigma^2}\right)}$, we will produce a new probability measure for our random variable X, which has a normal distribution with mean $\mu - \lambda$ and unchanged variance σ^2.[4] We will call this new measure $\mathbb{Q}$, and its density function is:

$$f_\mathbb{Q}(x) = \frac{1}{\sigma\sqrt{2\pi}}e^{-\frac{1}{2}\left(\frac{x-(\mu-\lambda)}{\sigma}\right)^2} = \xi(x)f_\mathbb{P}(x) = \frac{d\mathbb{Q}}{d\mathbb{P}}f_\mathbb{P}(x) \qquad (6.9)$$

Here, we have defined the Radon−Nikodym derivative for our change of measure:

$$\frac{d\mathbb{Q}}{d\mathbb{P}}(x) = \xi(x) = e^{\left(-\frac{\lambda x}{\sigma^2} + \frac{\mu\lambda}{\sigma^2} - \frac{\lambda^2}{2\sigma^2}\right)} \qquad (6.10)$$

This change of measure allows us to shift the mean of a normal distribution without affecting its variance. Next, we will perform a similar change of measure on a Brownian motion process.

6.2.7 Change of Brownian Motion

First, consider the process $X_t = Z_t + \mu t$, where Z_t is a standard Brownian motion and $\mu \geq 0$ is a constant or drift term. Denote $\mathbb{P}$ as the probability space for this process. We wish to find a change of measure that changes the mean μ while maintaining the variance of $\mathbb{P}$. Observe that for any fixed t, $X_t \sim N(\mu t, t)$. To simplify the notation and generalize the process, consider a random variable $X \sim N(\mu, \sigma^2)$ in probability space $\mathbb{P}$. We saw in the previous section that if $X \sim N(\mu, \sigma^2)$, then the change in measure:

$$\frac{d\mathbb{Q}}{d\mathbb{P}} = e^{\left(-\frac{\lambda x}{\sigma^2} + \frac{\mu\lambda}{\sigma^2} - \frac{\lambda^2}{2\sigma^2}\right)}$$

results in the random variable X having the distribution $\hat{X} \sim N(u - \lambda, \sigma^2)$ in probability space $\mathbb{Q}$. Note that we renamed the random variable X to be $\hat{X}$ in the new probability space $\mathbb{Q}$. Next, making the replacements: $X \to Z_t + \mu t$, $\sigma^2 \to t$, $\mu \to \mu t$, and $\lambda \to \lambda t$, suggests that the stochastic process $Z_t + \mu t$ in probability space $\mathbb{P}$ becomes the process $\hat{Z}_t + (\mu - \lambda)t$ in the probability space $\mathbb{Q}$ with the change of measure given by the Radon−Nikodym derivative at time t:

$$\xi_t = e^{\left[-\frac{\lambda t(Z_t + \mu t)}{t} + \frac{\mu t(\lambda t)}{t} - \frac{(\lambda t)^2}{2t}\right]} = e^{-\lambda Z_t - \frac{1}{2}\lambda^2 t}$$

$\hat{Z}_t$ is Brownian motion in the probability space $\mathbb{Q}$. Observe that this statement is not only referring to the resulting changes in the probability distributions, but is also referring to the entire resulting process ($\hat{Z}_t$), which has both continuity properties and independence properties. To prove the general result is beyond the level of this text, but this result is a special case of the Cameron−Martin−Girsanov theorem. Note that the change of measure has no effect on the volatility of the process, but it does shift the mean from μt to $(\mu - \lambda)t$. If we choose $\lambda = \mu$, then the process becomes standard Brownian motion $d\hat{Z}_t$ in probability space $\mathbb{Q}$.

This result can be generalized to any process of the form:

$$dX_t = dZ_t + \mu_t\, dt$$

where μ_t is a stochastic process that only depends upon time up to time t. This generalization is known as the Cameron−Martin−Girsanov theorem.

6.3 THE CAMERON–MARTIN–GIRSANOV THEOREM AND THE MARTINGALE REPRESENTATION THEOREM

In this and previous chapters, we discussed the change of probability measure from the physical space to the risk-neutral space. The *Cameron–Martin–Girsanov theorem* (Cameron and Martin, 1944; Girsanov, 1960) provides conditions under which probability measures for continuous-time stochastic processes can be converted to risk-neutral measures. More generally, the Cameron–Martin–Girsanov theorem can be used to convert one probability measure to another of the same type with the same variance but with a different drift. We start here by extending beyond our development of the Radon–Nikodym derivative for discrete processes to continuous processes.

6.3.1 Multiple Time Periods: Discrete Case to Continuous Case

In this section, we develop continuous-time processes from discrete processes. Recall our multiple period development of Radon–Nikodym derivatives in Section 6.2.4. Now, consider a continuous time–space stochastic process. Suppose X_t follows a stochastic process in continuous time t such that its differential satisfies $dX_t = dZ_t + \mu_t \, dt$ where Z_t is Brownian motion with respect to some probability measure $\mathbb{P}$ and μ_t is a continuous real-valued function of t. We solve for X_t as follows:

$$X_t - X_0 = \int_0^t (dZ_s + \mu_s \, ds)$$

which implies:

$$X_t = X_0 + Z_t + \int_0^t \mu_s \, ds$$

In this case, we say that the price X_t follows a Brownian motion process with variable drift. This continuous process can be approximated by a discrete process in the following way. Divide the time interval $[0,t]$ into n equal subintervals $[t_{i-1}, t_i]$ with $t_i - t_{i-1} = \Delta t = t/n$ for each $i = 1, 2, \ldots, n$. This implies that $t_i = t \times i/n$. Define $\Delta X_{t_i} = X_{t_i} - X_{t_{i-1}}$. This implies that:

$$X_t = X_0 + \Delta X_{t_1} + \Delta X_{t_2} + \ldots + \Delta X_{t_n}$$

For each ΔX_{t_i} we have:

$$\Delta X_{t_i} = Z_{t_i} - Z_{t_{i-1}} + \int_{t_{i-1}}^{t_i} \mu_s \, ds \approx Z_{t_i} - Z_{t_{i-1}} + \mu_{t_{i-1}} \Delta t$$

Thus, we can approximate the process X_t by:

$$X_t \approx X_0 + \sum_{i=1}^{n} (Z_{t_i} - Z_{t_{i-1}} + \mu_{t_{i-1}} \Delta t)$$

Since $Z_{t_i} - Z_{t_{i-1}} \sim N(0, \Delta t)$, $Z_{t_i} - Z_{t_{i-1}} + \mu_{t_{i-1}} \Delta t \sim N(\mu_{t_{i-1}} \Delta t, \Delta t)$. This means that the random variable $Z_{t_i} - Z_{t_{i-1}} + \mu_{t_{i-1}} \Delta t$ has the density function:

$$\frac{1}{\sqrt{2\pi \Delta t}} e^{-\frac{(x_i - \mu_{t_{i-1}} \Delta t)^2}{2\Delta t}}$$

Now, label the probability measure associated with this density function by $\mathbb{P}_i$. In Section 6.2.6, we demonstrated that using the Radon–Nikodym derivative:

$$\xi_t = e^{-\lambda Z_t - \frac{1}{2}\lambda^2 t}$$

changed the random variable $Z_t + \mu t$ to the random variable $\hat{Z}_t + (\mu - \lambda)t$. In this proof, we will only take into consideration the probability distribution features of the processes and not attempt to justify their additional continuity and independence properties. Making the replacements $Z_t \to Z_{t_i} - Z_{t_{i-1}}, \mu \to \mu_{t_{i-1}}, t \to \Delta t$, and $\lambda \to \lambda_{t_{i-1}}$, we conclude that the change in measure from $\mathbb{P}_{i-1,i}$ to $\mathbb{Q}_{i-1,i}$ with the Radon–Nikodym derivative:

$$\frac{d\mathbb{Q}_{i-1,i}}{d\mathbb{P}_{i-1,i}} = e^{-\lambda_{t_{i-1}}(Z_{t_i} - Z_{t_{i-1}}) - \frac{1}{2}\lambda_{t_{i-1}}^2 \Delta t}$$

will change the random variable $Z_{t_i} - Z_{t_{i-1}} + \mu_{t_{i-1}} \Delta t$ to the random variable $\hat{Z}_{t_i} - \hat{Z}_{t_{i-1}} + (\mu_{t_{i-1}} - \lambda_{t_{i-1}})\Delta t$ where $\hat{Z}_{t_i} - \hat{Z}_{t_{i-1}}$ is Brownian motion with respect to the probability measure $\mathbb{Q}_{i-1,i}$. $\{Z_{t_i} - Z_{t_{i-1}} + \mu_{t_{i-1}} \Delta t\}_{i=1}^n$ are independent random variables since $\{[t_{i-1}, t_i]\}_{i=1}^n$ are nonoverlapping intervals. Since the density function for a sum of independent random variables equals the product of their density functions, then the density function for:

$$\sum_{i=1}^{n}(Z_{t_i} - Z_{t_{i-1}} + \mu_{t_{i-1}} \Delta t)$$

is the product:

$$\prod_{i=1}^{n} \frac{1}{\sqrt{2\pi \Delta t}} e^{-\frac{(x_i - \mu_{t_{i-1}} \Delta t)^2}{2\,\Delta t}}$$

Next, label the probability measure associated with this density function as $\mathbb{P}_n$. Since $\{Z_{t_i} - Z_{t_{i-1}}\}_{i=1}^n$ are independent random variables, then the Radon–Nikodym derivative $\frac{d\mathbb{Q}_n}{d\mathbb{P}_n}$ to use to go from the random variable $\sum_{i=1}^{n}(Z_{t_i} - Z_{t_{i-1}} + \mu_{t_{i-1}} \Delta t)$ with respect to the probability measure $\mathbb{P}_n$ to the random variable $\sum_{i=1}^{n}[\hat{Z}_{t_i} - \hat{Z}_{t_{i-1}} + (\mu_{t_{i-1}} - \lambda_{t_{i-1}})\Delta t]$ with respect to the probability measure $\mathbb{Q}_n$ must be the product of the Radon–Nikodym derivatives $\frac{d\mathbb{Q}_{i-1,i}}{d\mathbb{P}_{i-1,i}}$ at each time increment from $t_{i-1,i}$ to t_i extending from 1 to n:

$$\frac{d\mathbb{Q}_n}{d\mathbb{P}_n} = \prod_{i=1}^{n} \frac{d\mathbb{Q}_{i-1,i}}{d\mathbb{P}_{i-1,i}} = \prod_{i=1}^{n} e^{-\lambda_{t_{i-1}}(Z_{t_i} - Z_{t_{i-1}}) - \frac{1}{2}\lambda_{t_{i-1}}^2 \Delta t} = e^{-\sum_{i=1}^{n} \lambda_{t_{i-1}}(Z_{t_i} - Z_{t_{i-1}}) - \frac{1}{2}\sum_{i=1}^{n} \lambda_{t_{i-1}}^2 \Delta t}$$

Referring to the definitions of stochastic integration in Section 6.1.2, in the limit as $n \to \infty$, we obtain:

$$\xi_t = e^{-\int_0^t \lambda_s \, dZ_s - \frac{1}{2} \int_0^t \lambda_s^2 \, ds}$$

This is the Radon–Nikodym derivative required for the change of measure in the Cameron–Martin–Girsanov theorem (see the following section) in the event that μ_t and λ_t are continuous real-valued functions. This result can be generalized to any integrable stochastic processes μ_t and λ_t, which is the statement of the Cameron–Martin–Girsanov theorem.

6.3.2 The Cameron–Martin–Girsanov Theorem

The *Cameron–Martin–Girsanov theorem* applies to stochastic processes X_t that are represented as Brownian motion with drift:

$$dX_t = dZ_t + \mu_t \, dt$$

under probability measure $\mathbb{P}$ with μ_t itself as a stochastic process and Z_t is standard Brownian motion. The theorem states that, under certain technical conditions (e.g., a finite variance), we can shift the drift so that $dX_t = d\hat{Z}_t + (\mu_t - \lambda_t)dt$ in an equivalent probability space $\mathbb{Q}$ using the Radon–Nikodym derivative

$$\xi_t = e^{-\int_0^t \lambda_s \, dZ_s - \frac{1}{2} \int_0^t \lambda_s^2 \, ds} \tag{6.11}$$

Essentially, this theorem states that we can change the drift component of a Brownian motion by changing the probability measure. In finance, this means that we can change the probability measure of a process from a physical measure in which investors are risk averse to a risk-neutral measure. As we shall see in the next two chapters, this shift is crucial to valuing options and bonds. It is also important that this shift is to the drift of the process, only its variance is not changed.

ILLUSTRATION: APPLYING THE CAMERON–MARTIN–GIRSANOV THEOREM

Consider the process $dX_t = dZ_t + 0.05 \, dt$ under some probability measure $\mathbb{P}$ with $X_0 = 10$. Solving for X_t gives:

$$X_t - X_0 = \int_0^t dX_s = \int_0^t (dZ_s + 0.05 \, ds) = Z_t + 0.05t$$

so that $X_t = Z_t + 0.05t + X_0$. Because of the drift term $0.05t$, X_t is not a martingale with respect to the probability measure $\mathbb{P}$. Using the Cameron–Martin–Girsanov Theorem, we will be able to find an equivalent probability measure $\mathbb{Q}$, so that the process X_t with respect to the probability measure $\mathbb{Q}$ will be a martingale. Choose $\lambda = 0.05$ in the Cameron–Martin–Girsanov theorem so that the Radon–Nikodym derivative for the change in measure will be:

$$\xi_t = e^{-\int_0^t 0.05 \, dZ_s - \frac{1}{2} \int_0^t (0.05)^2 \, ds} = e^{-0.05Z_t - 0.00125t}$$

By the Cameron–Martin–Girsanov theorem, this will result in the process X_t taking the form:

$$d\hat{X}_t = d\hat{Z}_t + (0.05 - 0.05)\, dt = d\hat{Z}_t$$

where $\hat{Z}_t$ is Brownian motion with respect to the probability measure $\mathbb{Q}$. Since obviously $\hat{X}_t = \hat{Z}_t$, then $\hat{X}_t$ is a martingale with respect to the probability measure $\mathbb{Q}$. The density function of X_t with respect to measure $\mathbb{P}$, the Radon–Nikodym derivative, and the density function of $\hat{X}_t$ with respect to $\mathbb{Q}$ are given as follows:

$$f_{\mathbb{P}}(x) = \frac{1}{\sqrt{2\pi t}}e^{-(x-\mu t)^2/2t} = \frac{1}{\sqrt{2\pi t}}e^{-(x-0.05t)^2/2t}$$

$$\xi_t(x) = e^{-0.05x+0.00125t}$$

$$f_{\mathbb{Q}}(x) = f_{\mathbb{P}}(x)\xi_t(x) = \frac{1}{\sqrt{2\pi t}}e^{-(x-0.05t)^2/2t}e^{-0.05x+0.00125t}$$

$$= \frac{1}{\sqrt{2\pi t}}e^{-\frac{1}{2t}x^2+0.05x-0.00125t-0.05x+0.00125t} = \frac{1}{\sqrt{2\pi t}}e^{-\frac{1}{2t}x^2}$$

6.3.3 The Martingale Representation Theorem

Suppose that M_t is a martingale process of the form $dM_t = \sigma_t\, dZ_t + \mu_t\, dt$ so that $\sigma_t \neq 0$ with probability 1. If N_t is another martingale with respect to the same probability measure, then there exists a stochastic process η_t such that N_t can be expressed in the form $dN_t = \eta_t\, dM_t$.

To gain a little more insight into this result, first consider an analogous problem involving real-valued continuously differentiable functions. Suppose $M_t = t^3$ and $N_t = t^2$. We calculate that $dM_t = 3t^2\, dt$ and $dN_t = 2t\, dt$. We wish to find η_t so that $dN_t = \eta_t\, dM_t$. The obvious choice for η_t is $\eta_t = (dN_t)/(dM_t) = (2t\, dt)/(3t^2\, dt) = 2/(3t)$. Let us check that it works:

$$\eta_t\, dM_t = \frac{2}{3t}(3t^2\, dt) = 2t\, dt = dN_t$$

as long as $t \neq 0$. You can see that it is important that the function $3t^2$ in the expression $dM_t = 3t^2\, dt$ is not equal to zero in order for η_t to be defined since η_t was obtained by the quotient: $(2t\, dt)/(3t^2\, dt) = 2/(3t)$.

Similarly, one needs $\sigma_t \neq 0$ with probability 1 (i.e., $\sigma_t = 0$ at worst on a set of probability measure 0) in order for the martingale result to also work. The martingale result is much deeper than the real-valued case, and it is beyond the level of this book to give a proof of the result. Nevertheless, a couple of examples can give us further insight. Observe that both processes must be martingales. Consider an example in which $M_t = \sigma_1 Z_t$ where σ_1 is any nonzero constant and Z_t is Brownian motion. Define $N_t = \sigma_2 Z_t$ where σ_2 is any constant. As we have already learned, both of these processes are martingales. In this case $dM_t = \sigma_1\, dZ_t$ and $dN_t = \sigma_2\, dZ_t$, so that if we choose $\eta_t = \sigma_2/\sigma_1$, then:

$$\eta_t\, dM_t = \frac{\sigma_2}{\sigma_1}\sigma_1\, dZ_t = \sigma_2\, dZ_t = dN_t$$

Contrast this example with one in which $M_t = \sigma_1 Z_t$ where σ_1 is any nonzero constant and Z_t is Brownian motion. But now, define $N_t = \sigma_2 Z_t + \mu_2 t$ where σ_2 is any constant and μ_2 is any nonzero constant. As we have seen, M_t is a martingale, while N_t is not a martingale. The differentials are $dM_t = \sigma_1\, dZ_t$ and $dN_t = \sigma_2\, dZ_t + \mu_2\, dt$. In this case, there does not exist any stochastic process η_t so that $dN_t = \eta_t\, dM_t$. If one tries the same simple approach used in the above example to find η_t, it would be:

$$\eta_t = \frac{dN_t}{dM_t} = \frac{\sigma_2\, dZ_t}{\sigma_1\, dZ_t + \mu_1\, dt} = \frac{\sigma_2}{\sigma_1 + \mu_1 \frac{dt}{dZ_t}}$$

Had Z_t been a reasonably nice real-valued function, the steps above would be well defined and η_t would have existed. One could have solved for the inverse function t as a function of Z_t and in many cases it would be differentiable, so that dt/dZ_t would have existed. However, Z_t is Brownian motion and the time variable t is not a differentiable function of Z_t.

6.4 ITÔ'S LEMMA

Itô's lemma is often regarded to be the fundamental theorem of stochastic calculus. Brownian motion processes are fractals that do not smooth as $\Delta t \to 0$. Newtonian calculus, which requires smoothing, cannot be used to differentiate or antidifferentiate Brownian motion functions. Hence, we will rely on Itô's lemma to analyze continuous-time stochastic processes. First, we will briefly review Taylor series expansions, originally covered in Chapter 1.

6.4.1 A Discussion on Taylor Series Expansions

As discussed in Chapter 1, we can use the Taylor series expansion to estimate the change in an infinitely differentiable function $y = y(t)$ as follows:

$$\Delta y = \frac{dy}{dt} \cdot \Delta t + \frac{1}{2} \cdot \frac{d^2 y}{dt^2} \cdot (\Delta t)^2 + \cdots$$

If Δt is small, higher-order terms (involving Δt raised to powers of 2 or greater) are negligible compared to terms just involving Δt. So we have the following approximation when Δt is small:

$$\Delta y = \frac{dy}{dt} \Delta t$$

This can also be expressed in differential notation:

$$dy = \frac{dy}{dt} dt$$

In particular, when evaluating integrals, terms involving Δt raised to powers of 2 or larger can be dropped. Thus, for example:

$$y(T) - y(0) = \int_0^T dy(t) = \int_0^T \frac{dy}{dt} dt$$

In this chapter, we will use the delta (Δ) notation when we are conducting Taylor series expansions to estimate changes in function. We will use the differential (d) notation for the case when we have dropped higher-order terms.

6.4.1.1 Taylor Series and Two Independent Variables

Now, suppose that $y = y(x, t)$; that is, y is a function of the independent variables x and t. The Taylor series expansion can be generalized to two independent variables as follows:

$$\Delta y = \frac{\partial y}{\partial x} \cdot \Delta x + \frac{\partial y}{\partial t} \cdot \Delta t + \frac{1}{2} \cdot \frac{\partial^2 y}{\partial x^2} \cdot (\Delta x)^2 + \frac{1}{2} \cdot \frac{\partial^2 y}{\partial t^2} \cdot (\Delta t)^2 + \frac{1}{2} \cdot \frac{\partial^2 y}{\partial x \, \partial t} \cdot (\Delta x \, \Delta t) + \cdots$$

If $x = x(t)$ is a differentiable function of time t, then we have the approximation $\Delta x = x'(t)\Delta t$. Ignoring higher-order terms in Δt, we obtain:

$$dy = \frac{\partial y}{\partial x} x'(t) \, dt + \frac{\partial y}{\partial t} dt$$

where we have expressed the result in differential form. This implies that:

$$y(T) - y(0) = \int_0^T dy(t) = \int_0^T \left(\frac{\partial y}{\partial t} x'(t) + \frac{\partial y}{\partial t} \right) dt$$

Note that for purposes of economy of notation we denoted $y(t) = y(x(t),t)$.

6.4.2 The Itô Process

Next, consider an Itô process, which means that the stochastic process X_t satisfies the equation:

$$dX_t = a(X_t, t)dt + b(X_t, t)dZ_t$$

The drift of the process is a, while b^2 is the instantaneous variance and dZ_t is a standard Brownian motion process. Taking Δt to be a small change in time and expressing $a = a(X_t, t)$ and $b = b(X_t, t)$ in order to economize the notation, we can write:

$$\Delta X_t = a\Delta t + b\Delta Z_t$$

where random variable $\Delta Z_t \sim N(0, \Delta t)$.

Now, suppose that $y = y(x, t)$ is an infinitely differentiable function with respect to the real variables x and t. Now, replace x with X_t so that $y = y(X_t, t)$. Thus, y itself becomes a stochastic process, since it is a function of a stochastic process X_t and time t. The Taylor series expansion above can be used to estimate Δy, the change in y, resulting from a change Δt in time:

$$\Delta y(X_t, t) = \frac{\partial y}{\partial t}\Delta t + \frac{\partial y}{\partial x}(a\Delta t + b\Delta Z_t) + \frac{1}{2} \cdot \frac{\partial^2 y}{\partial x^2}(a\Delta t + b\Delta Z_t)^2 + \frac{1}{2} \cdot \frac{\partial^2 y}{\partial t^2}(\Delta t)^2 + \frac{1}{2}$$

$$\cdot \frac{\partial^2 y}{\partial x \partial t}(a\Delta t + b\Delta Z_t)\Delta t + \cdots$$

$$= \frac{\partial y}{\partial t}\Delta t + \frac{\partial y}{\partial x}(a\Delta t + b\Delta Z_t) + \frac{1}{2} \cdot \frac{\partial^2 y}{\partial x^2}[a^2(\Delta t)^2 + 2ab\Delta t\Delta Z_t + b^2(\Delta Z_t)^2] + \frac{1}{2} \cdot \frac{\partial^2 y}{\partial t^2}(\Delta t)^2$$

$$+ \frac{1}{2} \cdot \frac{\partial^2 y}{\partial x \partial t}[a(\Delta t)^2 + b\Delta Z_t \Delta t] + \cdots$$

In the expansion above, we also economized the notation for the various partial derivatives evaluated at $(x,t) = (X_t,t)$ by denoting:

$$\frac{\partial y}{\partial x} = \frac{\partial y}{\partial x}(X_t, t), \frac{\partial y}{\partial t} = \frac{\partial y}{\partial t}(X_t, t), \frac{\partial^2 y}{\partial x^2} = \frac{\partial^2 y}{\partial x^2}(X_t, t), \frac{\partial^2 y}{\partial t^2} = \frac{\partial^2 y}{\partial t^2}(X_t, t), \frac{\partial^2 y}{\partial x \partial t} = \frac{\partial^2 y}{\partial x \partial t}(X_t, t)$$

When estimating Δy using the Taylor series expansion, all terms that are negligible compared to Δt as Δt approaches 0 can be dropped. Since $(Z_{t+\Delta t} - Z_t) \sim Z\sqrt{\Delta t}$ with $Z \sim N(0,1)$, the terms involving $(\Delta t)^2$, $\Delta t\, \Delta Z_t$, and higher-order terms can all be dropped. Our expansion simplifies to:

$$\Delta y = \frac{\partial y}{\partial t}\Delta t + \frac{\partial y}{\partial x}(a\Delta t + b\Delta Z_t) + \frac{b^2}{2} \cdot \frac{\partial^2 y}{\partial x^2}(\Delta Z_t)^2 \tag{6.12}$$

Since $(\Delta Z_t)^2 \sim Z^2\Delta t$, this term is not negligible. We now state a remarkable fact. We can actually replace $(\Delta Z_t)^2$ with Δt, where the random variable has seemingly disappeared. To show this, we will make use of the fact that Brownian motion increments are independent for disjoint intervals.

6.4.2.1 Demonstration that Δt Can Replace $(\Delta Z_t)^2$ in the Differential Δy

We begin by subdividing the interval $[t, t + \Delta t]$ into n smaller subintervals $[t + (i-1)\Delta t/n, t + i\Delta t/n]$ for $i = 1, 2, \ldots, n$. This means that the width of each subinterval equals $\Delta t/n$. In the Taylor series expansion above of Δy, the subintervals lead to $(\Delta Z_t)^2$ being replaced by $\sum_{i=1}^{n}(\Delta_i Z_t)^2$ with:

$$\Delta_i Z_t = Z_{t+\frac{i\Delta t}{n}} - Z_{t+\frac{(i-1)\Delta t}{n}}$$

Since:

$$\Delta_i Z_t = Z_{t+\frac{i\Delta t}{n}} - Z_{t+\frac{(i-1)\Delta t}{n}} \sim N\left(0, \frac{\Delta t}{n}\right)$$

the variance of the random variable $\Delta_i Z_t$ is:

$$E\left[(\Delta_i Z_t)^2\right] = Var(\Delta_i Z_t) = \frac{\Delta t}{n}$$

Note that this result gives us the expected value of the random variable $(\Delta_i Z_t)^2$. Next we calculate the variance of $(\Delta_i Z_t)^2$. From the second equation above, we see that:

$$\Delta_i Z_t \sim Z \sqrt{\frac{\Delta t}{n}}$$

with $Z \sim N(0,1)$, which implies:

$$(\Delta_i Z_t)^2 \sim Z^2 \frac{\Delta t}{n}$$

The solution to end-of-chapter Exercise 6.17.b verifies that $\text{Var}(Z^2 c) = 2c^2$ for any $c > 0$. By the last equation above and the fact that $\text{Var}(Z^2 c) = 2c^2$, we obtain:

$$\text{Var}\left[(\Delta_i Z_t)^2\right] = \frac{2(\Delta t)^2}{n^2}$$

The Brownian motion increments $\Delta_i Z_t$ are independent random variables since the corresponding time intervals are disjoint. Since the variance of a sum of independent random variables is the sum of each of their variances, then:

$$\text{Var}\left(\sum_{i=1}^n (\Delta_i Z_t)^2\right) = \sum_{i=1}^n \text{Var}\left((\Delta_i Z_t)^2\right) = \sum_{i=1}^n \frac{2(\Delta t)^2}{n^2} = \frac{2(\Delta t)^2}{n}$$

Since we showed above that $E\left[(\Delta_i Z_t)^2\right] = \frac{\Delta t}{n}$, we see that:

$$E\left[\sum_{i=1}^n (\Delta_i Z_t)^2\right] = \frac{\Delta t}{n} n = \Delta t$$

Thus, the quantity $\sum_{i=1}^n (\Delta_i Z_t)^2$ has an expected value of Δt and a variance $2(\Delta t)^2/n$ that approaches 0 as n approaches infinity. Since $(\Delta Z_t)^2$ must approach $\sum_{i=1}^n (\Delta_i Z_t)^2$ as $n \to \infty$ and as Δt gets arbitrarily small, this implies that $(\Delta Z_t)^2$ has an expected value of Δt and a variance approaching $2(\Delta t)^2/n \to 0$. But a random variable with variance 0 is simply equal to its expected value. So, we conclude that $(\Delta Z_t)^2 = \Delta t$.

6.4.2.2 Itô's Formula

Equation (6.12) above simplifies to:

$$\Delta y = \left(\frac{\partial y}{\partial t} + a \frac{\partial y}{\partial x} + \frac{1}{2} b^2 \frac{\partial^2 y}{\partial x^2}\right) \Delta t + b \frac{\partial y}{\partial x} \Delta Z_t$$

We can express this result, which becomes *Itô's formula* in differential form:

$$dy(X_t, t) = \left(\frac{\partial y}{\partial t} + a \frac{\partial y}{\partial x} + \frac{1}{2} b^2 \frac{\partial^2 y}{\partial x^2}\right) dt + b \frac{\partial y}{\partial x} dZ_t$$

where the partial derivatives are evaluated at $(x,t) = (X_t, t)$.

6.4.3 Itô's Lemma

The differential that we obtained above is the stochastic calculus version of the differential from ordinary calculus. This result is known as *Itô's lemma*:

Given a real-valued function: $y = y(x,t)$, define the stochastic process $y(X_t,t)$, where X_t is an Itô process $dX_t = a(X_t,t)\,dt + b(X_t,t)\,dZ_t$ with Z_t denoting standard Brownian motion. By Itô's formula, the differential of the process $y(X_t,t)$ satisfies the equation:

$$dy(X_t, t) = \left[a\frac{\partial y}{\partial x} + \frac{\partial y}{\partial t} + \frac{1}{2}b^2\frac{\partial^2 y}{\partial x^2} \right]dt + b\frac{\partial y}{\partial x}dZ_t$$

where the partial derivatives above are evaluated at $(x,t) = (X_t,t)$.

6.4.4 Applying Itô's Lemma

Evaluating stochastic integrals is generally trickier than evaluating ordinary real-valued integrals. We recommend the following three-step process to evaluate stochastic integrals:

1. As a first attempt, apply the form of the solution that mimics the solution for the analogous problem in ordinary calculus.
2. Invoke Itô's lemma to find the differential of the attempted solution.
3. Integrate both sides of the differential and rearrange the result to solve for the desired stochastic integral.

ILLUSTRATION: APPLYING ITÔ'S LEMMA

Suppose that we seek to evaluate $\int_0^T Z_t\,dZ_t$. Using the three-step technique described above, we evaluate $\int_0^T Z_t\,dZ_t$ as follows:

1. Attempt the ordinary calculus solution, which would suggest that $Y_T = \int_0^T Z_t\,dZ_t = \frac{1}{2}Z_T^2$.
2. Find the differential of our attempted solution $Y_T = \frac{1}{2}Z_T^2$ using Itô's lemma. Choose $y(x, t) = \frac{1}{2}x^2$ so that $Y_t = y(Z_t, t) = \frac{1}{2}Z_t^2$. With $dX_t = dZ_t = 0 \cdot dt + 1\,dZ_t$, and invoking Itô's lemma, we have:

$$dY_t = dy(Z_t, t) = \left[0 \cdot \frac{\partial y}{\partial x} + \frac{\partial y}{\partial t} + \frac{1}{2} \cdot 1^2 \cdot \frac{\partial^2 y}{\partial x^2} \right]dt + 1 \cdot \frac{\partial y}{\partial x}dZ_t$$

$$= \frac{\partial y}{\partial x}dZ_t + \frac{\partial y}{\partial t}dt + \frac{1}{2}\frac{\partial^2 y}{\partial x^2}dt$$

In Itô's lemma, we must evaluate the partial derivatives at $(x,t) = (Z_t,t)$. Since $\frac{\partial y}{\partial x} = x|_{x=Z_t} = Z_t$, $\frac{\partial y}{\partial t} = 0$ and $\frac{\partial^2 y}{\partial x^2} = 1$, we have:

$$dY_t = Z_t dZ_t + \frac{1}{2}dt$$

3. We will integrate both sides of this equation for $0 \le t \le T$. The left side of the following equation is based on our discussion concerning the integral of a stochastic differential at the start of Section 6.1.3. The right side of the following is based on the equation immediately above:

$$Y_T - Y_0 = \int_0^T dY_t = \int_0^T Z_t \, dZ_t + \int_0^T \frac{1}{2} dt$$

$$\frac{1}{2}(Z_T)^2 - \frac{1}{2}(Z_0)^2 = \int_0^T Z_t \, dZ_t + \int_0^T \frac{1}{2} dt$$

$$\frac{1}{2}(Z_T)^2 = \int_0^T Z_t \, dZ_t + \frac{1}{2} T$$

Solving for the desired integral results in:

$$\int_0^T Z_t \, dZ_t = \frac{1}{2}(Z_T)^2 - \frac{1}{2} T$$

Note that the result is a stochastic process. Also, observe that:

$$E\left[\frac{1}{2}(Z_T)^2 - \frac{1}{2} T\right] = \frac{1}{2}\text{Var}[Z_T] - \frac{1}{2} T = 0$$

which gives the correct expectation, and clearly, $\text{Var}[Z_T] = T$.

Intuitively, the reason why the Brownian motion integral and ordinary calculus integral result in different forms of solutions is because $x(t)$ is a continuously differentiable function while Z_t is not differentiable. More precisely, $dx(t) = x'(t) \, dt$ and $(dx(t))^2 = (x'(t))^2 (dt)^2$, while dZ_t has standard deviation equal to $\sqrt{dt}$ and $(dZ_t)^2 = dt$. This means that the second-order term in the Taylor expansion of $d\left[\frac{1}{2} Z_t^2\right]$ becomes important since $(dZ_t)^2 = dt$. In ordinary calculus, only the first-order term in the Taylor expansion of $d\left[\frac{1}{2}(x(t))^2\right]$ is important since $(dx(t))^2 = (x'(t))^2 (dt)^2$ and $(dt)^2$ is negligible insofar as integration is concerned.

6.4.5 Application: Geometric Brownian Motion

Geometric Brownian motion is an essential model for characterizing the stochastic process for a security with value S_t at time t:

$$dS_t = \mu S_t \, dt + \sigma S_t \, dZ_t \tag{6.13}$$

Recall that μ and σ are the geometric mean security return and the standard deviation of the security return per unit of time. We wish to determine the value of the security S_t as a function of time. First, rewrite the differential above in the form:

$$\frac{dS_t}{S_t} = \mu \, dt + \sigma \, dZ_t \tag{6.14}$$

The reason why this model is used extensively to model security prices S_t is that the instantaneous return on the security over the time interval $[t, t+dt]$ is the known drift $\mu \, dt$ plus a random Brownian motion component $\sigma \, dZ_t$. The drift is related to the time value of money and the Brownian motion term reflects the unknowable random factors affecting the security's price. Since Brownian motion has a normal distribution, we are, of course, assuming a particular type of random fluctuation. It is a subject of much debate in finance as to the merits of this choice to model securities.

Next, we will integrate both sides of Eq. (6.14) from 0 to T. To evaluate the integral of the left side, use our three-step procedure involving Itô's lemma. Step 1 is to evaluate the integral as though S_t is a real-valued function:

$$\int_0^T \frac{dS_t}{S_t} = \ln(S_t)\Big|_0^T = \ln(S_T) - \ln(S_0)$$

Since $\ln S_0$ is a constant, we can ignore it for the purposes of computing the differential of the right side function $\ln S_t - \ln S_0$. Step 2 is to take the differential of $\ln S_t$ using Itô's lemma applying it to the function $y = y(S_t, t) = \ln S_t$. In this formulation, dy from Itô's formula is $d(\ln S_t)$, a is μS_t, and b is σS_t. We substitute these values into Itô's formula as follows:

$$d(\ln S_t) = \left[\mu S_t \frac{\partial y}{\partial S} + \frac{\partial y}{\partial t} + \frac{1}{2}\sigma^2 S_t^2 \frac{\partial^2 y}{\partial S^2} \right] dt + \sigma S_t \frac{\partial y}{\partial S} dZ_t$$

Following standard rules for differentiation, the above equation is written:

$$d(\ln S_t) = \left[\mu S_t \frac{1}{S_t} + 0 - \frac{1}{2}\sigma^2 S_t^2 \frac{1}{S_t^2} \right] dt + \sigma S_t \frac{1}{S_t} dZ_t$$

which simplifies to:

$$d \ln(S_t) = \left[\mu - \frac{1}{2}\sigma^2 \right] dt + \sigma \, dZ_t$$

This equation would be the basis of a Monte Carlo simulation of geometric Brownian motion for stock price behavior should we wish to run one. Now, perform Step 3 by integrating both sides of this equation from $t = 0$ to $t = T$, which results in:

$$\int_0^T d \ln(S_t) = \int_0^T \left[\left(\mu - \frac{1}{2}\sigma^2 \right) dt + \sigma \, dZ_t \right]$$

$$\ln(S_t)\Big|_0^T = \left(\mu - \frac{1}{2}\sigma^2 \right) t + \sigma Z_t \Big|_0^T$$

$$\ln S_T - \ln S_0 = \ln\frac{S_T}{S_0} = \left(\mu - \frac{1}{2}\sigma^2 \right) T + \sigma Z_T$$

Exponentiating both sides of this equation and multiplying by S_0 yields:

$$S_T = S_0 e^{\left(\mu - \frac{1}{2}\sigma^2\right)T + \sigma Z_T} \tag{6.15}$$

Note that in the exponent of e, there is an extra factor of $-\sigma^2/2$ that would not appear in the solution to the classic problem $dS_t = \mu\, dt$. As we discussed in Chapter 4, the classic problem has the solution $S_T = S_0 e^{\mu T}$. Itô's formula shows that the effect of the Brownian motion term results in this increment to drift equal to $-\sigma^2/2$ in the solution for S_T. Also recall that the above solution was used to obtain the heuristic probabilistic derivation of the price of a European call option (Chapter 5), except that μ was replaced with the return r on a riskless bond. Next, we calculate the expected value of the logarithmic return:

$$E\left[\ln\frac{S_T}{S_0}\right] = E\left[\left(\mu - \frac{1}{2}\sigma^2\right)T + \sigma Z_T\right] = \left(\mu - \frac{1}{2}\sigma^2\right)T + \sigma E[Z_T] = \left(\mu - \frac{1}{2}\sigma^2\right)T = \alpha T \tag{6.16}$$

6.4.5.1 Returns and Price Relatives

The constant $\alpha = \mu - \sigma^2/2$ is known as the *mean logarithmic return* of the security per unit time. If we express the security price in the form:

$$S_T = S_0 e^{\alpha T + \sigma Z_T} \tag{6.17}$$

then $\ln(S_T/S_0) = \alpha T + \sigma Z_T$ is known as the *log of price relative* or the *logarithmic return*. It is also useful to calculate the variance of the logarithmic return:

$$\mathrm{Var}\left(\ln\frac{S_T}{S_0}\right) = E\left[(\ln\tfrac{S_T}{S_0} - \alpha T)^2\right] = E\left[\sigma^2 Z_T^2\right] = \sigma^2 E\left[Z_T^2\right] = \sigma^2 T \tag{6.18}$$

In comparison, let us examine the expected value and variance of the arithmetic return on S_T over time T. The *arithmetic return over time T* is defined as $r = S_T/S_0 - 1$. To derive expected arithmetic return over time T, we first observe that $E[\alpha T + \sigma Z_T] = \alpha T$ and $\mathrm{Var}[\alpha T + \sigma Z_T] = E[\sigma^2(Z_T)^2] = \sigma^2 T$. Since $\alpha T + \sigma Z_T \sim N(\alpha T, \sigma^2 T)$, by Eq. (2.26) in Section 2.6.4, we have:

$$E[r] = E\left[e^{\alpha T + \sigma Z_T} - 1\right] = e^{\alpha T + \frac{1}{2}\sigma^2 T} - 1 \tag{6.19}$$

We leave it as end-of-chapter Exercise 6.26 to verify that the variance of the arithmetic return is:

$$\mathrm{Var}[r] = \left(e^{\sigma^2 T} - 1\right)e^{2\alpha T + \sigma^2 T} \tag{6.20}$$

Because of the exponential nature of the arithmetic return in contrast to the linear nature of the logarithmic return, the expected value and variance of the arithmetic return grows exponentially with time T, while the expected value and variance of the logarithmic return grows linearly with time T.

6.4.5.2 Itô's Formula: Numerical Illustration

Suppose, for example, that the following Itô process describes the price path S_t of a given stock:

$$dS_t = 0.01 S_t\, dt + 0.015 S_t\, dZ_t$$

The differential for this process describes the infinitesimal change in the price of the stock. The expected rate of return per unit time is 0.01 and the standard deviation of the return per unit time is 0.015. The solution for this equation giving the actual price level at a point in time is given by:

$$S_T = S_0 e^{\left[\left(0.01-\frac{0.015^2}{2}\right)T+0.015Z_t\right]}$$

Suppose that one needed a return (or log of price relative) and variance over time T for the stock. The expected value and variance of the log of price relative are given by:

$$E\left[\ln\frac{S_T}{S_0}\right] = \alpha T = \mu T - \frac{1}{2}\sigma^2 T = \left(0.01 - \frac{0.015^2}{2}\right) \cdot 1 = 0.0098875$$

$$\text{Var}\left[\ln\frac{S_T}{S_0}\right] = \sigma^2 T = 0.015^2 \cdot 1 = 0.000225$$

The expected value and variance of the arithmetic return over a single period are given by:

$$E[r] = e^{\alpha T+\frac{1}{2}\sigma^2 T} - 1 = e^{\left(0.0098875\times 1+\frac{1}{2}0.015^2\times 1\right)} - 1 = 0.01005$$

and:

$$\text{Var}[r] = \left(e^{\sigma^2 T} - 1\right)e^{2\alpha T+\sigma^2 T} = \left(e^{0.015^2\times 1} - 1\right)e^{(2\times 0.0098875\times 1+0.015^2\times 1)} = 0.00022957$$

Figure 6.3 depicts a simulated geometric Wiener process over length of time 10,000, with $S0 = 100$, $\mu = 0.00001$, and $\sigma = 0.001$. This diagram was obtained from a simulation of a random process with 10,000 data points.

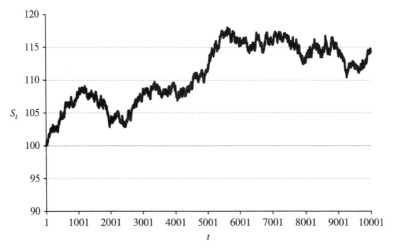

FIGURE 6.3   Geometric Brownian motion: $S_0 = 100$, $\mu = 0.00001$, $\sigma = 0.001$, $n = 10,000$.

6.4.6 Application: Forward Contracts

Let S_t be a stock price that follows a geometric Brownian motion:

$$dS_t = \mu S_t \, dt + \sigma S_t \, dZ_t$$

Let $F_{t,T}$ be the time t price of a forward contract that trades on that stock, settling at time T:

$$F_{t,T} = S_t e^{r(T-t)}$$

Recall Itô's formula:

$$dy = \left[a\frac{\partial y}{\partial x} + \frac{\partial y}{\partial t} + \frac{1}{2}b^2 \frac{\partial^2 y}{\partial x^2} \right] dt + b\frac{\partial y}{\partial x} dZ_t$$

Here, we choose $y(x,t) = x \, e^{r(T-t)}$ so that $F_{t,T} = y(S_t, t)$. In this case $a = \mu S_t$ and $b = \sigma S_t$. Note that $\frac{\partial y}{\partial x} = e^{r(T-t)}, \frac{\partial y}{\partial t} = -rS_t e^{r(T-t)}$, and $\frac{\partial^2 y}{\partial x^2} = 0$ evaluated at $(x,t) = (S_t,t)$. Applying Itô's formula gives:

$$dF_{t,T} = \left[\mu S_t e^{r(T-t)} - rS_t e^{r(T-t)} \right] dt + \sigma S_t e^{r(T-t)} dZ_t$$

Since $F_{t,T} = S_t \, e^{r(T-t)}$, the price of a forward contract on a single equity also follows a geometric Brownian motion process as follows:

$$dF_{t,T} = [\mu - r]F_{t,T} \, dt + \sigma F_{t,T} \, dZ_t$$

6.5 EXERCISES

6.1. Find the differential of the stochastic process $X_t = t^2 Z_t$.

6.2. Find the differential of the stochastic process $X_t = 0.06t + 100 + 0.02Z_t$.

6.3. Consider the following functions of a real-valued variable and of a stochastic process. Evaluate each and then contrast them:

   a. $\int_0^t 5 \, dX_s$ if $X_t = t^2$.

   b. $\int_0^t 5 \, dZ_s$ if Z_s is Brownian motion.

6.4. Suppose that X_t and Y_t are stochastic processes and C and D are real numbers. Verify that:

$$\int_a^b (C \, dX_t + D \, dY_t) = C(X_b - X_a) + D(Y_b - Y_a)$$

6.5. **a.** In Section 6.1.5, we claimed that for any integrable stochastic process f_t that is adapted to the same filtration $\{\mathscr{F}_t\}$ as a standard Brownian motion Z_t, we have $E\left[\int_s^T f_t \, dZ_t | \mathscr{F}_s \right] = 0$ whenever $s < T$. Verify this result using the tower property from Section 5.5.

   b. In our statement of the martingale theorem in Section 6.1.5, we claimed that if $dX_t = f_t \, dZ_t$, then X_t is a martingale, where f_t is an integrable stochastic process that is adapted to the same filtration $\{\mathscr{F}_t\}$ as a standard Brownian motion Z_t. Verify this result based on the result in part a.

6.6. Suppose that $S_{t-1} = 0$ and that potential outcomes S_t for a one-time period submartingale process are $+1$ with probability 0.6 and -1 with probability 0.4.

    a. Compute the expected value $E\left[S_t | S_{t-1}\right]$ of this process.

    b. Is the process in this problem a martingale?

    c. Compute the drift of the process.

6.7. In Section 6.2.2, we claimed that if the sample space Ω has at most a finite number of elements ω and $p(\omega) > 0$ for every $\omega \in \Omega$, then $\frac{d\mathbb{Q}}{d\mathbb{P}}(\omega) = \frac{q(\omega)}{p(\omega)}$ for every $\omega \in \Omega$. Verify this claim.

6.8. In Section 6.2.2, we claimed that if the probability measures $\mathbb{P}$ and $\mathbb{Q}$ are continuous with continuous density functions, say $p(x)$ and $q(x) > 0$, then the Radon–Nikodym derivative is $\frac{d\mathbb{Q}}{d\mathbb{P}}(x) = \frac{q(x)}{p(x)}$. Verify this claim.

6.9. Consider a probability measure $\mathbb{P}$ with $p_1 = 0.5$, $p_2 = 0.3$, and $p_3 = 0.2$. Consider a second measure $\mathbb{Q}$ with $q_1 = 0.5$, $q_2 = 0.5$, and $q_3 = 0$. Under what circumstances can values of the Radon–Nikodym derivative be calculated for a change of measure from $\mathbb{P}$ to $\mathbb{Q}$?

6.10. Consider a one-time period, two-potential outcome framework where there exists Company Q stock currently selling for $50 per share and a riskless $100 face value T-bill currently selling for $80. Suppose Company Q faces uncertainty, in that it will pay its owner either $30 or $70 in 1 year. Further assume that the physical probability that the stock price will drop is 0.2.

    a. List the risk-neutral probabilities for this payoff space.

    b. Compute values for the Radon–Nikodym derivative for this change of measure.

    c. Value call and put options on this stock, with exercise prices equal to $X = \$60$.

    d. Does put–call parity hold for this example?

6.11. Rollins Company stock currently sells for $12 per share and is expected to be worth either $10 or $16 in 1 year. The current riskless return rate is 0.125 and the physical probability that the stock price will increase is 0.75.

    a. List the risk-neutral probabilities for this payoff space.

    b. Compute values for the Radon–Nikodym derivative for this change of measure.

6.12. Suppose that we pay $S_0 = 0.7$ to purchase a security, with potential payoffs given as follows: $S_1 = 2$, $S_2 = 1$, and $S_3 = 0$ such that under physical probabilities, $E_\mathbb{P}[S] = 0.8$ and a variance equals $E_\mathbb{P}[S - E_\mathbb{P}[S]]^2 = 0.76$.

    a. Find the risk-neutral probabilities in measure $\mathbb{Q}$ based on the market price of the stock.

    b. Calculate Radon–Nikodym derivatives for this change of measure.

6.13. We know that if the changes in a security price over the different time intervals are independent of one another, then the Radon–Nikodym derivative $(d\mathbb{Q}_n/d\mathbb{P}_n)$ of the price changes from time 0 to n equals:

$$\frac{d\mathbb{Q}_n}{d\mathbb{P}_n} = \prod_{i=1}^{n} \frac{dQ_{i-1,i}}{dP_{i-1,i}}$$

Now, let the security price follow a binomial process.

    a. What is the Radon–Nikodym derivative for this binomial process at time n assuming a particular path with k increases to time n?

    b. What is the Radon–Nikodym derivative for this binomial process at time n assuming a particular event with k increases to time n?

6.14. Suppose that we wish to transform a normal density function for a probability measure $\mathbb{P}$ with mean $\mu = 0.1$ to an equivalent martingale $\mathbb{Q}$ with a mean of zero. The density functions for both measures will have standard deviations equal to 1.
   a. What is the Radon–Nikodym derivative for this transformation at $x = 0$?
   b. What is the density after this transformation at $x = 0$?
   c. What is the cumulative density after this transformation at $x = 0$?
   d. Recalculate parts a, b, and c assuming that both measures have standard deviations equal to 0.5.

6.15. Verify with appropriate algebra the following claim made near the end of Section 6.2.5:

$$f_{\mathbb{Q}}(x) = \frac{1}{\sigma\sqrt{2\pi}} e^{-\frac{1}{2}\left(\frac{x-(\mu-\lambda)}{\sigma}\right)^2} = f_{\mathbb{P}}(x)\xi(x)$$

where:

$$\xi(x) = e^{\left(-\frac{\lambda x}{\sigma^2} + \frac{\mu\lambda}{\sigma^2} - \frac{\lambda^2}{2\sigma^2}\right)}$$

6.16. Suppose that we have a random variable $X \sim N(\mu, 1)$.
   a. Write the density function $f_{\mathbb{P}}(x)$ under probability measure $\mathbb{P}$.
   b. Suppose that we change the probability measure of X to measure $\mathbb{Q}$, which is an equivalent martingale measure to $\mathbb{P}$. Write the density function $f_{\mathbb{Q}}(x)$ under probability measure $\mathbb{Q}$.
   c. Working from the Radon–Nikodym derivative for this change of measure that is:

$$\frac{d\mathbb{Q}}{d\mathbb{P}} = \frac{f_{\mathbb{Q}}(x)}{f_{\mathbb{P}}(x)}$$

use parts a and b to this question to write and simplify an expression for this Radon–Nikodym derivative.

6.17. In Section 6.4.2, we made the claim that the variance of the random variable $Z^2 c$ where $Z \sim N(0,1)$ and $c > 0$ equals $2c^2$.
   a. First, show that the variance of the random variable Z^2 where $Z \sim N(0,1)$ equals 2.
   b. Use the result of part a to show that the variance of the random variable $Z^2 c$ where $Z \sim N(0,1)$ and $c > 0$ equals $2c^2$.

6.18. Suppose a stock price S_t evolves according to $dS_t = tdt + \sigma dZ_t$ where Z_t is standard Brownian motion, $\sigma > 0$ is a constant, and S_0 is the initial price of the stock. Derive an equation to find the price of the stock at time T.

6.19. Suppose that the logarithmic return α on a stock follows a Wiener process (Brownian motion process) with drift, an expected value over 1 year equal to 5% and a variance equal to 0.09; that is, $\alpha \sim N(\mu,\sigma^2)$ with $\mu = 0.05$ and $\sigma^2 = 0.09$. Find the expected value and variance of the arithmetic return for the stock.

6.20. Suppose that the log of price relatives (instantaneous returns) for a stock follows a Wiener process with drift, an expected value equal to 6% per annum and a variance equal to 0.08 per annum. What are the expected value and variance of the arithmetic return r for the stock over 1 year?

6.21. Suppose that the following Itô process describes the price of a given stock after t weeks:

$$dS_t = 0.001 S_t \, dt + 0.02 S_t \, dZ_t$$

   a. What is the solution to this stochastic differential equation?
   b. Suppose that there are 52 periods in a year. What are the expected value and variance of the log of price relative for this stock over a 52-week period?

6.22. Suppose that X_t is a geometric Brownian motion process with $dX_t = \mu X_t\, dt + \sigma X_t\, dZ_t$. Consider a function Y_t of X_t with $Y_t = (X_t)^n$. Derive an expression for dY_t.

6.23. Suppose that a particular derivative instrument with price S_t satisfies the differential $dS_t = \mu(M - S_t)\, dt + \sigma(M - S_t)dZ_t$ and initial value S_0 with $0 < S_0 < M$. Find the solution for S_t that is valid as long as $0 < S_t < M$.

6.24. Suppose that $dX_t = \mu_t\, dt$ and $dY_t = \nu_t\, dt + \sigma_t\, dZ_t$ with μ_t, ν_t, and σ_t stochastic processes. In Section 6.1.1, we proved the special product rule in differential form. Provide a different proof of the same result based on Itô's lemma discussed in Section 6.4; that is, verify that $d(X_tY_t) = X_t dY_t + Y_t dX_t$. Hint: Use the identity: $X_tY_t = \frac{1}{2}(X_t + Y_t)^2 - \frac{1}{2}X_t^2 - \frac{1}{2}Y_t^2$.

6.25. Suppose that a stock price S_t satisfies the model $dS_t = \mu\, dt + Z_t\, dZ_t$ where Z_t is standard Brownian motion, $\mu > 0$ is a constant, and S_0 is the initial price of the stock. Derive an expression to price the stock at time t.

6.26. Derive the variance formula given by Eq. (6.20) for the arithmetic return of a security that follows a geometric Brownian motion of the form $S_T = S_0\, e^{\alpha T + \sigma Z_T}$.

NOTES

1. For the remainder of this chapter, we will normally use a slightly different notation for events with multiple outcomes. Subscripts of a quantity (example $f_t(\omega)$) will denote time (t). Symbols in the parentheses will refer to outcomes (ω), replacing subscripts as in $f_{t;\omega}$. For example, the probability at time 1 that a security would experience an upjump (u) at time 2 followed by a downjump (d) at time 3 will now be denoted by $p_{1,3}(ud)$. Since many of the probabilities are computed at time 0 for certain outcome ω by time t, we will write $p_t(\omega)$ rather than $p_{0,t}(\omega)$ for purposes of brevity. Thus, $p_2(du)$ denotes the probability at time 0 that a security experiences an downjump followed by an upjump at times 1 and 2.

2. The Radon–Nikodym derivative $\xi_1(\omega)$ with respect to $\mathscr{F}_1$ actually is defined for every element $\omega \in \Omega$. For example, a mathematical purist might write: $\xi_1(uu) = 1.25, \xi_1(ud) = 1.25, \xi_1(du) = 0.625, = 1.25, \xi_1(dd) = 0.625$. $\xi_1(uu) = \xi_1(ud)$ because uu and ud are indistinguishable at time 1; similarly $\xi_1(du) = \xi_1(dd)$. Recalculating $\mathbb{Q}(\phi)$ for the event $\phi = \{uu, ud\}$ more precisely would have: $\mathbb{Q}(\phi) = \xi_1(uu)p(uu) + \xi_1(ud)p(ud) = 1.25(0.6)(0.6) + 1.25(0.6)(0.4) = 0.75$. But we can support the usual expressions because the Radon–Nikodym derivative $\xi_1(\omega)$ remained constant when we only changed an outcome ω after time 1 ($\xi_1(uu) = \xi_1(ud) = 1.25$ and $\xi_1(du) = \xi_1(dd) = 0.625$), and we determined its values by only looking up to time 1. This justifies the simpler notation $\xi_1(u) = 1.25$ and $\xi_1(d) = 0.625$. This scenario is not unique to this example. Regardless of the probability spaces and the filtrations, every calculation of $\mathbb{Q}((\phi))$ for $\phi \in \mathscr{F}_t$ using the Radon–Nikodym derivative ξ_t can be reduced to only studying the probability spaces $\mathbb{P}$ and $\mathbb{Q}$ for distinguishable events up to time t.

3. Here, Riemann sums are calculated based on 20 rectangles from -3 to $+3$ where each has a width of 0.3. The height of each rectangle (except the first) is calculated based on the mean of each of its two vertical sides. The Riemann sum is an approximation for the cumulative density function; we cannot antidifferentiate the normal density function.

4. Review the beginning of Section 6.2.5 and work through Exercise 6.15 at the end of this chapter to verify this result.

References

Cameron, R.H., Martin, W.T., 1944. Transformation of wiener integrals under translations. Ann. of Math. 45, 386–396.
Girsanov, I., 1960. On transforming a certain class of stochastic processes by absolutely continuous substitution of measures. Theory Probab. Appl. 5 (3), 285–301.

7

Derivatives Pricing and Applications of Stochastic Calculus

7.1 OPTION PRICING INTRODUCTION

In Chapters 3, 4, and 5 of this book, we discussed several models for pricing options. The pure security methodology in Section 3.3 is particularly instructive because it highlights the importance of arbitrage in option pricing. That is, physical probabilities are not needed when capital markets are complete and where the Cameron−Martin−Girsanov theorem applies. We continued this discussion into Sections 3.4.2 and 5.2, where we discussed option pricing in binomial environments. The binomial model can be extended into a Black−Scholes continuous time environment with application of the central limit theorem from Chapter 2. In Section 5.4.1, paying only cursory attention to arbitrage and the importance of risk-neutral pricing, we heuristically derived the Black−Scholes model. In Chapter 6, we developed the mathematical methodology to rigorously derive from arbitrage-free perspectives the Black−Scholes model and many other models. We will apply this more rigorous methodology in this chapter.

We will first employ a probabilistic approach in this chapter to deriving the Black−Scholes model, in which we calculate a conditional expected discounted payoff given a filtration (information set) under a risk-neutral probability measure (Section 7.2.2.1). Then, we will employ the analytical approach of Black and Scholes themselves, where the option value is the solution to the appropriate boundary value problem. We will discuss estimating unobservable underlying security volatility, an essential input of the model, the model's sensitivities to its inputs, and then a variety of applications and extensions of the model.

7.2 SELF-FINANCING PORTFOLIOS AND DERIVATIVES PRICING

7.2.1 Introducing the Self-Financing Replicating Portfolio

In this section, we are going to introduce the self-financing replicating portfolio and how to use it to price derivatives. We will assume a market consisting of a stock and a riskless bond that will

P.M. Knopf & J.L. Teall: Risk Neutral Pricing and Financial Mathematics: A Primer.
DOI: http://dx.doi.org/10.1016/B978-0-12-801534-6.00007-2

be used to create the portfolio. For convenience, we will illustrate the concepts using a European call as our derivative, but the method will be the same for many types of derivatives. Let c_t denote the price of the call at time t. Consider a portfolio $(\gamma_{s,t}, \gamma_{b,t})$ combining $\gamma_{s,t}$ shares of stock at a per share price S_t at time t and $\gamma_{b,t}$ units of the riskless bond with a price per unit of B_t at time t. The value of the portfolio at time t is then $V_t = \gamma_{s,t}S_t + \gamma_{b,t}B_t$. Assume we purchase the call at time 0 and T is the call expiry and bond maturity date. On the expiry date, the value of the European call will be known: $c_T = \max(0, S_T - X)$, where X is the exercise price of the call.

We say that the portfolio $(\gamma_{s,t}, \gamma_{b,t})$ is a *self-financing replicating portfolio* for the call if and only if the following two properties are satisfied:

I. $dV_t = \gamma_{s,t}\, dS_t + \gamma_{b,t}\, dB_t$

and:

II. $c_T = \gamma_{s,T}S_T + \gamma_{b,T}B_T$

Property I is called the self-financing property. The interpretation of property I is that during every infinitesimal time interval $(t, t + dt)$ the change in the value of the portfolio is entirely due to the changes in the prices of the stock and bond. There is no net new investment in the portfolio. In other words, any infinitesimal purchases or sales of the stock and bond ($d\gamma_{s,t}$ and $d\gamma_{b,t}$) will offset each other so that the change in the value of the portfolio is entirely due to the changes in the value of the securities themselves. Property II states that the expiry date T value of the replicating portfolio will equal the price of the call at expiry date T. We will also assume the absence of arbitrage opportunities.

We will create a self-financing portfolio, consisting of a single T-period call, the underlying stock, and a T-period riskless bond. We will see that this arbitrage portfolio will have zero net investment from time 0 to time T, and it will be shown that its value is always equal to 0. This portfolio $(-1, \gamma_{s,t}, \gamma_{b,t})$ will consist of a short position in a single call along with positions in $\gamma_{s,t}$ shares of underlying stock and $\gamma_{b,t}$ units of the riskless bond. We shall see soon that, consistent with what we saw in Chapters 3 and 5, our short position in the call will be offset by a long position in the underlying stock ($\gamma_{s,t}$ will be positive), and a short position in $\gamma_{b,t}$ units of the riskless bond ($\gamma_{b,t}$ will be negative). Denote the values at time t of each unit of the call, stock, and bond by c_t (whose value is not yet known), S_t, and B_t, respectively. If P_t is the value of the portfolio, then:

$$P_t = -c_t + \gamma_{s,t}S_t + \gamma_{b,t}B_t \qquad (7.1)$$

Assume that we purchase the call at time 0 and T is the expiry date. On the expiry date, the value of the call will be known and the bond will mature. For example, a European call will be worth $c_T = \max(0, S_T - X)$, where X is the exercise price of the call. For our portfolio, the number of shares $\gamma_{s,T}$ of stock and the number of units of the bond $\gamma_{b,T}$ will be chosen so that the portfolio's expiry date value will be P_T:

$$P_T = -c_T + \gamma_{s,T}S_T + \gamma_{b,T}B_T = 0 \qquad (7.2)$$

We will also determine the values of $\gamma_{s,t}$ and $\gamma_{b,t}$ at every moment t so that during every infinitesimal time interval $(t, t + dt)$ there is zero net new investment in the portfolio. In other words, any infinitesimal purchases or sales of the stock and bond ($d\gamma_{s,t}$ and $d\gamma_{b,t}$) will offset each other so that the change in the value of the portfolio is entirely due to the changes in the value of the securities themselves. This can be expressed mathematically as:

$$dP_t = -dc_t + \gamma_{s,t}\,dS_t + \gamma_{b,t}\,dB_t \tag{7.3}$$

for any time t, $0 \le t \le T$. In a no-arbitrage market, conditions suggested by Eqs. (7.2) and (7.3) guarantee that the owner of the portfolio requires no capital at all to construct and maintain the portfolio. Such a portfolio is called a self-financing portfolio.[1] In an arbitrage-free market, this implies that the value of the portfolio P_t equals zero for all time t, $0 \le t \le T$. The reason is quite simple. Suppose at some point in time t the value of the portfolio was negative: $P_t = -c_t + \gamma_{s,t}S_t + \gamma_{b,t}B_t < 0$. By Eq. (7.2), a long position in this portfolio would certainly produce a riskless profit by time T, since the portfolio is constructed to have zero value at time T. Since the portfolio's current price is negative, its purchase would produce a positive time t cash flow. Since the portfolio satisfies the condition given by Eq. (7.3), we would not need to use any capital to maintain the portfolio from time t until the option expiry date T. Assuming an interest rate r, a guaranteed profit of $(\gamma_{s,t}S_t + \gamma_{b,t}B_t + c_t)e^{r(T-t)}$ with interest is locked in by the expiry date with no positive net investment. Of course, this would violate our no-arbitrage principle. Similarly, if the value of the portfolio were positive at any time t, we would short the portfolio to produce a time T arbitrage profit by simply reversing the positions taken by buying the portfolio. Once again, this would result in a guaranteed profit with no net expenditure on our part, violating the no-arbitrage requirement. Thus, we are able to conclude that:

$$P_t = -c_t + \gamma_{s,t}S_t + \gamma_{b,t}B_t = 0 \tag{7.4}$$

for all time t, $0 \le t \le T$. We can solve this equation for the price of the call at time t:

$$c_t = \gamma_{s,t}S_t + \gamma_{b,t}B_t \tag{7.5}$$

There is still a lot of work to do to obtain a numerical solution for the value of the call, but we have reduced the problem to ensuring that the call must satisfy Eqs. (7.2) and (7.3).

As we have seen, we are viewing the portfolio $(\gamma_{s,t}, \gamma_{b,t})$ of stock and bonds as replicating the derivative. Property I portrays the self-financing property of the replicating portfolio. In an arbitrage-free market, we showed that Properties I and II imply that the derivative value equals the value of the portfolio for all time t (Eq. (7.5)). For this reason, a portfolio $(\gamma_{s,t}, \gamma_{b,t})$ that satisfies Properties I and II is called a *self-financing replicating portfolio* for the derivative with value c_t. We have shown that if we can create a self-financing replicating portfolio for the derivative, then its value must equal the arbitrage-free price for the derivative. We are going to use two different approaches to derive the price of the derivative. The first approach will use martingales and the second approach will derive a partial differential equation known as the Black–Scholes equation that the derivative must satisfy.

7.2.2 The Martingale Approach to Valuing Derivatives

Following Harrison and Kreps (1979), Harrison and Pliska (1981) and Baxter and Rennie (1996), we use the expected arbitrage-free present value of the call payoff function to price a European call option. Finding the replicating self-financing portfolio will entail a several-step procedure making use of various results covered in Chapters 5 and 6 concerning martingales. First, we express prices in present value. We assume that the bond market has a constant riskless return rate $r \neq 0$, so that the time t price of a bond is $B_t = B_0 e^{rt}$. We note that a bond that has unit value at some future time T will have e^{-rT} as its present value.

We prefer to write the discounted value of the bond as e^{-rT} rather than 1 because we will also discount the other securities by e^{-rT} to eliminate their upward time value drift. This is important. Recall our efforts in Chapter 6 to implement changes of measure to eliminate drift. We will implement such changes of measure here as well. For example, we will shortly express the present value of a portfolio's future payoff at time T as $e^{-rT} V_T$, where V_t denotes the value of the portfolio at time t. Recall that we did something very similar in Section 3.2.1 to construct an arbitrage-free pricing model for bonds. Now, recall our stock pricing model $dS_t = \mu S_t\, dt + \sigma S_t\, dZ_t$. Create a variable Y_t to represent the present value of the underlying stock, but using the riskless return (or risk-neutral rate) as the discount rate:

$$Y_t = e^{-rt} S_t \tag{7.6}$$

Since S_t is a random variable, Y_t is a random variable. Since $d(e^{-rt}) = -r e^{-rt} dt$, showing that e^{-rt} has no volatility, by the special product rule for stochastic differentials in Section 6.1.1, we have:

$$\begin{aligned} dY_t &= e^{-rt}\, dS_t - r e^{-rt} S_t\, dt \\ &= e^{-rt} S_t(\mu\, dt + \sigma\, dZ_t) - r e^{-rt} S_t dt \\ &= e^{-rt} S_t[(\mu - r)dt + \sigma\, dZ_t] \\ &= Y_t[(\mu - r)\, dt + \sigma\, dZ_t] \end{aligned} \tag{7.7}$$

since $dS_t = S_t(\mu\, dt + \sigma\, dZ_t)$.

7.2.2.1 Implementing the Change of Measure

We will invoke the Cameron–Martin–Girsanov theorem, implementing a change of measure whereby $dY_t = \sigma Y_t\, d\hat{Z}_t$, where $\hat{Z}_t$ is standard Brownian motion with respect to the probability measure $\mathbb{Q}$. This is easily done. By Eq. (7.6) and (7.7), express:

$$\frac{dY_t}{\sigma Y_t} = \frac{e^{-rt} S_t[(\mu - r)dt + \sigma\, dZ_t]}{\sigma e^{-rt} S_t} = \frac{\mu - r}{\sigma}\, dt + dZ_t \tag{7.8}$$

By the Cameron–Martin–Girsanov theorem, there exists a change of measure from probability measure $\mathbb{P}$ to an equivalent probability measure $\mathbb{Q}$ so that $\frac{dY_t}{\sigma Y_t} = d\hat{Z}_t$, or equivalently, $dY_t = \sigma Y_t\, d\hat{Z}_t$. $\hat{Z}_t$ is standard Brownian motion with respect to the probability measure $\mathbb{Q}$. By the martingale theorem in Section 6.1.5, Y_t becomes a martingale with respect to the measure $\mathbb{Q}$. This is an important step. Recall that in earlier chapters, we priced a new security in terms of securities that were martingales with respect to an equivalent probability. Similarly, we will price a derivative in terms of the bond and stock now that we have expressed them as martingales.

The present value of the portfolio with future payoff V_T is simply $e^{-rT} V_T$. In order to price this portfolio for any time $t < T$, we will create a second martingale E_t with value $e^{-rT} V_T$ at time T. A simple way of doing this is to define:

$$E_t = E_{\mathbb{Q}}[e^{-rT} V_T | \mathscr{F}_t] \tag{7.9}$$

Intuitively, E_t would be the expected risk-neutral present value of the portfolio given the set of information as of time t. That is, E_t, an artificial construct, is the time 0 value of the portfolio expressed in terms of riskless bonds as the numeraire and with the information available at time t.

Observe that $E_T = E_{\mathbb{Q}}[e^{-rT}V_T|\mathcal{F}_T] = e^{-rT}V_T$ since $e^{-rT}V_T$ becomes known at time T. Furthermore, the process E_t is a martingale because:

$$E_{\mathbb{Q}}[E_{\mathbb{Q}}[e^{-rT}V_T|\mathcal{F}_t]|\mathcal{F}_u] = E_{\mathbb{Q}}[e^{-rT}V_T|\mathcal{F}_u]$$

for all $u < t$. (*Note*: We are using the subscript u here rather than s for time to avoid confusion with s as a subscript to refer to the underlying stock.) The equation above follows the tower property in Section 5.5. Since $\sigma Y_t \neq 0$, by the martingale representation theorem in Section 6.3.3, there is a stochastic process $\gamma_{s,t}$ such that $dE_t = \gamma_{s,t}\,dY_t$. This means that there is some number of units $\gamma_{s,t}$ of the stock of underlying shares that will replicate the dynamics of the portfolio over an instantaneous period with the riskless bond as the numeraire. Now, the desired self-financing portfolio can be constructed. Let the portfolio consist of $\gamma_{s,t}$ shares of stock and $\gamma_{b,t} = E_t - \gamma_{s,t}Y_t$ units of the bond. Then the value of the portfolio at time t is:

$$V_t = \gamma_{s,t}S_t + \gamma_{b,t}B_t = \gamma_{s,t}S_t + (E_t - \gamma_{s,t}Y_t)B_t = \gamma_{s,t}(S_t - Y_tB_t) + E_tB_t = E_tB_t \qquad (7.10)$$

because $Y_tB_t = Y_te^{rt} = S_t$ from Eq. (7.6) and our definition of B_t.

7.2.2.2 Demonstrating the Self-Financing Characteristic

We now demonstrate that the portfolio is self-financing. Since $dB_t = re^{rt}\,dt$ has no volatility, then by the special product rule for stochastic differentials in Section 6.1.1, and the fact that $S_t = Y_tB_t$, we have:

$$dS_t = Y_t\,dB_t + B_t\,dY_t$$

and:

$$dV_t = B_t\,dE_t + E_t\,dB_t = B_t\gamma_{s,t}\,dY_t + E_t\,dB_t \qquad (7.11)$$

where we used Eq. (7.10) and the result $dE_t = \gamma_{s,t}\,dY_t$ in Eq. (7.11) above. Since $E_t = (B_t)^{-1}V_t = \gamma_{s,t}Y_t + \gamma_{b,t}$, by Eqs. (7.6) and (7.10), then by Eq. (7.11) we have:

$$dV_t = \gamma_{s,t}B_t\,dY_t + (\gamma_{s,t}Y_t + \gamma_{b,t})dB_t = \gamma_{s,t}(B_t\,dY_t + Y_t\,dB_t) + \gamma_{b,t}\,dB_t = \gamma_{s,t}\,dS_t + \gamma_{b,t}\,dB_t$$

This shows that the portfolio replicating the derivative is self-financing, so V_t defines an arbitrage-free price of the derivative itself at any time t:

$$V_t = E_tB_t = e^{rt}E_{\mathbb{Q}}[e^{-rT}V_T|\mathcal{F}_t] = e^{-r(T-t)}E_{\mathbb{Q}}[V_T|\mathcal{F}_t] \qquad (7.12)$$

It is important to appreciate how far-ranging the implications are for the risk-neutral pricing methodology introduced here. V_T can represent the payoff function for many derivative instruments on the underlying security whose price is generated by $dS_t = \mu S_t\,dt + \sigma S_t\,dZ_t$. All we need is to find the self-financing replicating portfolio. We will apply our result above in the next section in a derivation of the Black–Scholes model for plain vanilla options and then later to certain other instruments.

7.2.3 Pricing a European Call Option and the Black–Scholes Formula

Now, we will be specific in our choice of derivatives; we will value a European call. First, we specify its payoff function at time T when it might be exercised. With exercise price X, the value of the option at expiry is $c_T = \text{Max}[S_T - X, 0]$. Based on Eq. (7.12) in the preceding section, the value in risk-neutral probability space of the call at time 0 is:

$$c_0 = \mathrm{e}^{-rT} E_{\mathbb{Q}}[V_T|\mathcal{F}_0] = \mathrm{e}^{-rT} E_{\mathbb{Q}}[c_T|\mathcal{F}_0] = \mathrm{e}^{-rT} E_{\mathbb{Q}}[\text{Max}(S_T - X, 0)]$$

We showed earlier that:

$$d(\mathrm{e}^{-rt}S_t) = \sigma \mathrm{e}^{-rt}S_t \, \mathrm{d}\hat{Z}_t$$

By the special product rule for stochastic differentials in Section 6.1.1, we have:

$$\mathrm{e}^{-rt} \, \mathrm{d}S_t - r\mathrm{e}^{-rt}S_t \, \mathrm{d}t = \sigma \mathrm{e}^{-rt}S_t \, \mathrm{d}\hat{Z}_t \tag{7.13}$$

Solving for $\mathrm{d}S_t$ gives:

$$\mathrm{d}S_t = rS_t \, \mathrm{d}t + \sigma S_t \, \mathrm{d}\hat{Z}_t$$

As we showed in Section 6.4.5, with the choice of $\mu = r$, the solution is:

$$S_T = S_0 \mathrm{e}^{\sigma \hat{Z}_T + (r - \frac{1}{2}\sigma^2)T}$$

Recall that we used a slight variation of this equation in Section 5.4.1 to value a call. Since $\hat{Z}_T \sim N(0, T)$, then $\hat{Z}_T = Z\sqrt{T}$ with $Z \sim N(0,1)$. Thus, the solutions S_T here and in Section 5.4.1 have identical probability distributions. Since the value of the call at time 0 is the same discounted expected value: $c_0 = e^{-rt} E_{\mathbb{Q}}[c_T] = e^{-rt} E_{\mathbb{Q}}[\text{Max}(S_T - X, 0)]$, this leads to the same result that we obtained in Section 5.4.1: $c_0 = -S_0 N(d_1) - X e^{-rt} N(d_2)$.

In Chapter 6 and earlier in this chapter, we have set forth the mathematics and pricing framework to value options and other derivatives. Our derivation of the Black–Scholes model in Section 5.4, while not incorrect, failed to show why we needed to change our probability measure to risk-neutral space and focus on the riskless return rather than the expected risk-adjusted return of the underlying stock. Now, it should be clear that our analysis and derivation need to focus on the expectation $E_{\mathbb{Q}}[\text{Max}(S_T - X, 0)]$, even though its computation is exactly as we demonstrated in Section 5.4.1. The key to this and the previous chapter is that we have demonstrated the critically important feature of this expectation in risk-neutral space: this expectation provides an arbitrage-free pricing of the call option or other derivative for which we can find a self-financing replicating portfolio. Since the portfolio is self-financing, it requires zero net investment to maintain the arbitrage position, and since it is replicating, its initial value must be the same as the call's.

7.3 THE BLACK–SCHOLES MODEL

7.3.1 Black–Scholes Assumptions

Black and Scholes (1973) set forth a rather strict set of assumptions for their model (the same assumptions apply to the martingale model in the previous section), but they are generally consistent with what we have been using throughout this book (save the distinction between discrete and continuous time). Most importantly, the model assumes that underlying share prices follow a geometric Brownian motion process and that investors can create hedged self-financing portfolios comprising calls, underlying shares, and riskless bonds. The set of assumptions on which the Black–Scholes model and its derivation are based are as follows:

1. There exist no restrictions on short sales of stock or writing of call options.
2. There are no transactions costs.
3. There exists continuous trading of stocks and options.
4. There exists a known constant riskless borrowing and lending rate r.
5. The underlying stock will pay no dividends or make other distributions during the life of the option.
6. The option can be exercised only on its expiration date; that is, it is a European option.
7. Shares of stock and option contracts are infinitely divisible.
8. Underlying stock prices follow a geometric Brownian motion process: $dS_t = \mu S_t \, dt + \sigma S_t \, dZ_t$.

It is important to note that, because of the Cox–Ross risk neutrality argument (change of measure to risk-neutral space), the following are not required as model inputs:

1. The expected or required return or risk premium on the stock or option and
2. Investor attitudes toward risk.

7.3.2 Deriving Black–Scholes

In Section 7.2.2, we priced a derivative using a self-financing replicating portfolio and martingales. Again in this section, we will price a derivative instrument using a self-financing replicating portfolio. However, rather than follow the methodology in Section 5.4, we will derive and solve the appropriate partial differential equation (known as the Black–Scholes differential equation) that the derivative instrument must satisfy. This partial differential equation can then be solved by standard techniques (as shown on the companion website) to obtain its price. In particular, we will derive the Black–Scholes option pricing model for a call, assuming that our all standard Black–Scholes assumptions hold. As we learned in Section 7.2, to price a call c_t at any time $t < T$, it is sufficient to construct a self-financing replicating portfolio $(\gamma_{s,t}, \gamma_{b,t})$ of stocks and bonds whose value equals the value of the call at time T. If we denote the value of the portfolio by

$$c_t = \gamma_{s,t} S_t + \gamma_{b,t} B_t \tag{7.14}$$

with c_T equal to the value of the call at expiry, and we require that the self-financing property is satisfied:

$$dc_t = \gamma_{s,t} \, dS_t + \gamma_{b,t} \, dB_t \tag{7.15}$$

It is convenient to rewrite the self-financing property (7.15) in the form:

$$\gamma_{b,t} \, dB_t = dc_t - \gamma_{s,t} \, dS_t \tag{7.16}$$

Since $B_t = e^{rt}$, $dB_t = re^{rt} \, dt = rB_t \, dt$, so Eq. (7.16) takes the form:

$$r\gamma_{b,t} B_t \, dt = dc_t - \gamma_{s,t} \, dS_t \tag{7.17}$$

Solving for $\gamma_{b,t} B_t = c_t - \gamma_{s,t} S_t$ in Eq. (7.14), and substituting this result into Eq. (7.17), Eq. (7.17) becomes:

$$r(c_t - \gamma_{s,t} S_t)\, dt = dc_t - \gamma_{s,t}\, dS_t \tag{7.18}$$

Recall that we assume that the stock price follows a geometric Brownian motion process:

$$dS_t = \mu S_t\, dt + \sigma S_t\, dZ_t$$

Now, we will invoke Itô's lemma from Chapter 6 to express the differential dc_t (Eq. (7.15)):

$$dc_t = \left(\frac{\partial c}{\partial t} + \mu S_t \frac{\partial c}{\partial S} + \frac{1}{2}\sigma^2 S_t^2 \frac{\partial^2 c}{\partial S^2} \right) dt + \sigma S_t \frac{\partial c}{\partial S}\, dZ_t \tag{7.19}$$

which we will use to replace dc_t in Eq. (7.18) and then simplify:

$$
\begin{aligned}
r(c_t - \gamma_{s,t} S_t)\, dt &= \left(\frac{\partial c}{\partial t} + \mu S_t \frac{\partial c}{\partial S} + \frac{1}{2}\sigma^2 S_t^2 \frac{\partial^2 c}{\partial S^2} \right) dt + \sigma S_t \frac{\partial c}{\partial S}\, dZ_t - \gamma_{s,t}\, dS_t \\[2mm]
&= \left(\frac{\partial c}{\partial t} + \mu S_t \frac{\partial c}{\partial S} + \frac{1}{2}\sigma^2 S_t^2 \frac{\partial^2 c}{\partial S^2} \right) dt + \sigma S_t \frac{\partial c}{\partial S}\, dZ_t - \gamma_{s,t}(\mu S_t\, dt + \sigma S_t\, dZ_t) \\[2mm]
&= \left(\frac{\partial c}{\partial t} + \frac{1}{2}\sigma^2 S_t^2 \frac{\partial^2 c}{\partial S^2} \right) dt + \mu S_t \left(\frac{\partial c}{\partial S} - \gamma_{s,t} \right) dt + \sigma S_t \left(\frac{\partial c}{\partial S} - \gamma_{s,t} \right) dZ_t
\end{aligned}
\tag{7.20}
$$

The instantaneous expected rate of return μ may reflect individual investor forecasts and risk preferences, and might even vary from investor to investor. It is unobservable and not very useful for most option valuation calculations. We encountered this situation before with physical probabilities. We dealt with this issue earlier by using Radon–Nikodym derivatives to transform physical probabilities into risk-neutral probabilities, which are consistent with eliminating arbitrage opportunities. If we choose $\gamma_{s,t} = \frac{\partial c}{\partial S}$, Eq. (7.20) will simplify to:

$$r\left(c_t - S_t \frac{\partial c}{\partial S} \right) dt = \left(\frac{\partial c}{\partial t} + \frac{1}{2}\sigma^2 S_t^2 \frac{\partial^2 c}{\partial S^2} \right) dt \tag{7.21}$$

This step is key. Note that this Eq. (7.21) makes no reference to μ, the instantaneous expected rate of return for the stock. This suggests our pricing model will not depend on a risk premium or investor risk preferences, as in fact we will see when we obtain the solution. Only the riskless return r is used. One can view the amount $c_t - S_t \frac{\partial c}{\partial S}$ as the value of a portfolio consisting of buying one call and shorting $\frac{\partial c}{\partial S}$ shares of the stock. This portfolio is known as a delta hedged portfolio. It is perfectly hedged to guarantee the riskless rate of return r. Equation (7.21) is divided by dt to become:

$$r\left(c_t - S_t \frac{\partial c}{\partial S} \right) = \frac{\partial c}{\partial t} + \frac{1}{2}\sigma^2 S_t^2 \frac{\partial^2 c}{\partial S^2}$$

Since the stock price is known at time t, the portfolio value c becomes a function of the real-valued variables S and t:

$$r\left(c - S\frac{\partial c}{\partial S}\right) = \frac{\partial c}{\partial t} + \frac{1}{2}\sigma^2 S^2 \frac{\partial^2 c}{\partial S^2} \tag{7.22}$$

or:

$$\frac{\partial c}{\partial t} = rc - rS\frac{\partial c}{\partial S} - \frac{1}{2}\sigma^2 S^2 \frac{\partial^2 c}{\partial S^2} \tag{7.23}$$

This partial differential equation is called the *Black–Scholes differential equation*. For the boundary value of the differential equation, we specify the value of the call at the time of expiration T. Of course, the value of the call $c(S,T)$ at expiry time T is a known function of the stock price S: c $(S,T) = \max(S - X,0)$. The Black–Scholes equation and the boundary condition guarantee that there is an arbitrage-free price for the call. For the interested reader, a derivation of the solution is on the text website. Of course, consistent with our derivations in Sections 7.2 and 5.4, the value of the call at time zero is:

$$c_0 = S_0 N(d_1) - X e^{-rT} N(d_2) \tag{7.24}$$

with:

$$d_1 = \frac{\ln\left(\frac{S_0}{X}\right) + \left(r + \frac{1}{2}\sigma^2\right)T}{\sigma\sqrt{T}} \quad \text{and} \quad d_2 = d_1 - \sigma\sqrt{T} \tag{7.25}$$

where $N(d^*)$ is the cumulative normal density function for (d^*).[2] From a computational perspective, one would first solve for d_1 and d_2 before c_0.

Again, it is important to note how far-ranging the implications are for the Black–Scholes differential equation. While we have focused on call valuation thus far, the equation can be used to value any derivative instrument on the underlying security whose price is generated by $dS_t = \mu S_t\, dt + \sigma S_t\, dZ_t$. We will discuss additional applications of the Black–Scholes differential equation to additional derivative instruments later in this chapter.

7.3.3 The Black–Scholes Model: A Simple Numerical Illustration

Consider the following example of a Black–Scholes model application where an investor can purchase a 6-month call option for $7.00 on a stock that is currently selling for $75. The exercise price of the call is $80 and the current riskless rate of return is 10% per annum. The variance of annual returns on the underlying stock is 16%. At its current price of $7.00, does this option represent a good investment? We will note the model inputs in symbolic form:

$$T = 0.5 \quad r = 0.10$$
$$X = 80 \quad \sigma^2 = 0.16$$
$$\sigma = 0.4 \quad S_0 = 75$$

Our first step is to find d_1 and d_2:

$$d_1 = \frac{\ln\left(\frac{75}{80}\right) + \left(0.10 + \frac{1}{2} \times 0.16\right) \times 0.5}{0.4\sqrt{0.5}} = \frac{\ln(0.9375) + 0.09}{0.2828} = 0.09$$

$$d_2 = 0.09 - 0.4\sqrt{0.5} = 0.09 - 0.2828 = -0.1928$$

Next, by either using a z-table (see end-of-book Appendix A.2) or using the polynomial estimating function above, we find cumulative normal density functions for d_1 and d_2:

$$N(d_1) = N(0.09) = 0.535864$$
$$N(d_2) = N(-0.1928) = 0.423549$$

Finally, we use $N(d_1)$ and $N(d_2)$ to value the call:

$$c_0 = 75 \times 0.536 - \frac{80}{e^{0.10 \times 0.5}} \times 0.424 = 7.958$$

Since the 7.958 value of the call exceeds its 7.00 market price, the call represents a good purchase. Next, we use the put–call parity relation to find the value of the put as follows:

$$p_0 = c_0 + Xe^{-rT} - S_0$$

$$p_0 = 7.958 + 80(0.9512) - 75 = 9.054$$

Sections, 7.4 and 7.5 will focus on issues related to the application of Black–Scholes to option pricing and to variance estimates. Afterward, we will begin to relax Black–Scholes assumptions to obtain additional applications of Black–Scholes.

7.4 IMPLIED VOLATILITY

Four of the five inputs required to implement the Black–Scholes model are easily observed. The option exercise price and term to expiry are defined by the option contract. The riskless return and underlying stock price are based on current quotes. Only the underlying stock return volatility during the life of the option cannot be observed. Instead, we often employ a traditional sample estimating procedure for return variance:

$$\sigma^2 = \text{Var}[r_t] = \text{Var}[\ln S_t - \ln S_{t-1}]$$

The difficulty with this procedure is that it requires that we assume that underlying security return variance is stable over time; more specifically, that future variances equal or can be estimated from historical variances. An alternative procedure first suggested by Latane and Rendleman (1976) is based on market prices of options that might be used to imply variance estimates. For example, the Black–Scholes option pricing model might provide an excellent means to estimate underlying stock variances if the market prices of one or more relevant calls and puts are known. Essentially, this procedure determines market estimates for underlying stock variance based on known market prices for options on the underlying securities. When we use this procedure, we assume that the market reveals its estimate of volatility through the market prices of options.

Consider the following example pertaining to a 6-month call currently trading for $8.20 and its underlying stock currently trading for $75:

$$T = 0.5 \quad r = 0.10 \quad c_0 = 8.20$$
$$X = 80 \quad S_0 = 75$$

If investors use the Black–Scholes option pricing model to value calls, the following should be expected:

$$8.20 = 75N(d_1) - 80e^{-0.1 \times 0.5}N(d_2)$$

$$d_1 = \frac{\ln\left(\dfrac{75}{80}\right) + (0.1 + 0.5 \times \sigma^2) \cdot 0.5}{\sigma\sqrt{0.5}}$$

$$d_2 = d_1 - \sigma\sqrt{0.5}$$

As we will demonstrate shortly, we find that this system of equations holds when $\sigma = 0.41147$. Thus, the market prices this call as though it expects that the standard deviation of anticipated returns for the underlying stock is 0.41147.

Unfortunately, the system of equations required to obtain an implied variance has no closed form solution. That is, we will be unable to solve this equation set explicitly for standard deviation; we must search, iterate, and substitute for a solution. One can substitute trial values for σ until one is found that solves the system. A significant amount of time can be saved by using one of several well-known numerical search procedures such as the method of bisection or the Newton–Raphson method.

7.4.1 The Method of Bisection

We seek to solve the above system of equations for σ. This is equivalent to solving for the root of:

$$f(\sigma*) = 0 = 75 \times N(d_1) - 80 \times e^{-0.1 \times 0.5} \times N(d_2) - 8.20$$

based on equations above for d_1 and d_2. There exists no closed form solution for σ. Thus, we will use the method of bisection to search for a solution. We first arbitrarily select endpoints for our range of guesses, such as $b_1 = 0.2$ and $a_1 = 0.5$ so that $f(b_1) = -4.46788 < 0$ and $f(a_1) = 1.860465 > 0$. Since these endpoints result in $f(\sigma)$ with opposite signs, our first iteration will be in the middle: $\sigma_1 = 0.5(0.2 + 0.5) = 0.35$. We find that this estimate for σ results in a value of -1.29619 for $f(\sigma)$. Since this $f(\sigma)$ is negative, we know that $\sigma*$ is in the segment $b_2 = 0.35$ and $a_2 = 0.5$. Moving to Row $n = 2$, we repeat the iteration process, finding after 16 iterations that $\sigma* = 0.41146$. Table 7.1 details the process of iteration.

7.4.2 The Newton–Raphson Method

The Newton–Raphson method can also be used to more efficiently iterate for an implied volatility. We will solve for the implied standard deviation in our illustration using the Newton–Raphson

TABLE 7.1   Using the Bisection Method to Estimate Implied Volatility

Equation for f: $S_0 N(d_1) - X e^{-rt} N(d_2) - c_0$

$a_1 = 0.5$			$b_1 = 0.2$		$\sigma_1 = 0.35$		$r = 0.1$		$S_0 = 75$		$X = 80$
$c_0 = 8.2$			$T = 0.5$								

	σ_n	$d_1(\sigma_n)$	$d_2(\sigma_n)$	$N(d_1)$	$N(d_2)$	$N(d_1)$	$N(d_2)$	$f(\sigma_n)$
$f(a_1) = 1.860465$	0.5	0.1356555	−0.2178978	0.553953	0.41375	0.553953	0.413754	1.860465
$f(b_1) = -4.46788$	0.2	−0.0320922	−0.1735135	0.487199	0.431122	0.487199	0.431124	−4.46788

n	a_n	b_n	σ_n	$d_1(\sigma_n)$	$d_2(\sigma_n)$	$N(d_1)$	$N(d_2)$	$N(d_1)$	$N(d_2)$	$f(\sigma_n)$
1	0.5	0.2	0.35	0.06499919	−0.1824882	0.525913	0.427597	0.525913	0.4276	−1.29619
2	0.5	0.35	0.425	0.10188237	−0.198638	0.540575	0.42127	0.540575	0.421273	0.284948
3	0.425	0.35	0.3875	0.08394239	−0.1900615	0.533449	0.424628	0.533449	0.424630	−0.50501
4	0.425	0.3875	0.40625	0.09302042	−0.1942417	0.537056	0.42299	0.537056	0.422993	−0.10987
5	0.425	0.40625	0.41562	0.09747658	−0.1964147	0.538826	0.42214	0.538826	0.422143	0.087583
6	0.41562	0.40625	0.41093	0.09525501	−0.1953217	0.537944	0.42256	0.537944	0.422571	−0.01113
7	0.41562	0.41093	0.41328	0.09636739	−0.1958666	0.538386	0.42235	0.538386	0.422357	0.038229
8	0.41328	0.41093	0.41210	0.09581161	−0.1955937	0.538165	0.42246	0.538165	0.422464	0.01355
9	0.41210	0.41093	0.41152	0.09553341	−0.1954576	0.538054	0.42251	0.538054	0.422517	0.00121
10	0.41152	0.41093	0.41123	0.09539424	−0.1953896	0.537999	0.42254	0.537999	0.422544	−0.00496
11	0.41152	0.41123	0.41137	0.09546383	−0.1954236	0.538027	0.42252	0.538027	0.422531	−0.00188
12	0.41152	0.41137	0.41145	0.09549862	−0.1954406	0.538041	0.42252	0.538041	0.422524	−0.00033
13	0.41152	0.41145	0.41148	0.09551602	−0.1954491	0.538048	0.42251	0.538048	0.422521	0.000438
14	0.41148	0.41145	0.41146	0.09550732	−0.1954449	0.538044	0.42251	0.538044	0.422522	0.000053
15	0.41146	0.41145	0.41145	0.09550297	−0.1954427	0.538042	0.42252	0.538042	0.422523	−0.00014
16	0.41146	0.41145	0.41146	0.09550514	−0.1954438	0.538043	0.42252	0.538043	0.422523	−0.00004

method to find the root of the equation: $f(\sigma) = S_0 N(d_1) - X e^{-rT} N(d_2) - c_0$. The Newton–Raphson method estimates the root of an equation $f(\sigma) = 0$ by using the formula:

$$\sigma_n = \sigma_{n-1} - \frac{f(\sigma_{n-1})}{f'(\sigma_{n-1})}$$

One starts with some initial rough estimate σ_0 for the root, then repeatedly applying the formula above, iterating until the desired accuracy is obtained.

For our example, we arbitrarily choose an initial trial solution of $\sigma = \sigma_0 = 0.6$. First, we need the derivative of $f(\sigma)$ with respect to the underlying stock return standard deviation σ.[3] Since we

TABLE 7.2 The Newton–Raphson Method and Implied Volatilities

Equation for f: $S_0 N(d_1) - Xe^{-rT} N(d_2)$

$r = 0.1$		$S_0 = 75$	$X = 80$	$c_0 = 8.20$	$T = 0.5$		$\sigma_0 = 0.6$
n	σ_n	$f'(\sigma_n)$	$d_1(\sigma_n)$	$d_2(\sigma_n)$	$N(d_1)$	$N(d_2)$	$f(\sigma_n)$
1	0.60000	20.82509	0.177864	−0.2464	0.5705853	0.402686	3.95012
2	0.41032	21.06194	0.094961	−0.19518	0.5378271	0.422626	−0.02415
3	0.41147	21.06085	0.095506	−0.19544	0.5380436	0.422522	0.00000

are treating c_0 as a given constant, differentiating this function is equivalent to differentiating the call function $c = S_0 N(d_1) - Xe^{-rT} N(d_2)$ with respect to σ. We leave as a homework exercise that:[2]

$$\frac{\partial c}{\partial \sigma} = \frac{S_0 \sqrt{T}}{\sqrt{2\pi}} e^{\frac{-d_1^2}{2}} > 0, \quad \text{Vega } \nu$$

with an arbitrarily selected initial trial solution of $\sigma_0 = 0.6$. We see from Table 7.2 that this standard deviation results in a value of $f(\sigma_0) = 3.95012$, implying a variance estimate that is too high. Substituting 0.6 for σ_0, we find that $f'(\sigma_0) = 20.82509$. Thus, our second trial value for σ is determined by $\sigma_1 \approx \sigma_0 - (f(\sigma_0) \div f'(\sigma_0)) = 0.6 - (3.95012 \div 20.82509) = 0.41032$. This process continues until we converge to a solution of approximately 0.41147. Note that the rate of convergence in this example is much faster when using the Newton–Raphson method than when using the method of bisection.

7.4.3 Smiles, Smirks, and Aggregating Procedures

We see that with an appropriate iteration methodology, solving for implied volatility is not a difficult matter. However, another difficulty arising with implied variance estimates results from the fact that there will typically be more than one option trading on the same stock. However, what if the short- and long-term uncertainty of a stock differs? Or, what if options with different strike prices disagree on the same underlying stock volatility? Both of these inconsistencies regularly occur. This latter effect in which implied volatilities vary with respect to option exercise prices is sometimes known as the smile or smirk effect. See Figure 7.1, which depicts a smile effect for five options on a single stock, and Figure 7.2, which depicts a smirk effect for a series of options on a second stock.[4] Each option's market price will imply its own underlying stock variance, and these variances are likely to differ. How might we use this conflicting information to generate the most reliable variance estimate? Each of our implied variance estimates is likely to provide some information, yet has the potential for having measured with error.

Settling these implied volatility problems is largely an empirical or practitioner issue.[5] There is an empirical literature that focuses on how to use the information implied by two or more options to determine the appropriate volatility forecast. For example, we can preserve much of the information from each of our estimates and eliminate some of our estimating error if we use for our own implied volatility a value based on an average of all of our estimates. However,

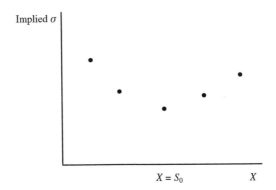

FIGURE 7.1   The volatility smile: implied volatility given S_0. Options with high or low exercise prices relative to the current underlying stock price produce higher implied volatilities.

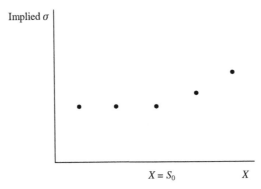

FIGURE 7.2   The volatility smirk: implied volatility given S_0. Options with high exercise prices relative to the current underlying stock price produce higher implied volatilities.

because volatility might be expected to vary over time, one should average only those variances implied by options with the same terms to expiration. Consider the following two methodologies for averaging implied standard deviation estimates:

1. Simple average: Here, the final standard deviation estimate is simply the mean of the standard deviations implied by the market prices of the calls.
2. Average based on price sensitivities to σ: Calls that are more sensitive to σ as indicated by $\partial c/\partial\sigma$ are more likely to imply a correct standard deviation estimate. Suppose we have n calls on a stock, and each call price c_j has an implied stock standard deviation σ_j. Each call price will have a sensitivity (vega; discussed in the next section) to its implied underlying stock standard deviation $\partial c_j/\partial\sigma_j$. The n option sensitivities can be summed, and a weighted average standard deviation estimate for the underlying stock based on its n options can be computed where the weight w_i associated with the implied standard deviation estimate for call option i is:

$$w_i = \frac{\frac{\partial c_{0,i}}{\partial \sigma_i}}{\sum_{j=1}^{n} \frac{\partial c_{0,j}}{\partial \sigma_j}} = \frac{S_{0,i}\sqrt{T}\frac{e^{\frac{-d_{1}^2}{2}}}{\sqrt{2\pi}}}{\sum_{j=1}^{n} S_{0,j}\sqrt{T}\frac{e^{\frac{-d_{1,j}^2}{2}}}{\sqrt{2\pi}}}$$

Thus, the final standard deviation estimate for a given stock based on all of the implied standard deviations from each of the call prices is:

$$\sigma = \sum_{i=1}^{n} w_i \sigma_i$$

While these aggregating procedures may provide useful information for volatility estimates, it might be useful to make entirely different sets of assumptions for option analysis, For example, while the Black–Scholes model assumes constant volatility, in Chapter 8, we will briefly discuss the assumption of stochastic volatility. More generally, we can allow volatility to be any function of time. In Chapter 2, we discussed jumps and Poisson processes. Black–Scholes assumes Brownian motion; this assumption can be relaxed to also allow for stock price jumps. We can also allow for time-varying interest rates, as we will discuss in Chapter 8. All of these assumption adjustments have the potential to explain or reduce the so-called smile or smirk effects.

7.5 THE GREEKS

Sources of sensitivity of the Black–Scholes model (see Black and Scholes, 1972) to each of its five inputs are known as the *Greeks*. The first of these Greeks is the option *delta*, or $N(d_1)$, which is the sensitivity of the option value to changes in the stock price:[6]

$$\frac{\partial c}{\partial S} = N(d_1) > 0 \quad \text{Delta} \; \Delta$$

This sensitivity means that the option's value will change by approximately $N(d_1)$ for each unit of change in the underlying stock price. This delta can be interpreted as the number of shares to short for every purchased call in order to maintain a portfolio that is hedged to changes in the underlying stock price. A delta-neutral portfolio means that the portfolio of options and underlying stock has a weighted average delta equal to zero so that its value is invariant with respect to the underlying stock price.

However, as the underlying stock's price changes over time, and, as the option's time to maturity diminishes, this hedge ratio will change. That is, because this delta is based on a partial derivative with respect to the share price, it holds exactly only for an infinitesimal change in the share price; it holds only approximately for finite changes in the share price. This delta only approximates the change in the call value resulting from a change in the share price because any change in the price of the underlying shares would lead to a change in the delta itself:

$$\frac{\partial^2 c}{\partial S^2} = \frac{\partial \Delta}{\partial S} = \frac{\partial N(d_1)}{\partial S_0} = \frac{e^{\left(-\frac{d_1^2}{2}\right)}}{\sqrt{2\pi T} S_0 \sigma} > 0 \quad \text{Gamma} \; \Gamma$$

This change in delta resulting from a change in the share price is known as *gamma*. However, again, this change in delta resulting from a finite share price change is only approximate.

Since each call and put option has a date of expiration, calls and puts are said to amortize over time. As the date of expiration draws nearer (T gets smaller), the value of the European call (and put) option might be expected to decline as indicated by a positive *theta*:[7]

$$\frac{\partial c}{\partial T} = rXe^{-rT}N(d_2) + \frac{S_0\sigma}{2\sqrt{2\pi T}}e^{-\frac{d_1^2}{2}} > 0 \quad \text{Theta } \theta$$

By convention, traders often refer to theta as a negative number since option values tend to decline as we move forward in time (T becomes smaller). In addition, many traders seek to maintain portfolios that are simultaneously neutral with respect to delta, gamma, and theta.

Vega, which actually is not a Greek letter, measures the sensitivity of the option price to the underlying stock's standard deviation of returns (vega is sometimes known as either *kappa* or *zeta*). One might expect the call option price to be directly related to the underlying stock's standard deviation:

$$\frac{\partial c}{\partial \sigma} = \frac{S_0\sqrt{T}}{\sqrt{2\pi}}e^{\left(-\frac{d_1^2}{2}\right)} > 0 \quad \text{Vega } \nu$$

Although the Black–Scholes model assumes that the underlying stock volatility is constant over time, in reality, as we discussed in the previous section, volatility can and does shift. Vega provides an estimate for the impact of a volatility shift on a particular option's value.

A trader should expect that the value of the call would be directly related to the riskless return rate and inversely related to the call exercise price:

$$\frac{\partial c}{\partial e^r} = TXe^{-rT}N(d_2) > 0 \quad \text{Rho } \rho$$

$$\frac{\partial c}{\partial X} = -e^{-rT}N(d_2) < 0$$

The option rho can be very useful in economies with very high or volatile interest rates, though most traders of "plain vanilla" options (standard options to buy or sell without complications) do not concern themselves much with it under typical interest rate regimes. Similarly, most traders of "plain vanilla" options ignore call value sensitivities to exercise prices.

ILLUSTRATION: GREEKS CALCULATIONS FOR CALLS

Here, we will calculate the Greeks for the call illustration from the previous Section 7.4 with the following parameters:

$$T = 0.5 \quad r = 0.10 \quad \sigma = 0.41147 \quad X = 80 \quad S_0 = 75 \quad c_0 = 8.20$$

$$\frac{\partial c}{\partial S} = \Delta = N(d_1) = N(0.0955) = 0.538$$

$$\frac{\partial^2 c}{\partial S^2} = \Gamma = \frac{\partial \Delta}{\partial S} = \frac{\partial N(d_1)}{\partial S} = \frac{e^{\left(\frac{-d_1^2}{2}\right)}}{S_0\sigma\sqrt{2\pi T}} = \frac{e^{\left(\frac{-0.0955^2}{2}\right)}}{75 \times 0.41147 \times \sqrt{\pi}} = 0.0182$$

$$\frac{\partial c}{\partial T} = \theta = rXe^{-rT}N(d_2) + S\frac{\sigma}{\sqrt{T}}\frac{e^{\left(\frac{-d_1^2}{2}\right)}}{2\sqrt{2\pi}} = 0.1 \times 80e^{-0.05}N(-0.1954) + 75\frac{0.41147\,e^{\left(\frac{-0.0955^2}{2}\right)}}{\sqrt{0.5}}\frac{1}{2\sqrt{2\pi}} = 11.882$$

$$\frac{\partial c}{\partial \sigma} = \nu = \frac{S_0\sqrt{T}}{\sqrt{2\pi}}e^{-\frac{d_1^2}{2}} = \frac{80 \times \sqrt{0.5}}{\sqrt{2\pi}}e^{-\frac{0.0955^2}{2}} = 21.06$$

$$\frac{\partial c}{\partial e^r} = \rho = TXe^{-rT}N(d_2) = 0.5 \times 80e^{-0.05}N(-0.1954) = 16.077$$

Note that traders sometimes change the sign for theta, actually using the derivative $\partial c/\partial t$ where T is replaced by $(T-t)$ with t approaching 0. More generally, a single unit increase in the relevant parameter changes the Black–Scholes estimated call value by an amount approximately equal to the associated "Greek."

ILLUSTRATION: GREEKS CALCULATIONS FOR PUTS

If put–call parity holds along with Black–Scholes assumptions given above, the Black–Scholes put value from our example in Section 7.5.1 is computed as follows:

$$p_0 = -S_0N(-d_1) + \frac{X}{e^{rT}}N(-d_2) = -75N(-0.0955) + \frac{80}{e^{0.05}}N(0.1954) = 9.298$$

In our numerical example above, the put would be worth 9.298 if its exercise terms are identical to those of the call. Put sensitivities formulas and calculations for our example are as follows:[8]

$$\frac{\partial p}{\partial S} = \Delta_p = N(d_1) - 1 = 0.538 - 1 = -0.462$$

$$\frac{\partial^2 p}{\partial S^2} = \Gamma_p = \frac{\partial \Delta_p}{\partial S} = \frac{\partial (N(d_1) - 1)}{\partial S} = \frac{e^{\left(\frac{-d_1^2}{2}\right)}}{S_0\sigma\sqrt{2\pi T}} = \frac{e^{\left(\frac{-0.0955^2}{2}\right)}}{75 \times 0.41447\sqrt{2\pi \times 0.5}} = 0.0182$$

$$\frac{\partial p}{\partial T} = \theta_p = rXe^{-rT}N(-d_2) - \frac{S_0\sigma}{2\sqrt{2\pi T}}e^{-\frac{d_1^2}{2}} = 0.1 \times 80e^{-0.05}N(0.1954) - \frac{75 \times 0.41147}{2\sqrt{\pi}}e^{-\frac{0.0955^2}{2}} = -4.27$$

$$\frac{\partial p}{\partial \sigma} = \nu_p = \frac{S_0\sqrt{T}}{\sqrt{2\pi}}e^{-\frac{d_1^2}{2}} = \frac{75 \times \sqrt{0.5}}{\sqrt{2\pi}}e^{-\frac{0.0955^2}{2}} = 21.06$$

$$\frac{\partial p}{\partial e^r} = \rho_p = -TXe^{-rT}N(-d_2) = -0.5 \times 80e^{-0.05}N(0.1954) = -21.97$$

Again, a single unit increase in the relevant parameter changes the Black–Scholes estimated put value by an amount approximately equal to the associated "Greek."

7.6 COMPOUND OPTIONS

A compound option (see Roll, 1977; Geske, 1979a,b) is simply an option on an option. While such options are useful in their own right as traded securities, their analyses have a variety of other applications as well. For example, limited liability equity gives shareholders the right to either pay off creditors in full and take control of the firm's assets or abandon their claim on the firm's assets. This limited liability corporate feature essentially results in equity being a call option on the firm's assets. Hence, a call option on limited liability equity is a compound option, a call on a call. American options have compound option features. For example, an American call might be exercised early just prior to the underlying stock's ex-dividend date. Alternatively, the owner of the call has the right to hold the call to its expiry, creating a call on a call, where the compound call is exercised by not exercising the American option just prior to the ex-dividend date. There are also a variety of real option applications (options on corporate securities and projects) for compound option pricing models.

A compound option has two exercise prices and two expiration dates, the first of each (X_1, T_1) that applies to the right to buy or sell the underlying option and a second of each (X_2, T_2) that applies to buying or selling the security that underlies the underlying option with $T_1 < T_2$. We will assume that the underlying option is a European option with exercise price X_2 and expiration date T_2. Denote the values of the underlying call and put at time t by $c_{u,t}$ and $p_{u,t}$, respectively. Assume that there is a European option on this underlying option with exercise price X_1 on expiration date T_1. Compound options come in four varieties with the following exercise date T_1 payoff functions, where $c_{u,T1}$ and $p_{u,T1}$ are exercise date underlying call and put values:

1. Call on call: $\text{Max}[c_{u,T1} - X_1, 0]$
2. Put on call: $\text{Max}[X_1 - c_{u,T1}, 0]$
3. Call on put: $\text{Max}[p_{u,T1} - X_1, 0]$
4. Put on put: $\text{Max}[X_1 - p_{u,T1}, 0]$

First, consider a call on a call. The compound call gives its owner the right to purchase an underlying call with value $c_{u,T1}$ at time T_1 for price X_1. This compound call is exercised at time T_1 only if the underlying call at time T_1 is sufficiently valuable, which will be the case if the stock underlying the underlying call is sufficiently valuable. That is, the call on the underlying call option is exercised at time T_1 only if the stock price S_{T1} exceeds some critical underlying stock price S_{T1}^* at the time. If the stock price S_{T1} exceeds this critical value S_{T1}^*, the underlying call value $c_{u,T1}$ will exceed the exercise price X_1 of the compound call. Thus, we first calculate this critical value S_{T1}^*, which solves:

$$c_{u,T1}(S_{T1}^*, T_1; r, T_2, X_2, \sigma) = X_1$$

To calculate the left side of this equation, we would use the Black–Scholes model equation (7.24), replacing S_0 with the variable S_{T1}^*, X with the number X_2, and T with the number $T_2 - T_1$. The numbers r and σ would remain the same. We will demonstrate shortly how to iterate to find the value of S_{T1}^* at time T_1 that would lead to optimal exercise of the compound call given the first exercise price X_1. There are several numerical and iterative techniques for solving this equality for S_{T1}^* (e.g., the methods of bisection and Newton–Raphson), but simple substitution and iteration will work.

7.6.1 Estimating Exercise Probabilities

Following the reasoning from Section 5.4.1, and assuming that the underlying stock price process is as we discussed in 5.4.1, the probability that the stock price at time T_1 will exceed the critical value $S_{T_1}^*$ is $N(d_2)$, calculated as follows:

$$d_1 = \frac{\left[\ln\left(\frac{S_0}{S_{T_1}^*}\right) + \left(r + \frac{1}{2}\sigma^2\right)T_1 \right]}{\sigma\sqrt{T_1}}$$

$$d_2 = d_1 - \sigma\sqrt{T_1}$$

where $N(d_2)$ is the cumulative normal density function for d_2.

Exercising the underlying call requires that the stock price S_{T_1} at time T_1 exceeds its critical value $S_{T_1}^*$ *and* that the stock price S_{T_2} exceeds the underlying option exercise price X_2 at time T_2. In other words, for both the compound call and its underlying call to be exercised, the value of the underlying stock must exceed $S_{T_1}^*$ at time T_1 (i.e., the value of the underlying call must exceed X_1) and the value of the underlying stock must exceed X_2 at time T_2. The probability of both occurring equals:

$$M\left(d_2, y_2; \sqrt{\frac{T_1}{T_2}}\right)$$

where:

$$y_1 = \frac{\left[\ln\left(\frac{S_0}{X_2}\right) + \left(r + \frac{1}{2}\sigma^2\right)T_2 \right]}{\sigma\sqrt{T_2}}$$

$$y_2 = y_1 - \sigma\sqrt{T_2}$$

and that the correlation coefficient between returns during our overlapping exercise periods equals $\rho = \sqrt{T_1/T_2}$.[9] The bivariate (multinomial) normal distribution function $M(*,**;\rho)$, discussed in Chapter 2, provides the joint probability distribution function that is used to calculate the probability that our two random variables, S_{T_1} and S_{T_2} exceed the time T_1 critical value $S_{T_1}^*$ and the time T_2 exercise price X_2, respectively.

7.6.2 Valuing the Compound Call

Relying on the logic presented in Section 5.4 (see Whaley, 1981, 1982), we can derive the value of a European compound call with exercise price X_1 and expiration date T_1, on an underlying call with exercise price X_2 and expiration date T_2 on a share of stock with current price S_0 with the following result:

$$c_{0,call} = S_0 M\left(d_1, y_1; \sqrt{\frac{T_1}{T_2}}\right) - X_2 e^{-rT_2} M\left(d_2, y_2; \sqrt{\frac{T_1}{T_2}}\right) - X_1 e^{-rT_1} N(d_2)$$

where:

$$d_1 = \frac{\left[\ln\left(\dfrac{S_0}{S_{T1}^*}\right) + \left(r + \dfrac{1}{2}\sigma^2\right)T_1\right]}{\sigma\sqrt{T_1}}$$

$$d_2 = d_1 - \sigma\sqrt{T_1}$$

$$y_1 = \frac{\left[\ln\left(\dfrac{S_0}{X_2}\right) + \left(r + \dfrac{1}{2}\sigma^2\right)T_2\right]}{\sigma\sqrt{T_2}}$$

$$y_2 = y_1 - \sigma\sqrt{T_2}$$

ILLUSTRATION: VALUING THE COMPOUND CALL

Consider a compound call with exercise price $X_1 = 3$, expiring in 3 months ($T_1 = 0.25$) on an equity call, with exercise price $X_2 = 45$, expiring in 6 months ($T_2 = 0.5$). The stock currently sells for $S_0 = 50$ and has a return volatility equal to $\sigma = 0.4$. Thus, the compound call gives its owner the right to buy an underlying call in 3 months for $3, which will confer the right to buy the underlying stock in 6 months for $45. The riskless return rate equals $r = 0.03$. What is the value of this compound call?

First, we calculate the critical value S_{T1}^* for underlying option exercise at time T_1. A search process reveals that this critical value equals 43.58191:

$$c_{u,T1} = S_{T1}^* \times N\left(\frac{\left[\ln\left(\dfrac{S_{T1}^*}{45}\right) + \left(0.03 + \dfrac{1}{2} \times 0.16\right) \times (0.5 - 0.25)\right]}{0.4 \times \sqrt{0.5 - 0.25}}\right)$$

$$- \frac{45}{e^{0.03 \times (0.5-0.25)}} N\left(\frac{\left[\ln\left(\dfrac{S_{T1}^*}{45}\right) + \left(0.03 - \dfrac{1}{2} \times 0.16\right) \times (0.5 - 0.25)\right]}{0.4 \times \sqrt{0.5 - 0.25}}\right) = X_1 = 3; \quad S_{T1}^* = 43.58191$$

We obtained this critical value by a process of substitution and iteration, assuming that the underlying call would be purchased for $X_1 = \$3$ and would confer the right to buy the underlying stock 3 months later for $X_2 = \$45$. The minimum acceptable value justifying the underlying call's exercise is $c_{u,T1} = X_1 = \$3$, which means that the underlying stock price must be at least $S_{T1}^* = 43.58191$ at time T_1 to exercise the right to purchase the underlying call.

Note that the correlation coefficient between our two random variables is $\rho = \sqrt{0.25/0.5} =$ 0.7071. We will use the reasoning from Section 5.4 and the bivariate normal distribution along with the formulas in Section 7.6.2 to calculate the compound call. But first, we make a few intermediate calculations:

$$d_1 = \frac{\left[\ln\left(\frac{50}{43.58191}\right) + \left(0.03 + \frac{1}{2} \times 0.16\right) \times 0.25\right]}{0.4\sqrt{0.25}} = 0.8244$$

$$d_2 = 0.8244 - 0.4 \times \sqrt{0.25} = 0.6244$$

$$y_1 = \frac{\left[\ln\left(\frac{50}{45}\right) + \left(0.03 + \frac{1}{2} \times 0.16\right) \times 0.5\right]}{0.4 \times \sqrt{0.5}} = 0.567$$

$$y_2 = 0.567 - 0.4 \times \sqrt{0.5} = 0.2841$$

Finally, we calculate the time 0 value of the compound call as follows:[10]

$$c_{0,\text{call}} = 50 \times M(0.8244, 0.567; 0.7071)$$
$$- 45e^{-(0.03 \times 0.5)} M(0.6244, 0.2841; 0.7071)$$
$$- 3e^{-(0.03 \times 0.25)} N(0.6244)$$
$$= 50 \times 0.6529 - 45 \times e^{-(0.03 \times 0.5)} \times 0.5507$$
$$- 3e^{-(0.03 \times 0.25)} \times 0.7338 = 6.0465$$

Thus, we find that the value of this compound call equals 6.0465. The bivariate probabilities were calculated using a spreadsheet-based multinomial cumulative distribution calculator (see Drezner, 1978; Hull, 2010, 2011, for details). There is an $N(d_2) = 0.7338$ probability that the compound call will be exercised to purchase the underlying call and a 0.6118 probability that the compound call *and* its underlying call will be exercised.

7.6.3 Put–Call Parity for Compound Options

Here, using the notation from above, we set forth pricing formulas for other compound options, including a call on a call (repeated from above), a call on a put, a put on a call, and a put on a put:

$$c_{0,\text{call}} = S_0 M\left(d_1, y_1; \sqrt{\frac{T_1}{T_2}}\right) - X_2 e^{-rT_2} M\left(d_2, y_2; \sqrt{\frac{T_1}{T_2}}\right) - X_1 e^{-rT_1} N(d_2)$$

$$p_{0,\text{call}} = X_2 e^{-rT_2} M\left(-d_2, y_2; -\sqrt{\frac{T_1}{T_2}}\right) - S_0 M\left(-d_1, y_1; -\sqrt{\frac{T_1}{T_2}}\right) + X_1 e^{-rT_1} N(-d_2)$$

$$c_{0,\text{put}} = X_2 e^{-rT_2} M\left(-d_2 - y_2; \sqrt{\frac{T_1}{T_2}}\right) - S_0 M\left(-d_2 - y_2; \sqrt{\frac{T_1}{T_2}}\right) - X_1 e^{-rT_1} N(-d_2)$$

$$p_{0,\text{put}} = S_0 M\left(d_1 - y_1; -\sqrt{\frac{T_1}{T_2}}\right) - X_2 e^{-rT_2} M\left(d_2 - y_2; -\sqrt{\frac{T_1}{T_2}}\right) + X_1 e^{-rT_1} N(d_2)$$

Let $c_{t,\text{call}}$, $p_{t,\text{call}}$, $c_{t,\text{put}}$, and $p_{t,\text{put}}$ denote the values of a call on a call, a put on a call, a call on a put, and a put on a put at time t, respectively. To obtain the formulas from our compound call pricing formula, first recall our compound option payoff functions from Section 7.6.1. We can demonstrate that the above pricing functions for our other compound options hold by first working with the exercise date T_1 payoff function for a portfolio, say, with a long position of one call on a call and a short position of one put on a call. At expiration date T_1, we have for our portfolio:

$$c_{T1,\text{call}} - p_{T1,\text{call}} = \text{Max}[c_{u,T1} - X_1, 0] - \text{Max}[X_1 - c_{u,T1}, 0] = c_{u,T1} - X_1$$

For any earlier time $t < T_1$, the value of the portfolio with time T_1 cash flow $c_{u,T1} - X_1$ must equal $c_{u,t} - X_1 e^{-r(T1-t)}$ since we must discount the cash associated with the exercise money X. In particular, at time 0, the value of the portfolio is:

$$c_{0,\text{call}} - p_{0,\text{call}} = c_{u,0} - X_1 e^{-rT_1}$$

This formula is the put–call parity relation for the options on the underlying call. Note that $c_{u,0}$ is the present value of the underlying call and $X_1 e^{-rT1}$ is the present value of the exercise money associated with the compound call.

Next, we demonstrate that the two relevant time 0 pricing formulas for the call on the call and the put on the call above are consistent with this portfolio value. We will demonstrate that our call on call value minus our put on call value yields the same value as our put–call formula for the options on the underlying call above:

$$S_0 M\left(d_1, y_1; \sqrt{\frac{T_1}{T_2}}\right) - X_2 e^{-rT_2} M\left(d_2, y_2; \sqrt{\frac{T_1}{T_2}}\right) - X_1 e^{-rT_1} N(d_2)$$

$$- \left[X_2 e^{-rT_2} M\left(-d_2, y_2; -\sqrt{\frac{T_1}{T_2}}\right) - S_0 M\left(-d_1, y_1; -\sqrt{\frac{T_1}{T_2}}\right) + X_1 e^{-rT_1} N(-d_2)\right] = c_{u,0} - X_1 e^{-rT_1}$$

Because the normal curve is symmetric, implying that the total area under the normal curve is $N(d_2) + N(-d_2) = 1$, this simplifies as follows:

$$S_0 M\left(d_1, y_1; \sqrt{\frac{T_1}{T_2}}\right) - X_2 e^{-rT_2} M\left(d_2, y_2; \sqrt{\frac{T_1}{T_2}}\right) - X_1 e^{-rT_1}$$

$$- \left[X_2 e^{-rT_2} M\left(-d_2, y_2; -\sqrt{\frac{T_1}{T_2}}\right) - S_0 M\left(-d_1, y_1; -\sqrt{\frac{T_1}{T_2}}\right)\right] = c_{u,0} - X_1 e^{-rT_1}$$

TABLE 7.3 Put–Call Parity Relation for Compound Options

Present value:

$c_{0,call} - p_{0,call} = c_{0,put} - p_{0,put} + S_0$

Time T_1 value:

$\text{Max}[c_{u,T1} - X_1, 0] - \text{Max}[X_1 - c_{u,T1}, 0] = \text{Max}[p_{u,T1} - X_1, 0] - \text{Max}[X_1 - p_{u,T1}, 0] + S_0 - X_2\,e^{-rT_2}$

$c_{u,T1} - X_1 = p_{u,T1} - X_1 + S_0 - X_2 e^{-rT_2}$

---

and with some minor rearrangement, is rewritten:

$$S_0\left[M\left(d_1, y_1; \sqrt{\frac{T_1}{T_2}}\right) + M\left(-d_1, y_1; -\sqrt{\frac{T_1}{T_2}}\right)\right]$$

$$- X_2 e^{-rT_2}\left[M\left(d_2, y_2; \sqrt{\frac{T_1}{T_2}}\right) + M\left(-d_2, y_2; -\sqrt{\frac{T_1}{T_2}}\right)\right] - X_1 e^{-rT_1} = c_{u,0} - X_1 e^{-rT_1}$$

Because $N(y_1) = M\left(d_1, y_1; \sqrt{\frac{T_1}{T_2}}\right) + M\left(-d_1, y_1; -\sqrt{\frac{T_1}{T_2}}\right)$, and

$N(y_2) = M\left(d_2, y_2; \sqrt{\frac{T_1}{T_2}}\right) + M\left(-d_2, y_2; -\sqrt{\frac{T_1}{T_2}}\right)$, we have:[11]

$$S_0 N(y_1) - X_2 e^{-rT_2} N(y_2) - X_1 e^{-rT_1} = c_{u,0} - X_1 e^{-rT_1}$$

$$S_0 N(y_1) - X_2 e^{-rT_2} N(y_2) = c_{u,0}$$

The final equality above is true because it is consistent with the European call value as derived in Section 7.2.4. This demonstrates that our put on call formula is consistent with our call on call formula, which was reasoned earlier. Similar manipulations can be used to verify our other compound option formulas and our compound option put–call parity formula depicted in Table 7.3. This simplifies to our standard plain vanilla put–call parity relation.

7.7 THE BLACK–SCHOLES MODEL AND DIVIDEND ADJUSTMENTS

The Black–Scholes model has an enormous number of extensions. Here, we will first consider stock options in scenarios in which certain standard Black–Scholes assumptions are violated. Relaxation of assumptions allows the model to be applied to options written under varying circumstances and will have a number of interesting implications. Here, we will discuss a few of the many extensions of the model.

7.7.1 Lumpy Dividend Adjustments

The Black–Scholes model allows for European options on nondividend-paying stocks only. However, what if underlying shares do pay dividends? How might the form of dividends affect option value?

7.7.1.1 The European Known Dividend Model

A *dividend-protected* call option allows for the option holder to receive the underlying stock and any dividends paid during the life of the option in the event of exercise. In practice, most options are not dividend protected. In effect, a dividend payment diminishes the value of the underlying stock by the value of the dividend on the ex-dividend date.[12] If a stock underlying such a European call option were to pay a known dividend of amount D, with ex-dividend date $t_D < T$, we might assume that the stock price S_t reduced by the value of the dividend at time t follows a geometric Brownian motion. With the boundary condition $c_T = \text{Max}[0, (S_T - De^{r(T-t_D)} - X)]$, the *European known dividend model* can be used to evaluate the option as follows:

$$c_0 = c_0\left[S_0 - De^{-rt_D}, T, r, \sigma, X\right] = (S_0 - De^{-rt_D})N(d_1) - \frac{X}{e^{rT}}N(d_2)$$

$$d_1 = \frac{\ln\left(\dfrac{S_0 - De^{-rt_D}}{X}\right) + \left(r + \dfrac{1}{2}\sigma^2\right)T}{\sigma\sqrt{T}}$$

$$d_2 = d_1 - \sigma\sqrt{T}$$

This is among the simplest of the dividend-adjusted Black–Scholes models. Multiple dividend payments produce the boundary condition $c_T = \text{Max}\left[0, (S_T - \sum_{t_{Di}=1}^{n} D_i e^{r(T-t_{Di})} - X)\right]$, where the firm makes a dividend payments to shareholders at each of n points in time t_{Di} prior to time T. Thus, generally, the analyst simply subtracts the present value of the known dividend payment (or payments) prior to the option expiry from the stock's price. However, this model still assumes that the call is of the European variety, so that the call will never be exercised early. Nevertheless, holders of nonprotected calls will not receive dividends if they obtain the underlying shares after the ex-dividend date. If this model should be used to value American calls on dividend-paying stocks, it will tend to undervalue these American calls that cannot be exercised on ex-dividend dates.

7.7.1.2 Modeling American Calls

An American call on a nondividend-paying stock will never be exercised before its expiration date T. Consider Table 7.4, which depicts a scenario in which an investor can choose whether to exercise his American call now prior to its expiration at time T. In effect, the investor can choose between two portfolios A and B where Portfolio A consists of one call that is not exercised before

TABLE 7.4   Exercising American Calls Early in the Absence of Dividends

Portfolio Value at Time T			
Portfolio	Current Value	$S_T \leq X$	$S_T > X$
A	$c_0 + Xe^{-rT}$	$0 + X$	$(S_T - X) + X$
B	S_0	S_T	S_T
Note that:		$V_{AT} > V_{BT}$	$V_{AT} = V_{BT}$

TABLE 7.5 Exercising American Calls Early in the Presence of Dividends

Portfolio Value at Time T			
Portfolio	Current Value	$S_T > X$	$S_T \leq X$
A	$c_0 + Xe^{-rT}$	$S_T - X + X$	X
B	$S_0 + De^{-rt_D}$	$S_T + De^{r(T-t_D)}$	$S_T + De^{r(T-t_D)}$
Note that:		$V_{AT} \leq V_{BT}$	$V_{AT} \gtrless V_{BT}$

expiry and the present value of X dollars Xe^{-rT} retained from not exercising the call. If the call is exercised before expiry, the portfolio, labeled as Portfolio B, will consist of one share of stock, which is purchased with the exercise money X. The last two columns of this table give expiration date portfolio values, depending on the underlying stock price S_T relative to the call exercise price X. Since the sum value of the call and exercise money is always either equal to or greater than the value of the stock on the expiration date ($V_{AT} \geq V_{BT}$), the call should never be exercised early. That is, Portfolio A is always preferred to Portfolio B.

However, Table 7.5 demonstrates that premature exercise of an American call might occur when its underlying stock pays a dividend, where t_D ($t_D \leq T$) is the time of the premature exercise. We will demonstrate shortly that this premature exercise will occur only on the ex-dividend date, and only when the dividend amount is sufficiently high relative to some critical ex-dividend date stock price S_{tD}^*.

7.7.1.3 Black's Pseudo-American Call Model

This model (Black, 1975) incorporates the European known dividend model with the choice of early exercise. Recall that an American option should never be exercised if the stock pays no dividend. Similarly, it can be shown that if an American call is to be exercised early, it will be exercised only at the instant (or immediately before) the stock goes ex-dividend. If, on the ex-dividend date t_D, the time value associated with exercise money, $X[1 - e^{-r(T-tD)}]$ is less than the dividend payment, D, the call is never exercised early. Otherwise it will be optimal to exercise early if the ex-dividend underlying stock price is sufficiently high. Black's model states that the call's value c_0 is determined by:

$$c_0 = \text{Max}[c_0^*, c_0^{**}]$$

$$c_0^* = c_0^*[S_0, t_D, r, \sigma, X]$$

$$c_0^{**} = c_0^{**}[S_0 - De^{-rt_D}, T, r, \sigma, X]$$

where c_0^* is the call's value assuming it is exercised immediately before the stock's ex-dividend date and c_0^{**} is the call's value assuming the option is held until it expires. Observe from the formula that the call's actual value, c_0, will be the larger of c_0^* or c_0^{**}. Since the call's value is the larger of c_0^* and c_0^{**}, and these values are both determined at time 0, the formula implies that the American call value is based on an exercise decision at time 0. That is, even though we do not decide whether to exercise early until the ex-dividend date, the formula implies that the call value is based on a decision today on whether to exercise on the ex-dividend date.

7.7.1.4 The Roll–Geske–Whaley Compound Option Formula

Since American options need to allow for early exercise of the option on dividend-paying stocks, they are significantly more difficult to price than their European counterparts. Their early exercise potential creates an inequality in the Black–Scholes differential equation, which makes it considerably more difficult to solve. In fact, even the standard put–call parity condition breaks down for American options.

Black's pseudo-American call model assumes ex-ante that the call on a dividend-paying stock may be exercised early, but it does not provide a probability for this early exercise. In effect, the option holder is assumed to decide at time 0 whether he prefers to exercise the call on the ex-dividend date to receive the stock and the dividend or wait until the call's expiry and consider exercising to receive the stock without the dividend. While this assumption is not true, the option value does reflect the assumption. Roll's (also attributed to Geske and Whaley) formula corrects for this early decision assumption, but is somewhat more complex.

The Roll–Geske–Whaley model (see Roll, 1977; Geske, 1979a,b; Whaley, 1981, 1982) requires that we first find that minimum stock price (S_{tD}^*) at the ex-dividend date (t_D) that will lead to early exercise of the option. That is, on the ex-dividend date, given a known dividend payment D, there is some minimum stock price S_{tD}^* at time t_D for which early exercise (on the ex-dividend date) is optimal. At this minimum price, or at a higher price, paying the exercise price and receiving the stock along with the dividend is worth more than retaining the call. Thus, it makes sense to exercise the call. All values of S_{tD} exceeding this price S_{tD}^* will lead to early option exercise. This minimum price is found by solving the following for S_{tD}^*:

$$c_{tD}(S_{tD}^*, t_D; r, T, X, \sigma) = S_{tD}^* + D - X \tag{7.26}$$

When dividends are zero, the call will never be exercised early. If dividends are small, the underlying stock price S_{tD}^* at time t_D will need to be very large to justify early exercise of the call. Thus, for any dividend amount $D > 0$, there is some minimum stock price S_{tD}^* on the ex-dividend date that would lead to early exercise of the American call. There are several numerical and iterative techniques for solving this equality for S_{tD}^* (e.g., the methods of bisection and Newton–Raphson), but simple substitution and iteration will work. Second, to find the call's current value c_0, solve the following:

$$c_0 = (S_0 - D \cdot e^{-r \cdot t_D}) N(y_1) + (S_0 - D \cdot e^{-r \cdot t_D}) M\left(d_1, -y_1; -\sqrt{\frac{t_D}{T}}\right)$$

$$- X e^{-r \cdot T} M\left(d_2, -y_2; -\sqrt{\frac{t_D}{T}}\right) - (X - D) e^{-r \cdot t_D} N(y_2) \tag{7.27}$$

where:

$$d_1 = \frac{\left[\ln\left(\frac{S_0 - D \cdot e^{-r \cdot t_D}}{X}\right) + \left(r + \frac{1}{2} \cdot \sigma^2\right)T\right]}{\sigma \sqrt{T}} \tag{7.28}$$

$$d_2 = d_1 - \sigma \cdot \sqrt{T} \tag{7.29}$$

$$y_1 = \frac{\left[\ln\left(\frac{S_0 - D \cdot e^{-r \cdot t_D}}{S_{tD}^*}\right) + \left(r + \frac{1}{2} \cdot \sigma^2\right)t_D\right]}{\sigma\sqrt{t_D}} \tag{7.30}$$

$$y_2 = y_1 - \sigma\sqrt{t_D} \tag{7.31}$$

$M\left(d(*), y(*); \sqrt{\frac{t_D}{T}}\right)$ is a cumulative multivariate (bivariate) normal distribution, where $\sqrt{\frac{t_D}{T}}$ is the correlation coefficient between random variables S_{tD} and S_T, assuming that stock returns are independent over nonoverlapping periods of time. The process of computing this cumulative probability is described in Section 2.8.

The intuition behind the Roll–Geske–Whaley model is fairly straightforward. An American call with a single ex-dividend date is essentially a compound option that can be replicated with the following portfolio:

a. A long position in a European call with expiration T and exercise price X.
b. A short position in a European compound call (call on the call in part a) with expiration t_D, and exercise price $S_{tD}^* + D - X$.
c. A long position in a European call with expiration t_D and exercise price S_{tD}^*.

This implies that the American call holder can hold his call to expiration as he would a European call unless it gets called away from him at time t_D by his exercising a second call at time t_D to take the underlying stock at price S_{tD}^*. The cash flows produced from this replicating portfolio on the ex-dividend date are given in Table 7.6. We see in Table 7.6 that if the dividend amount is sufficiently high such that $S_{tD} > S_{tD}^*$, the American call is exercised on the ex-dividend date. Note that Table 7.6 demonstrates that the payoff structure of the portfolio exactly duplicates the required payoff structure for an American call on the date t_D, thus demonstrating that the portfolio replicates the American call.

TABLE 7.6 Compound Call Option Payoffs on Ex-Dividend Date t_D

Position	$S_{tD} < S_{tD}^*$ (c Held)	$S_{tD} > S_{tD}^*$ (c Exercised)
European call (a)	$c(S_{tD}, t_D, r, \sigma, T, X)$	0 (called away by call (b))
Compound call (b)	0	$S_{tD}^* + D - X$ (exercise money)
European call (c)	0	$S_{tD} - S_{tD}^*$ (call is exercised)
Total	$c(S_{tD}, t_D, r, \sigma, T, X)$	$S_{tD}^* + D - X$

ILLUSTRATION: CALCULATING THE VALUE OF AN AMERICAN CALL ON DIVIDEND-PAYING STOCK

Suppose that we wish to calculate the value of a 6-month $X = \$45$ American call on a share of stock that is currently selling for $50. The stock is expected to go ex-dividend on a $5 payment in 3 months, and not again until after the option expires. The standard deviation of returns on the underlying stock is 0.4 and the riskless return rate is 3%. What is the call worth today assuming:

a. the European known dividend model?
b. Black's pseudo-American call model?
c. the Roll–Geske–Whaley model?

- Under the European known dividend model, the value of the call is \$5.39, calculated as follows:

$$c_0 = \left(50 - 5e^{-0.03 \times 0.25}\right) \times 0.5782 - \frac{45}{e^{0.03 \times 0.5}} \times 0.4660 = 5.39$$

where:

$$d_1 = \frac{\ln\left(\frac{50 - 5e^{-0.03 \times 0.25}}{45}\right) + \left(0.03 + \frac{1}{2} \times 0.16\right) \times 0.5}{0.4 \times \sqrt{0.5}} = 0.1974; N(d_1) = 0.5782$$

$$d_2 = 0.5782 - 0.4 \times \sqrt{0.5} = -0.0855; N(d_2) = 0.4660$$

- To use Black's pseudo-American call formula, we will first determine whether dividend exceeds the ex-dividend date time value associated with exercise money: $5 > 45[1 - e^{-0.03(0.5 - 0.25)}] = 0.336$. Since the dividend does exceed this value, we will calculate the value of the call assuming that we exercise it just before the underlying stock goes ex-dividend in 3 months, while it trades with dividend:

$$c_0 = 50 \times 0.7468 - \frac{45}{e^{0.03 \times 0.25}} \times 0.6788 = 7.02$$

where:

$$d_1 = \frac{\ln\left(\frac{50}{45}\right) + \left(0.03 + \frac{1}{2} \times 0.16\right) \times 0.25}{0.4 \times \sqrt{0.25}} = 0.6643; N(d_1) = 0.7468$$

$$d_2 = 0.6643 - 0.4 \times \sqrt{0.25} = 0.4643; N(d_2) = 0.6788$$

Under Black's pseudo-American call formula, we find that the value of the call is Max [5.39, 7.02], or 7.02.

- Using the Geske–Roll–Whaley model, we first calculate S_{tD}^*, the critical stock value on the ex-dividend date required for early exercise of the call:

$$c_{tD}^* = S_{tD}^* + D - X$$

$$c_{tD}(S_{tD}^*, 0.25, 0.03, 0.5, 45, 0.4) = S_{tD}^* + 5 - 45$$

Using simple substitution or an appropriate numerical or iterative technique, one finds that:

$$c_{tD}(42.4924, 0.25, 0.03, 0.5, 45, 0.4) = 2.4924 = 42.4924 + 5 - 45$$

$$S_{tD}^* = 42.4924$$

This means that if the stock is selling for $S_{tD}^* = \$42.4924$ in 3 months ($t_D = 0.25$), the call on the stock trading ex-dividend will have the same value as the stock, plus declared dividend minus the exercise money. Thus, as long as the ex-dividend value of the stock plus the dividend minus exercise money exceeds the ex-dividend call value, the call will be exercised early. The value of the American call is found to be 6.96, calculated from Eq. (7.27) as follows:

$$c_0 = (50 - 5 \cdot e^{-0.03 \times 0.25}) \cdot 0.6658$$
$$+ (50 - 5 \cdot e^{-0.03 \times 0.25}) \cdot 0.0798$$
$$- 45 \cdot e^{-0.03 \times \cdot 0.5} \cdot 0.07175$$
$$- (45 - 5) \cdot e^{-0.03 \times 0.25} \cdot 0.5903 = 6.96$$

where:

$$d_1 = \frac{\left[\ln\left(\frac{50 - 5 \cdot e^{-0.03 \times 0.25}}{45}\right) + \left(0.03 + \frac{1}{2} \cdot 0.16\right) \cdot 0.5\right]}{0.4 \cdot \sqrt{0.5}} = 0.1974$$

$$d_2 = 0.1974 - 0.4 \cdot \sqrt{0.5} = -0.0855$$

$$y_1 = \frac{\left[\ln\left(\frac{50 - 5 \cdot e^{-0.03 \times 0.25}}{42.4924}\right) + \left(0.03 + \frac{1}{2} \cdot 0.16\right) \cdot 0.25\right]}{0.4 \cdot \sqrt{0.25}} = 0.4283$$

$$y_2 = 0.4283 - 0.4 \cdot \sqrt{0.25} = 0.2283$$

- The Roll–Geske–Whaley model provided the correct price of $6.96 for the American call. Black's approximation ($7.02) provided a reasonable estimate, while the European known dividend model, as expected, undervalued the call ($5.39) relative to Black's model.

7.7.2 Merton's Continuous Leakage Formula

In some instances, dividends might be considered to be paid on a continuous basis. For example, many options on indices (index options) trade on portfolios whose *dividend leakage*, the rate at which continuous dividends are paid or received by a fund, can often be accounted for when assumed to occur continuously and without detectable seasonality given the time to option expiry. Other options can be traded on commodities or other assets that have costs of storage or other constant carry costs that are paid continuously over time. Here, we assume that the underlying stock follows the process below:

$$\frac{dS_t}{S_t} = (\mu - \delta)\,dt + \sigma\,dZ_t$$

where δ is the periodic dividend yield (see Merton, 1973). The self-financing portfolio in this case has the form:

$$dV_t = \gamma_{s,t}\,dS_t + r(V_t - \gamma_{s,t}S_t)\,dt + \delta\gamma_{s,t}S_t\,dt$$

The standard Black–Scholes differential equation for the continuous leakage model is as follows:

$$\frac{\partial c}{\partial t} = rc - (r - \delta)\frac{\partial c}{\partial S}S - \frac{1}{2}S^2\sigma^2\frac{\partial^2 c}{\partial S^2}$$

This is the continuous dividend-adjusted Black–Scholes differential equation. Its particular solution, subject to the *boundary condition* c_T = Max[0, S_T − X] for a European call, is given as follows:

$$c_0 = S_0 e^{-\delta T} N(d_1) - \frac{X}{e^{rT}}N(d_2)$$

where:

$$d_1 = \frac{\ln\left(\frac{S_0}{X}\right) + \left(r - \delta + \frac{1}{2}\sigma^2\right)T}{\sigma\sqrt{T}}$$

$$d_2 = d_1 - \sigma\sqrt{T}$$

$$p_0 = c_0 + Xe^{-rT} - S_0 e^{-\delta T}$$

ILLUSTRATION: CONTINUOUS DIVIDEND LEAKAGE

Return to our earlier example where an investor is considering 6-month options on a stock that is currently priced at $75. The exercise price of the call and put are $80 and the current riskless rate of return is 10% per annum. The variance of annual returns on the underlying stock is 16%. However, this stock will pay a continuous annual dividend at a rate of 2%. What are the values of these call and put options on this stock? Our first step is to find d_1 and d_2:

$$d_1 = \frac{\ln\left(\frac{75}{80}\right) + \left(0.10 - 0.02 + \frac{1}{2} \times 0.16\right) \times 0.5}{0.4\sqrt{0.5}} = 0.0547$$

$$d_2 = 0.0547 - 0.4\sqrt{0.5} = -0.2282$$

With $N(d_1) = 0.522$ and $N(d_2) = 0.41$, we value the call and put as follows:

$$c_0 = 75e^{-0.02 \times 0.5} \times 0.522 - \frac{80}{e^{0.10 \times 0.5}} \times 0.41 = 7.56$$

$$p_0 = 7.56 + 80(0.9512) - 75e^{-0.02 \times 0.5} = 9.41$$

7.8 BEYOND PLAIN VANILLA OPTIONS ON STOCK

The Black–Scholes differential equation and model have a huge number of widely varying applications, particularly when it is extended by relaxing assumptions. Relaxation of assumptions has many useful applications and interesting implications. Here, we will discuss a few of the many extensions of the model.

7.8.1 Exchange Options

In this section, we discuss a variation of the Black–Scholes model provided by Margrabe (1978) for the valuation of an option to exchange one risky asset for another. Suppose that the prices S_1 and S_2 of two assets follow geometric Brownian motion processes with σ_1 and σ_2 as standard deviations of logarithms of price relatives (or returns) for each of the two securities:

$$\frac{dS_1}{S_1} = \mu_1\, dt + \sigma_1\, dZ_1$$

$$\frac{dS_2}{S_2} = \mu_2\, dt + \sigma_2\, dZ_2$$

where $E[dZ_1 \times dZ_2] = \rho_{1,2}\, dt$, where $\rho_{1,2}$ is the correlation coefficient between logarithms of price relatives $\ln(S_1/S_2)$ between the two securities. Furthermore, the variance of logarithms of price relatives of the two assets relative to one another is σ^2:[13]

$$\sigma^2 = \sigma^2_{\ln\left(S_1/S_2\right)} = \sigma^2_{\ln(S_1)} + \sigma^2_{\ln(S_2)} - 2\rho_{1,2}\sigma_{\ln(S_1)}\sigma_{\ln(S_2)}$$

That is, σ is the anticipated standard deviation of $[\ln(S_1/S_2)]$ over the life of the option. We will use the standard deviation of logs of relative returns σ^2 for our option valuation. Note that asset 1 is potentially given up in the exchange and asset 2 is potentially received.

7.8.1.1 Changing the Numeraire for Pricing the Exchange Call

Using our notation from Section 7.2.3, the exchange option is valued in risk-neutral probability space as:

$$E_Q[V_T|\mathscr{F}_0] = E_Q[c_T|\mathscr{F}_0] = E_Q[\mathrm{Max}[S_{2,T} - S_{1,T}, 0]|\mathscr{F}_0]$$

However, we can change our numeraire to stock 1. Now, the exchange option can be valued in a Black–Scholes environment with stock 1 as the numeraire:

$$\frac{c_0}{S_{1,0}} = E_Q\left[\mathrm{Max}\left[\frac{S_{2,T}}{S_{1,T}} - 1, 0\right]|\mathscr{F}_0\right]$$

$$\frac{c_0}{S_{1,0}} = \frac{S_{2,0}}{S_{1,0}}N(d_1) - N(d_2)$$

$$d_1 = \frac{\ln\left(\frac{S_{2,0}}{S_{1,0}}\right) + \frac{1}{2}\sigma^2 T}{\sigma\sqrt{T}}$$

$$d_2 = d_1 - \sigma\sqrt{T}$$

Thus, ignoring dividends, we calculate the value of a single exchange call (c_0) as follows:

$$c_0 = S_{2,0}N(d_1) - S_{1,0}N(d_2)$$

$$d_1 = \frac{\ln\left(\frac{S_{2,0}}{S_{1,0}}\right) + \frac{1}{2}\sigma^2 T}{\sigma\sqrt{T}}$$

$$d_2 = d_1 - \sigma\sqrt{T}$$

It is interesting to note that not even the riskless return rate plays a role in valuing the exchange option in the absence of dividends; in effect, the riskless return is embedded in the price of stock 1 in the risk-neutral environment. Recall that we used fairly minor variations of this equation in Sections 5.4.1 and 7.2.4 to value plain vanilla calls (just assume that asset 1 is riskless with value X). Following the procedure set forth in Section 7.2.4 leads to the solution for c_0 given above for the exchange option.

7.8.1.2 Exchange Option Illustration

Suppose that the Predator Company has announced its intent to acquire the Prey Company through an exchange offer. That is, the Predator Company has extended an offer to Prey Company shareholders to exchange one share of its own stock for each share of Prey Company stock. This offer expires in 90 days (0.25 years). Shares of stock in the two companies follow geometric Brownian motion processes with a variance of $\sigma^2_{\ln(S1)} = 0.16$ for Predator (Predator is company 1) and $\sigma^2_{\ln(S2)} = 0.36$ for Prey, and zero correlation $\rho_{1,2} = 0$ between the two. Predator Company stock is currently selling for $40 and Prey shares are selling for $50. The current riskless return rate is 0.05. What is the value of the exchange option associated with this tender offer? First, we calculate σ^2, the variance of changes (actually, logs thereof) of stock prices relative to each other as follows:

$$\sigma^2 = \sigma^2_{\ln\left(\frac{S_1}{S_2}\right)} = \sigma^2_{\ln(S_1)} + \sigma^2_{\ln(S_2)} - 2\rho_{1,2}\sigma_{\ln(S_1)}\sigma_{\ln(S_2)} = 0.16 + 0.36 - 2\times 0\times 0.4\times 0.6 = 0.52$$

The value of this exchange option is calculated as 12.61 as follows:

$$c_0 = 50N(d_1) - 40N(d_2) = 50\times 0.7879 - 40\times 0.6695 = 12.61$$

$$d_1 = \frac{\ln\left(\frac{50}{40}\right) + \left(\frac{1}{2}\times 0.7211^2\right)\times 0.25}{0.7211\times\sqrt{0.25}} = 0.79917; \quad N(d_1) = 0.7879$$

$$d_2 = 0.79917 - 0.7211\sqrt{0.25} = 0.4386; \quad N(d_2) = 0.6695$$

With put–call parity, which is unchanged from the usual plain vanilla scenario, we find that the put is worth 2.61 in the Black–Scholes environment. Note that a long position in one call plus a short position in one put has a combined value equal to 10, the difference between the prices of the two shares. A long position in the exchange call and a short position in the exchange put essentially mean a one-for-one exchange of shares produces a profit of 10 to the portfolio holder.

Creating a call option to "buy" one share of Predator Company with one share of Prey Company is identical to creating a put option to "sell" one share of Prey Company for one share of Predator Company. Working through the Black–Scholes equations above, reversing the prices of stock 1 and stock 2, we find that the exchange put is worth 12.61, just as was the call price earlier. The exchange call in this reversed scenario is worth 2.61.

7.8.2 Currency Options

In this section, we discuss a variation of the Black–Scholes model provided by Garman and Köhlagen (1983), Biger and Hull (1983), and Grabbe (1983) for the valuation of currency options. This option is much like the exchange option presented above. The stock option pricing model is based on the assumption that investors price calls as though they expect the underlying stock to earn the risk-free rate of return. However, when we value currency options, we acknowledge that interest rates may differ between the foreign and domestic countries. Hence, we quote two interest rates $r(f)$ and $r(d)$, one each for the foreign and domestic currencies. The standard differential equation for currency options is as follows:

$$\frac{\partial c}{\partial t} = r(d)c - \left[r(d) - r(f)\right]\frac{\partial c}{\partial s}s - \frac{1}{2}s^2\sigma^2\frac{\partial^2 c}{\partial s^2}$$

where s is the exchange rate. Note that this differential equation is identical to the Merton dividend leakage model above, where the dividend yield δ is replaced by the foreign interest rate $r(f)$ and $r(d)$ is simply the domestic riskless rate. Its particular solution, subject to the *boundary condition* $c_T = \text{Max}[0, s_T - X]$, is given as follows:

$$c_0 = \frac{s_0}{e^{r(f)T}}N(d_1) - \frac{X}{e^{r(d)T}}N(d_2) = c_0\left[s_0, T, r(f), r(d), \sigma, X\right]$$

$$d_1 = \frac{\ln\left(\frac{s_0}{X}\right) + \left(r(d) - r(f) + \frac{1}{2}\sigma^2\right)T}{\sigma\sqrt{T}}$$

$$d_2 = d_1 - \sigma\sqrt{T}$$

where $r(d)$ is the domestic riskless rate (the rate for the currency that will be given up if the option is exercised), $r(f)$ is the foreign riskless rate (the rate for the currency that may be purchased). The spot rate for the currency is s_0. The standard deviation of proportional changes in the currency value to be purchased in terms of the domestic currency is σ. The exercise price of the option, X, represents the number of units of the domestic currency to be given up for the foreign currency.

7.8.2.1 *Currency Option Illustration*

Consider the following example where call options are traded on Brazilian real (BRL). One US dollar is currently worth two Brazilian real ($s_0 = 0.5$). We wish to evaluate a 2-year European call and put on BRL with exercise prices equal to $X = 0.45$. US and Brazilian interest rates are 0.03 and 0.08, respectively. The annual standard deviation of exchange rates is 0.15. We calculate $d_1 = 0.1313$ and $d_2 = -0.0808$; $N(d_1)$ and $N(d_2)$ are 0.5522 and 0.4678, respectively. Thus the call is valued at $c_0 = 0.037$ and the put is valued at $p_0 = 0.035$.

7.9 EXERCISES

7.1. Demonstrate that the definition given for self-financing in property (7.7) is equivalent to the condition that over any infinitesimal time interval from t to $t + dt$, the change in the value of the portfolio resulting purely from the transactions $d\gamma_{s,t}$ and $d\gamma_{b,t}$ that are executed over this time interval must equal zero.

7.2. Suppose that a self-financing portfolio includes a single short position in a call option with exercise price X. This single short position remains constant over time t. The portfolio also has a long position in $\gamma_{s,t}$ shares of stock that varies over time t along with $\gamma_{b,t}$ short positions in bonds (face value $= X$) that also vary over time. The bonds will be paid off by the exercise money realized if and when the call is exercised at time T.

    a. Write an expression that gives the portfolio value V_T at time T.

    b. Suppose that at time T, the value of the stock exceeds the exercise price X of the call. What will be $\gamma_{s,T}$ and $\gamma_{b,T}$?

    c. Suppose that at time T, the value of the stock is less than the exercise price X of the call. What will be $\gamma_{s,T}$ and $\gamma_{b,T}$?

7.3. Assuming a Black–Scholes environment, evaluate calls and puts for each of the following European stock option series:

Option 1	Option 2	Option 3	Option 4
$T = 1$	$T = 1$	$T = 1$	$T = 2$
$S = 30$	$S = 30$	$S = 30$	$S = 30$
$\sigma = 0.3$	$\sigma = 0.3$	$\sigma = 0.5$	$\sigma = 0.3$
$r = 0.06$	$r = 0.06$	$r = 0.06$	$r = 0.06$
$X = 25$	$X = 35$	$X = 35$	$X = 35$

7.4. Evaluate each of the European options in the series on ABC Company stock assuming a Black–Scholes environment. Current market prices for each of the options are listed in the table. Determine whether each of the options in the series should be purchased or sold at the given market prices. The current market price of ABC stock is 120, the August options expire in 9 days, September options in 44 days, and October options in 71 days. The stock variances prior to expirations are projected to be 0.20 prior to August, 0.25 prior to September, and 0.20 prior to October. The Treasury bill rate is projected to be 0.06 for each of the three periods prior to expiration. Convert the number of days given to fractions of 365-day years, as we shall assume that trading occurs 365 days per year.

Calls				
X	Aug	Sep	Oct	
110	9.500	10.500	11.625	$\sigma = 0.20$ for Aug
115	4.625	7.000	8.125	$\sigma = 0.25$ for Sep
120	1.250	3.875	5.250	$\sigma = 0.20$ for Oct
125	0.250	2.125	3.125	$r = 0.06$
130	0.031	0.750	1.625	$S = 120$

(Continued)

(Continued)

Puts			
X	Aug	Sep	Oct
110	0.031	0.750	1.500
115	0.375	1.750	2.750
120	1.625	6.750	4.500
125	5.625	6.750	7.875
130	10.625	10.750	11.625

Exercise prices for 15 calls and 15 puts are given in the left columns. Expiration dates are given in column headings and current market prices are given in the table interiors.

7.5. Use put–call parity and the Black–Scholes call pricing model to verify the following in a Black–Scholes environment:

$$p_0 = Xe^{-rT}N(-d_2) - S_0N(-d_1)$$

7.6. Emu Company stock currently trades for $50 per share. The current riskless return rate is 0.06. Under the Black–Scholes framework, what would be the standard deviations implied by 6-month (0.5-year) European calls with current market values based on each of the following striking prices? That is, with market prices of calls taken as given and equal to Black–Scholes estimates, what standard deviation estimates in Black–Scholes models would yield call values equal to market values in each of the following scenarios?
   a. $X = 40$; $c_0 = 11.50$
   b. $X = 45$; $c_0 = 8.25$
   c. $X = 50$; $c_0 = 4.75$
   d. $X = 55$; $c_0 = 2.50$
   e. $X = 60$; $c_0 = 1.25$

7.7. As discussed in Section 7.4, the sensitivity of the call's price to changes in the underlying stock return standard deviation is often known as "vega," which is calculated from the following:

$$\frac{\partial c}{\partial \sigma} = \frac{S_0\sqrt{T}}{\sqrt{2\pi}}e^{\frac{-d_1^2}{2}} > 0 \quad \text{Vega } \nu$$

Derive this equation above for vega, using the identity:

$$S_0\frac{\partial N(d_1)}{\partial d_1} = Xe^{-rT}\frac{\partial N(d_2)}{\partial d_2}$$

which is derived in the companion website.

7.8. Cannondale Company stock is currently selling for $40 per share. Its historical standard deviation of returns is 0.5. The 1-year Treasury bill rate is currently 5%. Assume that all of the standard Black–Scholes option pricing model assumptions hold.

   a. What is the value of a put on this stock if it has an exercise price of $35 and expires in 1 year?

   b. What is the implied probability that the value of the stock will be less than $30 in 1 year?

7.9. What is the gamma of a long position in a single futures contract? Why?

7.10. In Section 7.5.2, we stated the delta and gamma for a put. Derive both.

7.11. a. As the time to maturity T of a call option increases, the call value will increase. Find and simplify the derivative of the Black–Scholes option pricing model c_0 with respect to time to maturity (T). As we discussed in Section 7.5, this derivative is often known as the option theta. Hint: Make use of the product and chain rules from calculus and pay close attention to the simplification procedure described in the companion website.

   b. As the call option approaches maturity, its value will diminish. That is, as we move forward through time (t), the value of the call will tend to decline as its expiration draws closer. Based on part a of this problem, find the rate of change of the value for the call with respect to time.

   c. The Black–Scholes option pricing model is derived from the Black–Scholes differential equation:

$$\frac{\partial c}{\partial t} = rc - r\frac{\partial c}{\partial S}S - \frac{1}{2}S^2\sigma^2\frac{\partial^2 c}{\partial S^2}$$

   Using your results from part b of this problem, verify that the Black–Scholes pricing model is a valid solution to the Black–Scholes differential equation.

7.12. Tiblisi Company stock currently sells for 30 per share and has an anticipated volatility (annual return standard deviation) equal to 0.6. Three-month (0.25-year) call options are available on this stock with an exercise price equal to 25. The current riskless return rate equals 0.05.

   a. Calculate the call's value.

   b. Calculate the call's delta, gamma, theta, vega, and rho.

   c. What is the Black–Scholes implied probability that the stock price will exceed 25 three months from now?

   d. What is the value of a put with the same exercise terms as the call?

   e. What are the Greeks for the put in part d?

7.13. Consider a compound call on a call with exercise price $X_1 = 5$, expiring in 6 months ($T_1 = 0.5$) on an equity call, with exercise price $X_2 = 75$, expiring in 1 year ($T_2 = 1$). The stock currently sells for $S_0 = 70$ and has a return volatility equal to $\sigma = 0.4$. The riskless return rate equals $r = 0.05$. What is the value of the compound call?

7.14. Consider 2-year options on a stock that is currently priced at $20. The exercise price of the call and put are $10 and the current riskless rate of return is 4% per annum. The standard deviation of annual returns on the underlying stock is 0.8. However, this stock will pay a continuous annual dividend at a rate of 5%. What are the values of these call and put options on this stock assuming continuous dividend leakage?

7.15. An investor has the opportunity to purchase a 6-month call option for $2.00 on a stock that is currently selling for $30. The stock is expected to go ex-dividend in 3 months on a declared dividend of $2. The exercise price of the call is $35 and the current riskless rate of return is 3% per annum. The variance of annual returns on the underlying stock is 16%.

a. Based on the European known dividend model, and its current price of $2.00, does this option represent a good investment?

b. What is the value of a European put on this stock?

c. Value the American call using Black's pseudo-American call model.

d. Value the American call using the Geske–Roll–Whaley compound call model.

7.16. Suppose that S_1 and S_2 are random variables. In our exchange option discussion in Section 7.8.1, we claimed that we could measure the variance of logarithmic of price relatives of the two assets relative to one another with a particular variant of σ^2:

$$\sigma^2_{\ln\left(S_1/S_2\right)} = \sigma^2_{\ln(S_1)} + \sigma^2_{\ln(S_2)} - \rho_{1,2}\sigma_{\ln(S_1)}\sigma_{\ln(S_2)}$$

a. Demonstrate more generally that $\mathrm{Var}[S_1 - S_2] = \mathrm{Var}[S_1] + \mathrm{Var}[S_2] - \mathrm{Cov}[S_1,S_2]$.

b. Based on your demonstration for part a, demonstrate the following:

$$\sigma^2_{\ln\left(S_1/S_2\right)} = \sigma^2_{\ln(S_1)} + \sigma^2_{\ln(S_2)} - 2\rho_{1,2}\sigma_{\ln(S_1)}\sigma_{\ln(S_2)}$$

7.17. Six-month currency options on Thai baht (THB) denominated in US dollars with exercise prices equal $0.03. The US interest rate is 0.03 and the Thai rate is 0.10. The current exchange rate is $0.032/THB and the standard deviation associated with the exchange rate is 0.01. What is the value of this FX call? What is the value of a put with the same exercise terms?

7.18. Owners of many commodities must make payments to store and maintain their inventories. In addition, many of these inventories suffer from depreciation or depletion. Suppose that the combined costs of storage and depletion for a given amount of a particular commodity (say a bushel of a particular variant of corn) is qS_t, where q is the proportional storage and depletion cost per unit of the commodity per unit time.

a. Write the Black–Scholes differential equation defining the option price path for this unit of corn.

b. Write a variation of the Black–Scholes option pricing model for a call on a single bushel of corn.

7.19. A *chooser option* (or *as you like it option*) is sometimes held to allow its owner to benefit from underlying security volatility, as in 1991, when the Persian Gulf War led to high levels of price volatility in oil markets. The *standard chooser option* allows its owner at a *choice date* $0 < t_c < T$ to choose between the option taking the form of a plain vanilla call or a put. Thus, on the choice date t_c, the owner of the standard chooser option decides if he wishes the option to be a plain vanilla call or a put for the remainder of its life.

a. Write a function that gives the choice date t_c payoff for the standard chooser option in terms of the plain vanilla options from which the owner can choose.

b. Making use of the put–call relation for plain vanilla options, evaluate as of time t_c the plain vanilla put component of the chooser option.

c. Making use of your solution to part b, rewrite the function from part a that gives the choice date t_c payoff for the standard chooser option in terms of the plain vanilla options from which the owner can choose. Comment on your findings.

   d. Under what circumstances does the owner of the chooser option select the call on the choice date t_c? Under what circumstances does the owner of the chooser option select the put on the choice date t_c?

   e. Devise a model to evaluate the chooser option in a Black–Scholes environment.

   f. An investor needs to evaluate a 1-year chooser option on a nondividend-paying stock in a Black–Scholes environment. The choice date is in 6 months. The riskless return rate is 0.05 and the standard deviation of underlying stock returns is 0.4. Both the current stock price and the exercise price of the chooser option are 50. What is the current value of the chooser option?

NOTES

1. Many authors only require condition 3 for the definition of a self-financing portfolio.
2. Our replicating portfolio $(-1, \gamma_{s,t}, \gamma_{b,t})$ requires a long position $\gamma_{s,t} = N(d_1)$ in stock and a short position $\gamma_{b,t} = -Xe^{-rT}N(d_2)$ in bonds.
3. See Section 7.5 for additional details and discussion concerning vega, one of the "Greeks," and Exercise 7.7 for its derivation.
4. The "smile effect" pertains to the empirical finding that options with very high and very low exercise prices relative to current underlying security prices produce high implied volatilities relative to options trading at or near the money. The "smirk effect" occurs when only either high or low exercise prices produce higher implied volatilities.
5. There are a small number of studies that suggest that historical volatilities contain useful information not contained in implied volatilities, and might even, in some instances, be better predictors of future volatility (see, for example, Canina and Figlewski, 1993).
6. The companion website provides a derivation deriving the delta and gamma expressions.
7. This expression is derived in end-of-chapter Exercise 7.11.a. More importantly, its relation to the all-important Black–Scholes differential equation (7.11) in Section 7.3.2 is described in Exercise 7.11.c. Many traders refer to theta by its negative value, emphasizing that the value of the option decays through the passage of time. Theta is also known as amortization. Many traders will also divide the annual theta by 252, the number of trading days in a year in order to reflect the daily amortization.
8. The put delta and gamma are derived in Exercise 7.10.
9. The processes are perfectly correlated during their overlapping period T_1, and are independent over their nonoverlapping period $T_2 - T_1$. Thus, T_1/T_2 is the proportion of variability in one random variable explained by the other and the r-square value between the two processes is T_1/T_2. Thus, the correlation coefficient is $\rho = \sqrt{T_1/T_2}$.
10. See Section 2.7.3 for the procedure to calculate bivariate normal distributions.
11. It is fairly straightforward to verify these relationships by using a multinomial normal distribution calculator and a z-table.
12. A shareholder holding the stock on the ex-dividend date receives the dividend. Shareholders obtaining the stock after the ex-dividend date do not. Also, when underlying stock returns volatility is computed, dividends are excluded from the calculations.
13. End-of-chapter Exercise 7.16 provides a verification of this equality.

References

Baxter, M.W., Rennie, A.J.O., 1996. Financial Calculus: An Introduction to Derivative Pricing. Options Markets. Cambridge University Press, Cambridge, UK.

Biger, N., Hull, J., 1983. The valuation of currency options. Financ. Manage. 12, 24–28.

Black, F., 1975. Fact and fantasy in the use of options. Financ. Analysts J. 31, 36–41, 61–72.

Black, F., Scholes, M., 1972. The valuation of options contracts and a test of market efficiency. J. Financ. 27, 399–417.

Black, F., Scholes, M., 1973. The pricing of options and corporate liabilities. J. Pol. Econ. 81, 637–654.

Canina, L., Figlewski, S., 1993. The informational content of implied volatility. Rev. Financ. Stud. 6, 659–681.

Cox, J.C., Rubinstein, M., 1985. Options Markets. Prentice Hall, Englewood Cliffs, NJ.

Drezner, Z., 1978. Computation of the bivariate normal integral. Math. Comput. 32, 277–279.

Garman, M., Köhlagen, S., 1983. Foreign currency option values. J. Int. Money Financ. 2, 231–237.

Geske, R., 1979a. The valuation of compound options. J. Financ. Econ. 7, 63–81.

Geske, R., 1979b. A note on an analytic valuation formula for unprotected American call options on stocks with known dividends. J. Financ. Econ. 7, 375–380.

Grabbe, J.O., 1983. The pricing of call and put options on foreign exchange. J. Int. Money Finance. 2, 239–253.

Harrison, J.M., Kreps, D.M., 1979. Martingales and arbitrage in multi-period securities markets. J. Econ. Theory. 381–408.

Harrison, J.M., Pliska, S., 1981. Martingales and stochastic integrals in the theory of continuous trading. Stoch. Processes Their Appl. 11, 215–260.

Hull, J.C., 2010. DerivaGem, version 2.01, online software package. <http://www.rotman.utoronto.ca/~hull/software/bivar.xls> (accessed 17.08.14.).

Hull, J.C., 2011. Futures, Options and Other Derivatives. eighth ed. Prentice Hall, Englewood Cliffs, NJ.

Latane, H., Rendleman, R., 1976. Standard deviations of stock price ratios implied by option prices. J. Financ. 31, 369–381.

Margrabe, W., 1978. The value of an option to exchange one asset for another. J. Financ. 33, 177–186.

Merton, R.C., 1973. Theory of rational option pricing. Bell J. Econ. Manage. Sci. 4 (1), 141–183.

Roll, R., 1977. An analytic valuation formula for unprotected American call options on stocks with known dividends. J. Financ. Econ. 5, 251–258.

Whaley, R.E., 1981. On the valuation of American call options on stocks with known dividends. J. Financ. Econ. 9, 207–211.

Whaley, R.E., 1982. Valuation of American call options on dividend-paying stocks: empirical tests. J. Financ. Econ. 10, 29–58.

Mean-Reverting Processes and Term Structure Modeling

8.1 SHORT- AND LONG-TERM RATES

As discussed in Chapters 3 and 4, the yield curve depicts varying spot rates over associated bond terms to maturity. Spot rates are interest rates on loans originated at time 0 (e.g., now). The yield curve is typically constructed from the yields of benchmark, highly liquid (where possible), default-free fixed income, and/or zero-coupon instruments. Debt instruments, including Treasury bills and Treasury bonds, issued by national governments are often used as benchmarks as they are generally considered to be the least likely to default and have the greatest liquidity. We calculate spot rates from cash flows yielded by bonds (e.g., coupon payments and principal repayment) and their market prices. The term structure is crucial because yields on securities with fixed maturities are the backbone for obtaining discount rates and riskless return rates needed for valuing bonds, stocks, derivatives, and other instruments.

Many stock pricing models assume that stock returns follow constant volatility Brownian motion processes with drift, based on underlying assumptions that prices tend to maintain somewhat predictable proportional drifts over time and that unanticipated price changes are caused by random and continuous arrivals of new information. However, the returns and prices of many assets, particularly assets with cyclical returns, high transactions costs, and assets with finite lives, are less likely to follow Brownian motion processes. For example, the price of a $1000 zero-coupon bond will tend to drift toward its maturity value of $1000. Similarly, some rates, such as exchange rates and interest rates, have either "natural levels" or are managed by governments. Consider, for example, the case of bonds whose prices are affected by short-term (technically, *instantaneous short rates*) and long-term interest rates. Instantaneous rates, which are unobservable because they do not actually exist in reality, might be proxied by overnight rates. Empirical observation suggests that short rates tend to drift toward somewhat predictable or mean rates over long periods of time.

Understanding the stochastic processes that drive security prices, interest rates, and other rates is essential to valuing and hedging the securities, their derivative securities, and portfolios related to them. In this chapter, we analyze alternatives to Brownian motion processes for security and

P.M. Knopf & J.L. Teall: Risk Neutral Pricing and Financial Mathematics: A Primer.
DOI: http://dx.doi.org/10.1016/B978-0-12-801534-6.00008-4

247

rate process, with a particular focus on interest rates and debt instruments. We will focus on single factor models, where the entire yield curve and its related bond prices are driven by a single factor, being short-term interest rates.

In Chapters 3 and 4, we made contradictory assumptions about the nature of the yield curve. In Chapter 3, we allowed for non-flat yield curves that did not shift. In our duration and convexity models from Chapter 4, we allowed only for flat yield curves that experienced shocks, leading to parallel yield curve shifts. Later in Chapter 4, we described term structure with non-flat yield curves. In this chapter, we will analyze term structures and price bonds and calculate non-flat yield curves that are driven by short-term interest rates, fluctuate continuously, and shift in a non-parallel manner. More important, we are interested in term structure modeling, which describes how the yield curve itself evolves over time.

8.1.1 Rates and Arithmetic Brownian Motion

The major issue in term structure modeling is that future interest rates are not known with certainty. One of our problems is to model this uncertainty. The Merton (1973) term structure model prices bonds based on the assumption that short-term interest rates follow an arithmetic Brownian motion process:

$$dr_t = \mu \, dt + \sigma_r \, dZ_t$$

An attractive characteristic of rates following arithmetic Brownian motion is that they are normally distributed, and normal distributions are often very easy to work with. On the other hand, the expected change in interest rates is normally unrelated to how they compared to historical or long-term rates. This would mean that the directional move for short-term interest rates cannot be reasonably well predicted based on available information, particularly when μ is low compared to σ. The range of potential interest rate changes is from negative infinity to positive infinity. Negative interest rates are possible under Brownian motion, but seem less likely in practice. Furthermore, the arithmetic Brownian motion process implies a constant volatility for interest rates, regardless of whether the short-term rate is high or close to zero. Again, in practice, we generally observe higher interest rate volatility when rates are high. More discussion on this model and its pricing implications is provided in Section 8.4.1.

8.1.2 Rates and Geometric Brownian Motion

The problems associated with the arithmetic Brownian motion model that can produce negative interest rates and constant interest rate volatility can be alleviated with the Rendleman and Bartter (1980) geometric Brownian motion process model:

$$dr_t = \mu r_t \, dt + \sigma_r r_t \, dZ_t$$

In this model, the long-term interest rate is lognormally distributed. As interest rates approach zero, the rate volatility $\sigma_r r_t$ approaches zero, and rates will not become negative. Volatility will increase as rates increase from zero in this geometric Brownian motion process. Nevertheless, this model does not capture the tendency for short-term interest rates to revert toward some long-term mean as we often observe in actual financial markets.

8.2 ORNSTEIN–UHLENBECK PROCESSES

Unlike returns for stocks, short-term interest rates have a generally accepted empirically verified tendency to drift toward some long-term mean rate μ (the empirical evidence for stock drift is rather mixed, though weak at best). This long-term mean interest rate is not the same as the long-term interest rate, but the mean of short-term rates over an extended period of time. For example, when the short-term rate exceeds the long-term mean rate $(r_t > \mu)$, the drift might be expected to be negative so that the short-term rate drifts down toward the long-term mean rate. We might say that interest rates are currently high in this scenario, and we expect for them to drop toward the long-term mean. When the short-term rate r_t is less than the long-term mean rate $(r_t < \mu)$, the drift might be expected to be positive. When the short-term rate r_t equals the long-term mean (or average) rate μ, the drift might be expected to be zero, though random changes might still occur. Thus, the short-term rate has a tendency to revert to its long-term mean μ, whose value might be the value justified by economic fundamentals such as capital productivity, long-term monetary policy, etc. Similar scenarios might exist with respect to currency exchange rates and arbitrage portfolio returns. We will briefly discuss such scenarios at the end of this chapter.

The Ornstein–Uhlenbeck process (Uhlenbeck and Ornstein, 1930) is a form of Markov process in which the random variables indexed by time have a tendency to revert toward a given constant or long-term mean μ. We will examine this process, first by characterizing its path, then solving the Ornstein–Uhlenbeck differential equation and finally providing illustrations. In Section 8.3, we will discuss the Vasicek bond model, which prices bonds based on this process.

8.2.1 The Ornstein–Uhlenbeck Path

Short-term interest and exchange rates frequently exhibit a mean-reverting behavior as do many commodity prices and self-financing portfolio values. Define the term $0 < \lambda < 1$ to be a constant that reflects the speed of the mean-reverting adjustment for the instantaneous short-term rate r_t toward its constant long-term mean rate μ; that is, λ is the mean reversion factor, sometimes called a "pullback factor." This pullback factor is typically estimated or calibrated based on a statistical analysis of historical data. Let $\sigma_r \, dZ_t$ represent random shocks or disturbances to r_t, where σ_r is a constant. The following defines the Ornstein–Uhlenbeck mean-reverting process:

$$dr_t = \lambda(\mu - r_t)dt + \sigma_r \, dZ_t$$

This is a stochastic differential equation where dZ_t is standard Brownian motion. The Ornstein–Uhlenbeck process is a stationary diffusion-type Markov process. The Ornstein–Uhlenbeck process is sometimes called an elastic random walk, and has been used in physics to model, among other phenomena, the behavior of springs.

Recall the illustration on mean-reverting interest rates from Section 4.2.1. In our discussion there, the short rate diverged from the long-term mean, because there was no mechanism to accomplish this divergence. Here, we will see that the Ornstein–Uhlenbeck process has two components, the mean reversion component $\lambda(\mu - r_t)$ and the Brownian motion component $\sigma_r \, dZ_t$. The Brownian motion component is the disturbance factor that causes the short-rate r_t to diverge from the long-term mean rate μ. The mean reversion component draws the short-term rate r_t back toward the long-term mean rate μ. A higher value for λ implies a faster reversion by the short-term rate r_t to the long-term mean rate μ.

If λ were greater than 1, the process would no longer be *convergent*, and would ultimately "explode." This means that the short-term rate would ultimately move away from its long-term mean at an increasing rate. If λ were zero, there would be no mean reversion and the process would be Brownian motion. In an interest rate environment, the Ornstein–Uhlenbeck process is often known as the *Vasicek model*. One drawback to the Vasicek interest rate model is that interest rate shifts have a normally distributed component, leading to the unfortunate result that it is possible for the interest rate to become negative. Obviously, this creates an arbitrage opportunity when cash is available for investors to hold. We will focus on this difficulty later in this chapter.

Figures 8.1 and 8.2 depict simulations of Ornstein–Uhlenbeck processes. In each figure, the process is depicted over length of time 200, with $r_0 = 0.05$ and $\sigma_r = 0.02$. In Figure 8.1, the mean reversion is more significant, with $\lambda = 0.9$: λ is only 0.1 in Figure 8.2. Notice that as λ gets closer to 1, the process becomes more mean reverting. This means that the pullback factor has a greater propensity to pull the rate back toward its long-term mean. As λ approaches 0, the Ornstein–Uhlenbeck process approaches Brownian motion. Also notice on Figure 8.2 that the process can drop below zero.

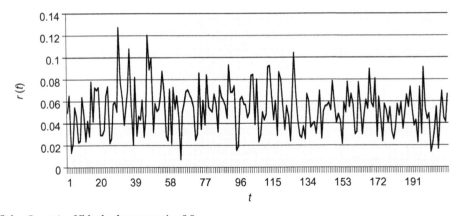

FIGURE 8.1   Ornstein–Uhlenbeck process: $\lambda = 0.9$.

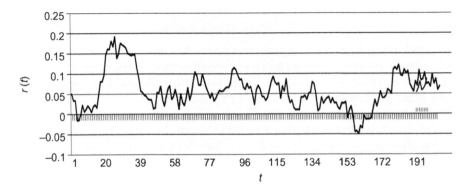

FIGURE 8.2   Ornstein–Uhlenbeck process: $\lambda = 0.1$.

8.2.2 Solving the Ornstein–Uhlenbeck Stochastic Differential Equation

Suppose that r_t follows the following Ornstein–Uhlenbeck process:

$$dr_t = \lambda(\mu - r_t)\,dt + \sigma_r\,dZ_t$$

We solve the Ornstein–Uhlenbeck stochastic differential equation by first multiplying both sides by $e^{\lambda t}$ then rewriting as follows:

$$e^{\lambda t}\,dr_t = \lambda(e^{\lambda t}\mu - e^{\lambda t}r_t)\,dt + e^{\lambda t}\sigma_r\,dZ_t$$

$$e^{\lambda t}\,dr_t + \lambda e^{\lambda t}r_t = \lambda e^{\lambda t}\mu\,dt + e^{\lambda t}\sigma_r\,dZ_t$$

The reason for rewriting the equation is so that we can apply the special product rule from Section 6.1.1 to the left-hand side, in order to rewrite it as $d(e^{\lambda t}r_t)$. Note that the left-hand side maintains its value when the product rule is applied to $d(e^{\lambda t}r_t)$:

$$e^{\lambda t}\,dr_t + \lambda e^{\lambda t}r_t = d(e^{\lambda t}r_t)$$

$$= \lambda e^{\lambda t}\mu\,dt + e^{\lambda t}\sigma_r\,dZ_t$$

Changing the variable from t to s, and integrating from 0 to t, we find that:

$$e^{\lambda t}r_t - r_0 = \mu\int_0^t \lambda e^{\lambda s}\,ds + \sigma_r\int_0^t e^{\lambda s}dZ_s$$

We evaluate the first integral and then solve algebraically for r_t. Rewriting further, we express the solution so as to distinguish the predictable part of r_t from the Brownian motion aspect of it. Following these steps gives:

$$r_t = e^{-\lambda t}\left[r_0 + \mu(e^{\lambda t} - 1) + \sigma_r\int_0^t e^{\lambda s}\,dZ_s\right] = \mu + e^{-\lambda t}(r_0 - \mu) + \sigma_r\int_0^t e^{\lambda(s-t)}\,dZ_s \qquad (8.1)$$

Conditional on the current instantaneous (short) rate r_0, the expected instantaneous rate $E[r_t]$ at time t is:

$$E[r_t] = \mu + e^{-\lambda t}(r_0 - \mu) \qquad (8.2)$$

which is implied by the martingale theorem in Section 6.1.5. This equation gives us the conditional expected time t instantaneous interest rate. This instantaneous rate is the periodic (usually annualized) rate for a loan that matures in a single instant after its origination. Note that as t gets large, the expected rate $E[r_t]$ approaches the long-term mean rate μ. The conditional short rate at time t given the current level is:

$$r_t = \mu + e^{-\lambda t}(r_0 - \mu) + \sigma_r\int_0^t e^{\lambda(s-t)}dZ_s = E[r_t] + \sigma_r\int_0^t e^{\lambda(s-t)}dZ_s$$

To find the variance σ_t^2 of r_t, first observe that:

$$\sigma_t^2 = \mathrm{Var}(r_t) = E\left[(r_t - E[r_t])^2\right] = E\left[\sigma_r^2\left(\int_0^t e^{\lambda(s-t)}dZ_s\right)^2\right]$$

Using the Itô isometry from Section 6.1.5, we can determine the variance:

$$\sigma_t^2 = \sigma_r^2 \int_0^t e^{2\lambda(s-t)} \, ds = \sigma_r^2 \left(\frac{1 - e^{-2\lambda t}}{2\lambda} \right) \tag{8.3}$$

Invoking the last theorem in Section 6.1.5, r_t must have a normal distribution; thus, $r_t \sim N(E[r_t], \sigma_t^2)$. Because r_t is normally distributed, it can take on negative values. Also, as t increases without bound, from Eq. (8.3) we see that the variance of the instantaneous rate r_t approaches $\sigma_r^2/(2\lambda)$.

In Section 4.2.1, we studied the analogous deterministic model in which $\sigma_r = 0$. We derived the solution Eq. (8.1) to be: $r_t = \mu + Ke^{-\lambda t}$. If we assume an initial interest rate of r_0, it is easy to check that the solution takes the form: $r_t = \mu + (r_0 - \mu)e^{-\lambda t}$. For this deterministic model, the interest rate r_t converges to the long-term mean rate of μ. For the Ornstein–Uhlenbeck process with $\sigma_r > 0$, as the time t increases without bound, the interest rates r_t have their expected values approach μ as we can see from Eq. (8.2). However, as we observed above, r_t maintains a normal distribution so that its variance approaches $\sigma_r^2/(2\lambda)$ as time t increases without bound. Nevertheless, there is a mean reversion in the sense that the initial interest rate r_0 at time 0 becomes irrelevant as time t becomes large, and the interest rates r_t will tend to fluctuate randomly about their long-term mean of μ.

ILLUSTRATION: CONDITIONAL EXPECTED SHORT RATES AND THE VASICEK MODEL

In this illustration, we present a simplified scenario to help understand our yield curve derivations later in this chapter. Here we will focus on various expected short rates and their evolution over time as a function of the instantaneous spot rate. Suppose that short-term (instantaneous) interest rates follow an Ornstein–Uhlenbeck (Vasicek model). The long-term mean interest rate μ equals 0.06, the short-term (instantaneous) rate r_0 currently equals 0.02, the pullback factor λ equals 0.1, and the instantaneous standard deviation σ_r of short rates equals 0.01:

$$dr_t = \lambda(\mu - r_t) \, dt + \sigma_r \, dZ_t = 0.1(0.06 - r_t) \, dt + 0.01 \, dZ_t$$

8.2.2.2 Expected Short Rates and Variances

First, we will focus on the instantaneous rates associated with the process above, and later in this chapter, we will discuss yields and prices associated with bonds. Instantaneous rates apply for a single instant. Using our expected short rate expression above, we calculate instantaneous forward rates as below and in Table 8.1, which also include calculations for instantaneous variances:

$$E[r_t] = \mu + e^{-\lambda t}(r_0 - \mu) = r_0 e^{-\lambda t} + \mu\left(1 - e^{-\lambda t}\right)$$

$$\sigma_t^2 = \sigma_r^2 \left(\frac{1 - e^{-2\lambda t}}{2\lambda} \right)$$

TABLE 8.1 Sampling of Expected Instantaneous Rates and Variances

t	$E[r_t]$	σ_t^2	t	$E[r_t]$	σ_t^2
1	0.023807	0.000091	10	0.045285	0.000432
2	0.027251	0.000165	20	0.054587	0.000491
3	0.030367	0.000226	30	0.058009	0.000499

For example, for 1 year, we have:

$$E[r_1|r_0 = 0.02] = 0.06 + e^{-0.1 \times 1}(0.02 - 0.06) = 0.023807$$

$$\sigma_1^2 = 0.0001 \left(\frac{1 - e^{-2 \times .1 \times 1}}{2 \times 0.1} \right) = 0.000091$$

We see in Table 8.1 that the expected short rate is anticipated to increase to 2.3807% in 1 year, 2.7251% in 2 years, 5.8009% in 30 years, and ultimately approaching the long-term mean rate of 6% as t approaches ∞. Another interpretation of this expected rate $E[r_t]$ after t years is that it is a weighted average of the current short rate r_0 and the long-term mean value μ. The weight on the short rate declines exponentially with t and the pullback factor λ, while the weight on the long term mean rate increases at an offsetting rate. The variance of the short rate also increases with time, asymptotically approaching its stationary long-term level $\sigma_r^2/2\lambda$. Notice from Table 8.1 that, because of the normal range of the pullback factor, $0 < \lambda < 1$, the variance of the process increases at a rate less than time t.

8.3 SINGLE RISK FACTOR INTEREST RATE MODELS

Interest rate models are important because interest rates are not constant over time as we sometimes assume and because of interest rate products that are used to hedge this uncertainty require inputs to estimate risk. In the next two sections, we assume that interest rates are driven by a single source of risk. The short-term interest rate r_t will be expressed as a stochastic differential equation with a term involving Brownian motion as the single source of risk. We will devote particular attention to the Vasicek model. The Vasicek model is, in some ways, more reasonable and consistent with reality than Brownian motion or Rendleman–Bartter processes in that these two processes lack the tendency to revert to mean values. Here, we will seek to value bonds in stochastic interest rate environments, and to estimate spot and forward rates of interest over different terms to maturity so as to construct yield curves.

8.3.1 Pricing a Zero-Coupon Bond

Valuing a zero-coupon bond with face value F in an environment with continuously discounted rates r_s following a continuous stochastic process is accomplished with the following very general form:

$$B_0 = E \left[Fe^{\left(-\int_0^T r_s \, ds \right)} | r_0 \right] \qquad (8.4)$$

This implies that the time 0 bond value, B_0, is the expected value of the time T payment F, where the instantaneous discount rate r_s varies over time s. In effect, and in intuitive terms, at each instant, the face value F is discounted at rate r_s, and these discount functions are summed over all time periods from 0 to T. Individual payments associated with coupon bonds would be valued separately as though each were zero-coupon bonds, then individual cash flow present values would be summed into portfolios.

Before we derive the formulas for pricing zero-coupon bonds, we will derive the partial differential equation that the bond price must satisfy. This equation is the analogue to the Black–Scholes partial differential equation that was used to price options. We will give two derivations. Both derivations assume that the instantaneous short interest rate r_t is described by any stochastic differential equation (the Ornstein–Uhlenbeck process being a special case). However, the first derivation is heuristic since it is based on the assumption that the instantaneous expected return on the bond equals the instantaneous short-term interest rate.

8.3.1.1 *The Interest Rate Process and Itô's Lemma*

We will assume that the instantaneous interest rate r_t existing in physical probability space $\mathbb{P}$ can be characterized by the following stochastic differential equation:

$$dr_t = a\,dt + b\,dZ_t \qquad (8.5)$$

where $a = a(r_t,t)$ and $b = b(r_t,t)$ are themselves stochastic processes.[1] Note that while the state variable r_t takes on values in the physical probability space, as in previous chapters, valuation will usually occur in risk-neutral probability space $\mathbb{Q}$. This is consistent with option pricing models that we discussed earlier. Assume that the face value of the bond $B_T = F$ for some positive constant F. The time t price of a bond $B_t = B(t, T, r_t)$ is a function of time $t \le T$, the bond's maturity date T and the time t instantaneous (short) rate of interest r_t. Since the interest rate r_t is a function of time t, we can also view the bond price as a function of just t and T; that is, $B(t, T)$. By Itô's lemma, the change in the price of the bond from time t to time $t + dt$ is:

$$dB = \frac{\partial B}{\partial r}\,dr + \frac{\partial B}{\partial t}\,dt + \frac{1}{2}\frac{\partial^2 B}{\partial r^2}(dr)^2$$

Based on our assumption that the instantaneous interest rate r_t follows a stochastic differential equation given by Eq. (8.5) that enables us to write and simplify our bond price process, and recalling that $(dZ_t)^2 = dt$, we have:

$$dB = \frac{\partial B}{\partial r}(a\,dt + b\,dZ_t) + \frac{\partial B}{\partial t}\,dt + \frac{1}{2}\frac{\partial^2 B}{\partial r^2}(a\,dt + b\,dZ_t)^2$$

$$dB = \left(a\frac{\partial B}{\partial r} + \frac{\partial B}{\partial t} + \frac{1}{2}b^2\frac{\partial^2 B}{\partial r^2}\right)dt + b\frac{\partial B}{\partial r}\,dZ_t \qquad (8.6)$$

Now, we will divide Eq. (8.6) by B so as to express the return over the infinitesimal time change dt:

$$\frac{dB}{B} = \frac{1}{B}\left(a\frac{\partial B}{\partial r} + \frac{\partial B}{\partial t} + \frac{1}{2}b^2\frac{\partial^2 B}{\partial r^2}\right)dt + \frac{b}{B}\frac{\partial B}{\partial r}\,dZ_t$$

We will split this equation into its predictable and Brownian motion components, rewriting it as follows:

$$\frac{dB}{B} = E[r_B]\,dt + \sigma_B\,dZ_t \qquad (8.7)$$

based on the following definitions:

$$E[r_B] = \frac{1}{B}\left(a\frac{\partial B}{\partial r} + \frac{\partial B}{\partial t} + \frac{1}{2}b^2\frac{\partial^2 B}{\partial r^2}\right) \tag{8.8}$$

and:

$$\sigma_B = \frac{b}{B}\frac{\partial B}{\partial r} \tag{8.9}$$

Heuristically, one can think of $E[r_B]$ as the expected instantaneous rate of return for the bond and σ_B as its instantaneous volatility. This follows from the non-rigorous step of taking the expectation on both sides of Eq. (8.7):

$$E\left[\frac{dB}{B}\right] = E[E[r_B]\,dt] + E[\sigma_B\,dZ_t] = E[r_B]\,dt$$

We can rewrite Eq. (8.8) in the form:

$$a\frac{\partial B}{\partial r} + \frac{\partial B}{\partial t} + \frac{1}{2}b^2\frac{\partial^2 B}{\partial r^2} - BE[r_B] = 0 \tag{8.10}$$

subject to the bond's terminal payoff function, $B(t, T) = F$. However, as the increment of time approaches zero, if we assume that the expected instantaneous bond return approaches the instantaneous market interest rate r, then our pricing differential equation becomes:

$$a\frac{\partial B}{\partial r} + \frac{\partial B}{\partial t} + \frac{1}{2}b^2\frac{\partial^2 B}{\partial r^2} - rB = 0 \tag{8.11}$$

The next scenario is more general, where we allow for the possibility that the expected instantaneous rate of return on the bond may be different than the instantaneous short-term interest rate. The argument is also more rigorous, where we avoid taking the expectation of an infinitesimal.

8.3.1.2 Setting the Self-Financing Portfolio Combinations

Valuing bonds when their underlying interest rates are uncertain is, in many respects, similar to valuing a stock option when the underlying stock price is risky. However, a complication that we encounter when valuing bonds in an uncertain interest rate environment is that the source of risk is not an underlying security such as a stock, but instead an interest rate r. This means that we cannot hedge the risk in the same way. Without an underlying hedging security, hedging risk as in the Black–Scholes and binomial option pricing models requires a different mindset. Instead of hedging with an underlying security, interest rate risk is hedged by combining bonds of different maturities into a self-financing portfolio.

With our single factor model, we will be able to demonstrate a common risk premium that applies to all time periods and all bonds in our market. The market will be complete as long as all of the bonds are driven by the same interest rate model. Similar to the derivation of the Black–Scholes differential equation, we will apply a no-arbitrage argument to construct the differential equation for the bond price. This differential equation can then be solved. Thus, our pricing mechanism is quite similar to those discussed in Chapter 7.

Consider the price dynamics of two bonds $B_1(t, T_1, r_t)$ and $B_2(t, T_2, r_t)$ in the same economy (that is, there is a single source of risk affecting both bonds) with different maturity dates:

$$dB_1 = \left(a\frac{\partial B_1}{\partial r} + \frac{\partial B_1}{\partial t} + \frac{1}{2}b^2\frac{\partial^2 B_1}{\partial r^2} \right)dt + B_1\sigma_{B1}\,dZ_t \tag{8.12a}$$

and

$$dB_2 = \left(a\frac{\partial B_2}{\partial r} + \frac{\partial B_2}{\partial t} + \frac{1}{2}b^2\frac{\partial^2 B_2}{\partial r^2} \right)dt + B_2\sigma_{B2}\,dZ_t \tag{8.12b}$$

where the short-term interest rate r_t satisfies the stochastic differential Eq. (8.5) (the source of risk). Equations (8.12a) and (8.12b) were already derived (Eq. 8.6). The terms σ_{B1} and σ_{B2} represent standard deviations of instantaneous returns for bonds 1 and 2. For brevity of notation in the calculations to follow denote:

$$E[r_{B1}] = \frac{1}{B_1}\left(a\frac{\partial B_1}{\partial r} + \frac{\partial B_1}{\partial t} + \frac{1}{2}b^2\frac{\partial^2 B_1}{\partial r^2} \right) \tag{8.13a}$$

and:

$$E[r_{B2}] = \frac{1}{B_2}\left(a\frac{\partial B_2}{\partial r} + \frac{\partial B_2}{\partial t} + \frac{1}{2}b^2\frac{\partial^2 B_2}{\partial r^2} \right) \tag{8.13b}$$

Notice that the pricing dynamics of both bonds draw from the same underlying process, but that their parameters $(E[r_{B1}], E[r_{B2}], \sigma_{B1}$ and $\sigma_{B2})$ differ.

To derive the partial differential equation for pricing a bond, we construct a portfolio that consists of buying γ_1 units of a bond B_1 and shorting γ_2 units of a bond B_2 so that the total value of the self-financing portfolio P is:

$$P = \gamma_1 B_1 - \gamma_2 B_2$$

The self-financing portfolio requires that the change in the value of the portfolio is due only to changes in the prices of the bonds themselves. This requires:

$$dP = \gamma_1\,dB_1 - \gamma_2\,dB_2$$

Rewrite, substituting Eqs. (8.12a,b) and (8.13a,b) from above:

$$dP = \left(\gamma_1\left(\frac{\partial B_1}{\partial r}\lambda(\mu - r_t) + \frac{\partial B_1}{\partial t} + \frac{1}{2}\frac{\partial^2 B_1}{\partial r^2}\sigma_r^2 \right) - \gamma_2\left(\frac{\partial B_2}{\partial r}\lambda(\mu - r_t) + \frac{\partial B_2}{\partial t} + \frac{1}{2}\frac{\partial^2 B_2}{\partial r^2}\sigma_r^2 \right) \right)dt$$

$$+ \left(\gamma_1\sigma_{B1} - \gamma_2\sigma_{B2} \right)dZ_t$$

$$dP = \left(\gamma_1 E[r_{B1}]B_1 - \gamma_2 E[r_{B2}]B_2 \right)dt + \left(\gamma_1\sigma_{B1}B_1 - \gamma_2\sigma_{B2}B_2 \right)dZ_t$$

We will hedge away portfolio risk by choosing γ_1 and γ_2 so that $\gamma_1\sigma_{B1}B_1 = \gamma_2\sigma_{B2}B_2$, eliminating the Brownian motion component from the hedging portfolio dynamic. Thus:

$$dP = \left(\gamma_1 E[r_{B1}]B_1 - \gamma_2 E[r_{B2}]B_2 \right)dt$$

We have selected γ_1 and γ_2 so that portfolio risk has been hedged away, in a manner very similar to our Black–Scholes hedge in Chapter 7. Our self-financing portfolio value and hedge of portfolio risk leads to the following two linear equation systems:

$$P = \gamma_1 B_1 - \gamma_2 B_2$$

$$0 = \gamma_1 \sigma_{B1} B_1 - \gamma_2 \sigma_{B2} B_2$$

Now, we calculate the portfolio strategy. Multiply the first of these two equations by σ_{B1} and subtract from the second, which yields:

$$-\sigma_{B1} P = \gamma_2 B_2 (\sigma_{B1} - \sigma_{B2})$$

$$\gamma_2 = \frac{\sigma_{B1} P}{B_2(\sigma_{B2} - \sigma_{B1})}$$

Similarly, we have:

$$\gamma_1 = \frac{\sigma_{B2} P}{B_1(\sigma_{B2} - \sigma_{B1})}$$

These portfolio transactions provide for a dynamic replication strategy analogous to that used in the Black–Scholes formula derivation, and ensure that our self-financing portfolio is riskless over the interval dt. The ratio of γ_1 to γ_2, as with the Black–Scholes hedge ratio, must be updated continuously to maintain the hedge. For our purposes, the importance of this hedge ratio is to enable us to obtain arbitrage-free pricing of the bonds. We will see shortly that this will lead to a risk premium that is common to all of the bonds in the arbitrage-free market. This will enable us to eliminate the individual drifts of the separate bonds for purposes of pricing, similar to the manner in which arbitrage-free pricing of options led to eliminating the individual drifts μ of the stocks. Furthermore, just as in the case of options, in the risk-neutral probability space $\mathbb{Q}$, the discounted bond becomes a martingale. Thus, we seek to value bonds in a risk-neutral environment just as we did options in Chapter 7.

Our next issue is to show that we can select B_1 and B_2 so that the bond portfolio is self-financing. If we substitute the portfolio strategy γ_1 and γ_2 from the equations above into the expression for dP and then divide by P we obtain:

$$\frac{dP}{P} = \frac{\sigma_{B2}}{\sigma_{B2} - \sigma_{B1}} \frac{dB_1}{B_1} - \frac{\sigma_{B1}}{\sigma_{B2} - \sigma_{B1}} \frac{dB_2}{B_2}$$

We see from this expression that changes in portfolio value result entirely from changes in component bond prices. This means that we have a differential expression for the stochastic process P that represents the price path of the self-financing portfolio. It is a matter of determining P so that this differential expression is satisfied. We do not need to know the actual solution to this equation, we just need to know that the solution exists. As you may know from an elementary course in differential equations, a first-order differential equation with real-valued variables and sufficiently continuous terms is always guaranteed of having a solution. Similarly, the first-order stochastic differential above is guaranteed to have a solution for P. Thus, our portfolio is both self-financing and riskless.

8.3.1.3 *The Market Price of Risk and the Bond Pricing Differential Equation*

Bonds of varying levels of maturity have different levels of risk due to the volatile nature of interest rates. Investors normally require higher returns for higher levels of risk. We will derive an expression for the risk premium of a security that provides how much additional return investors will require per additional unit of risk. This risk premium may vary with time, but all of the bonds in this market are driven by the same short-term interest rate model. Now, we use our portfolio combinations to write our self-financing portfolio dynamics from above as follows:

$$dP = \left(\gamma_1 E[r_{B1}]B_1 - \gamma_2 E[r_{B2}]B_2\right)dt = \left(\frac{\sigma_{B2}P}{(\sigma_{B2} - \sigma_{B1})}E[r_{B1}] - \frac{\sigma_{B1}P}{(\sigma_{B2} - \sigma_{B1})}E[r_{B2}]\right)dt$$

Simplify slightly, and in the absence of arbitrage opportunities, the self-financing portfolio's return must equal the riskless rate:

$$\frac{dP}{P} = \left(\frac{\sigma_{B2}}{(\sigma_{B2} - \sigma_{B1})}E[r_{B1}] - \frac{\sigma_{B1}}{(\sigma_{B2} - \sigma_{B1})}E[r_{B2}]\right)dt = r_t dt$$

The absence of arbitrage opportunities means that our self-financing portfolio must earn the riskless rate. Long-term bonds face risk even when they are default risk-free. Now we rewrite our self-financing portfolio return dynamic in terms of risk premiums by multiplying each expression by $(\sigma_{B2} - \sigma_{B1})$ then subtracting $(\sigma_{B2} - \sigma_{B1})r_t \, dt$ from each of the expressions and rearranging to obtain:

$$\frac{E[r_{B1}] - r_t}{\sigma_{B1}} = \frac{E[r_{B2}] - r_t}{\sigma_{B2}}$$

More generally, we can write this ratio for any bond as follows:

$$\frac{E[r_B] - r_t}{\sigma_B} = \Theta(r_t, t) \tag{8.14}$$

This ratio is referred to as the risk premium or *Sharpe ratio*, and it suggests a similarity between the model presented here and the Capital Asset Pricing Model. The implications of this equality are profound. First, it implies that the excess return (risk premium) at time t required by the market for each unit of risk is the same for all default-free bonds at all maturities, as long as their returns are driven by the same short-term interest rate model. That is, investors require additional return for risk bearing, and that risk premium is always proportional to the level of risk σ_B.[2] Thus, $\Theta(r_t, t)$ is the market price per unit of risk σ at time t. Rearranging and expressing in terms of bond expected instantaneous return, we have:

$$E[r_B] = r_t + \Theta \sigma_B$$

Observe that this equation shows that the drift $E[r_B]$ of an individual bond B may depend on its volatility σ_B, but this drift must have the same linear relationship with respect to its volatility as all other bonds that follow the same underlying interest rate process r_t.

Now, we substitute our definitions of expected instantaneous return and volatility of the bond from Eqs. (8.8) and (8.9) above, multiply both sides by B and rearrange as follows:

$$\frac{1}{B}\left(a\frac{\partial B}{\partial r} + \frac{\partial B}{\partial t} + \frac{1}{2}\frac{\partial^2 B}{\partial r^2}\sigma_r^2\right) = r_t + \Theta\frac{\partial B}{\partial r}\frac{b}{B}$$

$$(a - \Theta b)\frac{\partial B}{\partial r} + \frac{\partial B}{\partial t} + \frac{1}{2}\frac{\partial^2 B}{\partial r^2}\sigma_r^2 - Br_t = 0 \qquad (8.15)$$

This is the differential equation for pricing a bond in probability space $\mathbb{P}$, with boundary condition $B(T, T) = F$. The solution to this pricing differential equation is obtained on the companion website. Alternatively, we can use the martingale approach to price the bond in a risk-neutral environment, as we will discuss next.

8.3.1.4 *The Martingale Approach*

To price bonds and other interest rate instruments, we need to express the present value of some future certain cash flow. Suppose, for example, the price of a long-term bond is driven by some continuous instantaneous interest rate r_t. Here, we will first approximate the continuous model with a discrete model. Divide the interval of time $[0, t]$ into n subintervals of equal width $\Delta t = (t - 0)/n = t/n$. Recall the derivation in Chapter 4 of the present value of a bond that follows an interest rate $r(t)$ for the case of a real-valued non-stochastic model. We derive the present value PV_t of a bond with face value equal to 1. Using a similar argument, one obtains the same form for its approximate value for the present situation in which $r(t)$ is a stochastic process:

$$PV_t \approx (1)(1 + r_{t_0}\Delta t)^{-1}(1 + r_{t_1}\Delta t)^{-1}\cdots(1 + r_{t_{n-1}}\Delta t)^{-1}$$

Taking logs on both sides, using the fact that the log of a product is a sum of the logs, and expressing the result in summation notation, we obtain:

$$\ln PV_t \approx -\sum_{i=1}^{n}\ln(1 + r_{t_{i-1}}\Delta t)$$

Similar to the derivation in Chapter 4 for bond pricing, we have the approximation for each time t_{i-1}:

$$\ln\left(1 + r_{t_{i-1}}\Delta t\right) \approx r_{t_{i-1}}\Delta t$$

The approximation for $\ln PV_t$ becomes:

$$\ln PV_t \approx -\sum_{i=1}^{n}r_{t_{i-1}}\Delta t$$

Next, we exponentiate both sides of the equation above:

$$PV_t \approx \exp\left(-\sum_{i=1}^{n}r_{t_{i-1}}\Delta t\right)$$

In the limit as n approaches infinity, the approximation becomes an equality:

$$\mathrm{PV}_t = \lim_{n \to \infty} \exp\left(-\sum_{i=1}^{n} r_{t_{i-1}} \Delta t\right) = e^{\left(-\int_0^t r_s \, ds\right)}$$

To price an option, recall that we discounted the value of the option at expiration time T with e^{-rT}. Similarly, we discount the face value F of the bond at maturity T as follows:

$$Fe^{\left(-\int_0^t r_s \, ds\right)}$$

Suppose that the original probability measure is some physical probability measure $\mathbb{P}$. In the martingale approach to pricing an option, recall that the value of the option at time 0 is the expected value taken at time 0 of its discounted expiration value with respect to the risk-neutral equivalent probability measure $\mathbb{Q}$. Similarly, the value of the long-term bond at time 0 is the expected value of the bond with respect to the risk-neutral equivalent probability measure $\mathbb{Q}$ and is:

$$B(0, T) = E_{\mathbb{Q}}\left[Fe^{\left(-\int_0^T r_s \, ds\right)} | \mathscr{F}_0\right] \tag{8.16}$$

Compare this result with the price obtained for the bond derived by the PDE approach in on the companion website, where we show that:

$$B(0, T) = E_{\mathbb{P}}\left[F\exp\left(-\int_0^T r_s \, ds - \frac{1}{2}\int_0^T \Theta^2(s, r_s) \, ds - \int_0^T \Theta(s, r_s) \, dZ_s\right) | \mathscr{F}_0\right] \tag{8.17}$$

where $\Theta(s, r_s)$ is a per unit risk premium. This suggests that the Radon–Nikodym derivative from the Cameron–Martin–Girsanov theorem to change from the probability space $\mathbb{P}$ to the equivalent risk-neutral probability measure $\mathbb{Q}$ should be:

$$\xi_T = \exp\left(-\frac{1}{2}\int_0^T \Theta^2(s, r_s) ds - \int_0^T \Theta(s, r_s) \, dZ_s\right) \tag{8.18}$$

This will enable us to rewrite Eq. (8.17) as:

$$B(0, T) = E_{\mathbb{P}}\left[Fe^{\left(-\int_0^T r_s \, ds\right)} \xi_T | \mathscr{F}_0\right] = E_{\mathbb{Q}}\left[Fe^{\left(-\int_0^T r_s \, ds\right)} | \mathscr{F}_0\right]$$

thus showing that Eqs. (8.16) and (8.17) give equivalent pricing for the bond. In a manner analogous to option pricing, the discounted face value of the bond is a martingale with respect to the risk-neutral probability space $\mathbb{Q}$. If Eq. (8.18) is the Radon–Nikodym derivative in the Cameron–Martin–Girsanov theorem, then we know that it will shift the drift by a factor $-\Theta(t, r_t)$ for a process of the form $dX_t = \mu_t \, dt + dZ_t$ in probability space $\mathbb{P}$ to $dX_t = (\mu_t - \Theta(t, r_t)) \, dt + d\hat{Z}_t$ in probability space $\mathbb{Q}$. More generally, a process of the form $dX_t = \mu_t + \sigma_t \, dZ_t$ will be shifted to $dX_t = (\mu_t - \Theta(t, r_t) \, \sigma_t) \, dt + \sigma_t \, d\hat{Z}_t$. We will, in fact, shortly show that for the Vasicek model and for the case that the risk premium is a constant that the short-term interest rate process r_t is shifted in precisely this manner.

8.3.1.5 *The Risk-Neutral Probability Space and the Vasicek Model*

Now we discuss the shift to the risk-neutral probability space to show its effect on the Vasicek bond pricing model. We show on the companion website that if we assume that the market price of risk Θ is constant (in the probability space $\mathbb{P}$), then the bond price at time 0 is:

$$B(0, T) = Fe^{\left[\frac{1}{\lambda}(1-e^{-\lambda T})(r_\infty - r_0) - r_\infty T - \frac{\sigma_r^2}{4\lambda^3}(1-e^{-\lambda T})^2\right]} \tag{8.19}$$

with face value F and where the infinitely long rate is:

$$r_\infty = \mu - \frac{\sigma_r \Theta}{\lambda} - \frac{\sigma_r^2}{2\lambda^2}$$

Suppose that our hypothesis is that the risk premium is zero in risk-neutral probability space $\mathbb{Q}$. Based on observations we made earlier concerning the shift in the drift arising from the Cameron–Martin–Girsanov theorem, we would expect that the short-term interest rate process in probability space $\mathbb{P}$ to follow:

$$dr_t = \lambda(\mu - r_t)dt + \sigma_r\,dZ_t$$

and in probability space $\mathbb{Q}$, to follow:

$$dr_t = \lambda\left(\mu - \frac{\sigma_r \Theta}{\lambda} - r_t\right)dt + \sigma_r\,d\hat{Z}_t$$

where $\sigma_r \Theta$ is the risk premium associated with σ_r units of risk. This would mean that in risk-neutral probability space $\mathbb{Q}$, we can write the new risk premium to be $\hat{\Theta} = 0$ if the Vasicek model has the form:

$$dr_t = \lambda(\hat{\mu} - r_t)\,dt + \sigma_r\,d\hat{Z}_t$$

with:

$$\hat{\mu} = \mu - \frac{\sigma_r \Theta}{\lambda}$$

This is, in fact, exactly what happens. If we replace Θ with $\hat{\Theta}$ and μ with $\hat{\mu}$ in the solution (8.19) to obtain:

$$B(0, T) = Fe^{\left[\frac{1}{\lambda}(1-e^{-\lambda T})(r_\infty - r_0) - r_\infty T - \frac{\sigma_r^2}{4\lambda^3}(1-e^{-\lambda T})^2\right]}$$

where:

$$r_\infty = \hat{\mu} - \frac{\sigma_r \hat{\Theta}}{\lambda} - \frac{\sigma_r^2}{2\lambda^2}$$

Since $\hat{\Theta} = 0$ and $\hat{\mu} = \mu - \sigma_r \Theta/\lambda$, then:

$$r_\infty = \mu - \frac{\sigma_r \Theta}{\lambda} - \frac{\sigma_r^2}{2\lambda^2}$$

The solutions are identical. Let us summarize what we have just shown. Suppose we first price a bond with short-term interest rates following the Vasicek model with long-term mean μ and with a constant risk premium equal to Θ in probability space $\mathbb{P}$. Then, we obtain the same pricing of the bond if in a probability space $\mathbb{Q}$ short-term interest rates follow the Vasicek model with shifted long-term mean $\hat{\mu} = \mu - \sigma_r \Theta / \lambda$ so that the risk premium becomes $\hat{\Theta} = 0$.

8.3.1.6 The Bond Pricing Formula

In this section, we will price bonds in a risk-neutral environment under probability measure $\mathbb{Q}$, enabling us to set the risk premium equal to zero as we justified above. For convenience, we will label the long-term mean interest rate by μ rather than $\hat{\mu}$. By Eq. (8.19), the bond's time 0 value B_0 can be expressed in the following way, where F is the face value of the bond:

$$B_0 = Fe^{\left[\kappa(r_\infty - r_0) - r_\infty T - \frac{(\sigma_r \kappa)^2}{4\lambda}\right]} \tag{8.20a}$$

where[3]:

$$r_\infty = \mu - \frac{\sigma_r^2}{2\lambda^2}$$

$$\kappa = \frac{1 - e^{-\lambda T}}{\lambda}$$

Equation (8.20a) is useful for computational purposes. If we furthermore define:

$$\Lambda = \left[\mu - \frac{\sigma_r^2}{2\lambda^2}\right][\kappa - T] - \frac{\sigma_r^2 \kappa^2}{4\lambda} = r_\infty[\kappa - T] - \frac{(\sigma_r \kappa)^2}{4\lambda}$$

then we can also express the bond price in the form:

$$B_0 = Fe^{\Lambda - r_0 \kappa} = Fe^{-r_{0,T} T} \tag{8.20b}$$

The term Λ is the log of the bond price when $r_0 = 0$ (the vertical intercept term in a graph of B_0 against r_0) and κ represents the rate of change in the bond price (per unit of the bond) as the instantaneous spot rate r_0 increases. That is, κ is a slope term for the single factor for B_0 against r_0. In effect, κ tells us the impact on bond prices of a change in the overnight federal funds rate. The term $r_{0,T}$ is the yield, which we will discuss shortly.

We earlier defined in Eq. (8.9) the instantaneous volatility of a bond. We use the derivative of our bond pricing equation with respect to the instantaneous rate above to rewrite the absolute value of the volatility:

$$|\sigma_B| = \left|\frac{\partial B}{\partial r}\frac{\sigma_r}{B}\right| = \kappa \sigma_r = \sigma_r \frac{1 - e^{-\lambda T}}{\lambda}$$

Note that this volatility is increasing and concave down with respect to the length of time T to maturity. Also notice that as the maturity date T approaches 0, the bond's volatility approaches zero. This should seem reasonable because the bond is considered default risk-free, meaning that its value approaches certainty as its maturity approaches. Thus, we should always expect that the volatility of a default-free bond will approach zero as it approaches its maturity.

8.3.2 The Yield Curve

As we discussed in Chapter 4, the market spot rate of interest, $r_{0,T}$, implied by the bond's price is given in general by:

$$r_{0,T} = -\frac{1}{T}\ln\left(\frac{B_0}{F}\right)$$

In the more specific Vasicek model discussed here, we see from Eqs. (8.20a) and (8.20b) that:

$$r_{0,T} = r_\infty - \frac{\kappa}{T}(r_\infty - r_0) + \frac{(\sigma_r \kappa)^2}{4\lambda T}$$

Note that as $T \to \infty$, the spot rate $r_{0,T}$ will asymptotically approach the infinite long-rate $r_\infty = \mu - 0.5\sigma_r^2/\lambda^2$ in the risk-neutral environment where $\Theta = 0$.

As we have seen above, the yield curve produced by the Vasicek model starts (intercepts) at the current instantaneous rate r_0 and approaches the infinite long-rate $r_\infty = \mu - 0.5\sigma_r^2/\lambda^2$. The following conditions produce upward sloping, downward sloping, and humped yield curves:

$$r_0 \leq r_\infty - \frac{\sigma_r^2}{4\lambda^2} \quad \text{(upward sloping)}$$

$$r_0 \geq r_\infty + \frac{\sigma_r^2}{2\lambda^2} \quad \text{(downward sloping)}$$

$$r_\infty + \frac{\sigma_r^2}{2\lambda^2} \geq r_0 \geq r_\infty - \frac{\sigma_r^2}{4\lambda^2} \quad \text{(humped)}$$

ILLUSTRATION: MEAN-REVERTING INTEREST RATES AND BOND PRICING

Suppose that short-term rates follow the Vasicek model in a risk-neutral environment. The long-term mean interest rate μ equals 0.04, the short-term (instantaneous) rate r_0 currently equals 0.015, the pullback factor λ equals 0.5, and the instantaneous standard deviation σ_r of short rates equals 0.02:

$$dr_t = \lambda(\mu - r_t)\, dt + \sigma_r\, dZ_t = 0.05(0.04 - r_t)\, dt + 0.02\, dZ_t$$

What is the current value of a 1-month zero-coupon bond B_0 with face value $F = 1$? What is the 1-month $(T = 1/12)$ spot rate $r_{0,1/12}$? Finally, plot out the yield curve over 25 years. The following are our calculations:

$$\kappa = \frac{1 - e^{-\lambda T}}{\lambda} = \frac{1 - e^{-0.5/12}}{0.5} = 0.0816211,$$

$$r_\infty = 0.04 - \frac{0.02^2}{2 \times 0.5^2} = 0.0392$$

$$B_0 = e^{\left[0.0816211(0.0392-0.015)-0.0392/12-\frac{(0.02 \times 0.0816211)^2}{4 \times 0.5}\right]} = 0.998708$$

$$r_{0,1/12} = -\ln\left(\frac{B_0}{F}\right)/T = -\frac{\ln\left(\frac{0.998708}{1}\right)}{0.08333333} = 0.015513$$

Thus, we find that the 1-month, zero-coupon \$1 bond is currently worth $B_0 = 0.998708$, implying a 1-month continuously compounded spot rate equal to $r_{0,1/12} = 0.015513$. Table 8.2 provides spot rates to map points on the yield curve over 25 years. The yield curve is plotted in Figure 8.3. By Eq. (8.2), the expected instantaneous forward rate $E[r_{1/12}]$ in 1 month is:

$$E[r_{1/12}] = \mu + e^{-\lambda T}(r_0 - \mu) = 0.04 + e^{-0.5/12}(0.015 - 0.04) = 0.01602$$

TABLE 8.2   Yield Curve Data; Spot Rates When $\mu = 0.04$, $r_0 = 0.015$, $\lambda = .5$, $\sigma_r = 0.02$

t	$r_{0,t}$	t	$r_{0,t}$
0.08333	0.015513225	7	0.032601995
0.25	0.016495583	8	0.033357180
0.5	0.017866202	9	0.033968889
0.75	0.019124042	10	0.034471537
1	0.020279938	11	0.034890116
2	0.024062513	12	0.035243001
3	0.026827440	14	0.035803048
4	0.028887086	16	0.036225981
5	0.030449394	20	0.036820106
6	0.031655336	25	0.037296007

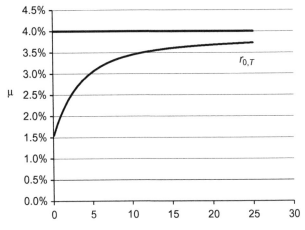

FIGURE 8.3   Yield curve; $\mu = 0.04$, $r_0 = 0.015$, $\lambda = 0.5$, $\sigma_r = 0.02$.

8.3.3 The Problem with Vasicek Models

The Vasicek yield curve model has a number of desirable characteristics. The model captures the empirical tendency for interest rates to revert toward some sort of mean rate. The model is driven off short-term interest rates, much as actual interest rates might be impacted by the federal funds rate, the "overnight" bank-to-bank controlled by the central bank (the Fed). However, there are a number of problems with the Vasicek model in characterizing the behavior of the yield curve:

1. The Vasicek model is likely to apply only in reasonably "normal" scenarios. For example, in situations involving crises such as hyperinflation, mean reversion is not likely to characterize the behavior of interest rates.
2. Even in more typical circumstances, empirical tests reveal that the Vasicek model does not characterize observed interest rate structures well.
3. Because it is based on Brownian motion, the Vasicek model does not allow for discreet jumps in the interest rate process.
4. The Vasicek model produces the result that all short- and long-term rates are perfectly correlated.
5. Related to the difficulty put forth just above, the Vasicek model assumes only a single underlying risk factor when, in fact, there is significant evidence that there may well be multiple factors. For example, sometimes the yield curve can "twist"; that is, long- and short-term rates can move in opposite directions. Multiple risk factors can often explain such "twisting."
6. Finally, the Vasicek model allows for the possibility of negative interest rates, even for negative real interest rates, a phenomenon that we should expect to observe rarely, if at all. We discussed a solution to this problem earlier (Bartter and Rendleman) and will introduce the Cox–Ingersoll–Ross model shortly as an alternative.

Why work with an interest rate model that presents all of these difficulties? As with most other financial models, we simply balance realism and ease of model building. The Vasicek model does capture some of the characteristics of a reasonable interest rate process and it is rather easy to work with, particularly in terms of parameter calibration. In addition, it is useful and sometimes very straightforward to adapt this framework into more realistic alternative depictions of interest rate processes.

8.4 ALTERNATIVE INTEREST RATE PROCESSES

In Section 8.1, we discussed arithmetic and geometric short-rate processes. Geometric processes never lead to negative short rates and neither allow for mean reversion. The Vasicek model does allow for mean reversion, but might result in negative short-term interest rates. We offer an alternative next. We do not seek to provide comprehensive mathematical discussions of models here; we only introduce basic models that can be studied more carefully and applied elsewhere.

8.4.1 The Merton Model

The short-rate dynamics, pricing differential equation, and bond pricing model produced by Merton (1973) are given by the next three equations:

$$dr_t = \mu \, dt + \sigma_r \, dZ_t$$

where μ and σ_r are constants. Substituting $a = \mu$ and $b = \sigma_r$ into Eq. (8.15), we obtain the partial differential equation governing the bond price for the Merton model:

$$\frac{\partial B}{\partial r}(\mu - \Theta\sigma_r) + \frac{\partial B}{\partial t} + \frac{1}{2}\sigma_r^2\frac{\partial^2 B}{\partial r^2} - rB = 0$$

Assuming a constant risk premium Θ, the following solution to this differential equation is derived by making use of Eq. (8.17)[4]:

$$B_0 = Fe^{\left[-r_0 T - \frac{(\mu-\Theta\sigma_r)T^2}{2} + \frac{\sigma_r^2 T^3}{6}\right]} \tag{8.21}$$

The corresponding term structure (equation for the yield curve) is:

$$r_{0,T} = -\frac{1}{T}\ln\left(\frac{B_0}{F}\right) = r_0 + \frac{(\mu - \Theta\sigma_r)T}{2} - \frac{\sigma_r^2 T^2}{6} \tag{8.22}$$

Note here that a shift in the short-term rate r_0 will result in a parallel shift in the yield curve; that is, the shift in $r_{0,T}$ will be the same for all T. Such a parallel shift is inconsistent with the empirical observation that short-term rates are more volatile than are long-term rates. Note also that the yield curve can never be monotonically increasing: it is either humped (when $\mu > \Theta\sigma_r$) or downward sloping. However, often when it is humped, the upward sloping portion of the hump can extend beyond the period being analyzed. Thus, the yield curve can be upward sloping over the "relevant" period.

Recall our numerical illustration from the previous section with the long-term mean interest rate $\mu = 0.04$, $r_0 = 0.015$, $\lambda = 0.5$, $F = 1$, and $\sigma_r = 0.02$. In a risk-neutral Merton (1973) environment, we value a 1-month zero-coupon bond and obtain its yield to maturity as follows:

$$dr_t = \mu\, dt + \sigma_r\, dZ_t$$

$$B_0 = e^{\left[-0.083333 \times 0.015 - \frac{0.04 \times 0.083333^2}{2} + \frac{0.02^2 \times 0.083333^3}{6}\right]} = 0.99861$$

$$r_{0,T} = \frac{-1}{0.083333}\ln\left(\frac{0.99861}{1}\right) = 0.01669$$

In comparison, one can check that a 30-year bond would produce a yield of 0.0376 and a bond price equal to 0.3276.

8.4.2 The Cox, Ingersoll, and Ross Process

A simple "square root correction" in the Cox–Ingersoll–Ross model (CIR, Cox, Ingersoll and Ross (1985a)) eliminates the potential problem of negative interest rates in the mean reverting Vasicek model (see Figure 8.4):

$$dr_t = \lambda(\mu - r_t)dt + \sigma_r\sqrt{r_t}\, dZ_t$$

A simulation of this model is depicted in Figure 8.4, with $\mu = r_0 = 0.05$, $\lambda = 0.1$, and $\sigma = 0.02$. This model has two components, the mean reversion component $\lambda(\mu - r_t)$ and the scaled Brownian motion component $\sigma_r\sqrt{r_t}\, dZ_t$. As in the Vasicek model, the mean reversion component draws the short-term rate back toward the long-term rate. However, notice that the

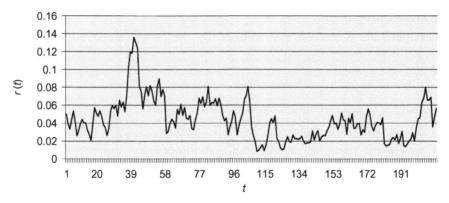

FIGURE 8.4 Cox, Ingersoll and Ross process simulation over 200 periods: $\lambda = 0.1$.

nominal (not proportional) volatility of rates is not constant; it declines as the interest rate declines. That is, as the short-term rate r_t approaches zero, the random component of interest rate changes $\sigma_r \sqrt{r_t}\, dZ_t$ will also approach zero. First, this characteristic is more realistic for interest rates, as the volatility of rates tends to be reduced as interest rates approach zero. Second, this process ensures that rates will not drop to or below zero, eliminating arbitrage opportunities for investors holding cash. This represents a second advantage of CIR over the Vasicek model.

8.4.3 Bond Pricing and the Yield Curve with CIR

The fundamental pricing differential equation and bond pricing model produced by Cox, Ingersoll, and Ross (1985b) are:

$$\left(\lambda(\mu - r) - \Theta' \sigma_r \sqrt{r}\right)\frac{\partial B}{\partial r} + \frac{\partial B}{\partial t} + \frac{1}{2}r\sigma_r^2 \frac{\partial^2 B}{\partial r^2} - rB = 0$$

$$B_0 = \left(Fe^{\frac{2\lambda\mu}{\sigma_r^2}\ln\left[\frac{2\varrho e^{(\lambda+\Theta'\,+\,\varrho)T/2}}{2\varrho+(\lambda+\Theta'\,+\,\varrho)(e^{\varrho T}-1)}\right]}\right)\left(e^{-r_0 \times \frac{2(e^{\varrho T}-1)}{2\varrho+(\lambda+\Theta'\,+\,\varrho)(e^{\varrho T}-1)}}\right) = Fe^{\Lambda-r_0\kappa} = Fe^{-r_{0,T}T} \qquad (8.23)$$

where F is the face value of the bond and Θ' is the CIR risk premium ($\Theta' r$ is the covariance between changes in interest rates and changes in market portfolio value) and:

$$\varrho = \sqrt{(\lambda+\Theta')^2 + 2\sigma_r^2}$$

$$\kappa = \frac{2(e^{\varrho T}-1)}{2\varrho + (\lambda+\Theta'+\varrho)(e^{\varrho T}-1)}$$

$$\Lambda = \frac{2\lambda\mu}{\sigma_r^2}\ln\left[\frac{\kappa\varrho e^{(\lambda+\Theta'+\varrho)T/2}}{e^{\varrho T}-1}\right]$$

For computational purposes, it is desirable to use: $B_0 = Fe^{\Lambda-r_0\kappa}$ as defined above in Eq. (8.23).

8.4.3.1 *The Yield Curve*

The market spot rate of interest, $r_{0,T}$, implied by the bond's price, is given by either of the following:

$$r_{0,T} = -\frac{1}{T}\ln\left(\frac{B_0}{F}\right)$$

$$r_{0,T} = -\frac{\Lambda}{T} + r_0\frac{\kappa}{T}$$

Note that as $T \to \infty$, the spot rate $r_{0,T}$ will asymptotically approach the infinite long-rate $r_\infty = 2(\mu\lambda)/(\lambda + \Theta' + \varrho)$.

8.4.3.2 *Numerical Illustration*

As we did in our numerical illustration of the Vasicek pricing model, suppose that short-term rates follow a CIR stochastic process in a risk-neutral environment. The long-term mean interest rate μ equals 0.04, the short-term (instantaneous) rate r_0 currently equals 0.015, the pullback factor λ equals 0.5, and the instantaneous standard deviation σ_r of short rates equals 0.02:

$$dr_t = \lambda(\mu - r_t)\, dt + \sigma_r\sqrt{r_t}\, dZ_t = 0.5(0.04 - r_t)\, dt + 0.02\sqrt{r_t}\, dZ_t$$

What is the current value of a 1-year zero-coupon bond B_0 with face value $F = 1$? What is the 1-year spot rate $r_{0,1}$? Finally, plot out the yield curve over 25 years. The following are our calculations:

$$\varrho = \sqrt{(\lambda+\Theta')^2 + 2\sigma_r^2} = \sqrt{0.5^2 + 2\times 0.02^2} = 0.5008$$

$$\kappa = \frac{2(e^{\varrho T} - 1)}{2\varrho + (\lambda + \Theta' + \varrho)(e^{\varrho T} - 1)} = \frac{2\times(0.65004)}{1.0016 + (1.0008)(0.65004)} = 0.7869$$

$$\Lambda = \frac{2\lambda\mu}{\sigma_r^2}\ln\left[\frac{\kappa\varrho e^{(\lambda+\Theta'+\varrho)T/2}}{e^{\varrho T} - 1}\right] = 100\times\ln\left[\frac{0.7869\times 0.5008 e^{(1.0008)\times 1/2}}{0.65004}\right] = -0.00852$$

$$B_0 = 1\times e^{-0.00852 - 0.015\times 0.7869} = 0.97988$$

8.4.3.3 *The Yield Curve*

The market spot rate of interest, $r_{0,T}$, implied by the bond's price is given by either of the following:

$$r_{0,T} = -\frac{1}{T}\ln\left(\frac{B_0}{F}\right) = -\ln\left(\frac{0.97988}{1}\right) = 0.02033$$

$$r_{0,T} = -\frac{\Lambda}{T} + r_0\frac{\kappa}{T} = -\frac{-0.00852}{1} + 0.015\frac{0.7869}{1} = 0.02033$$

TABLE 8.3 CIR Yield Curve Data; Cir Spot Rates When $\mu = 0.04$, $r_0 = 0.015$, $\lambda = 0.5$, $\sigma_r = 0.02$

t	$r_{0,t}$	t	$r_{0,t}$
1	0.02033	10	0.03501
2	0.02419	15	0.03665
3	0.02705	20	0.03747
4	0.02918	25	0.03797
5	0.03081	30	0.03831

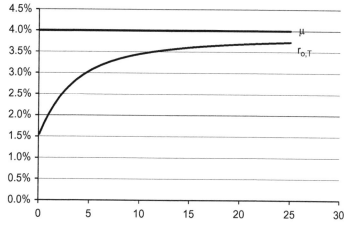

FIGURE 8.5 CIR yield curve: $\mu = 0.04$, $r_0 = 0.015$, $\lambda = 0.5$, $\sigma_r = 0.02$.

Thus, we find that the 1-year, zero-coupon \$1 bond is currently worth $B_0 = 0.97988$, implying a 1-year continuously compounded spot rate equal to $r_{0,1} = 0.02033$. Table 8.3 provides spot rates to map points on the yield curve over 25 years as in Figure 8.5.

8.4.4 Yield Curve Models: Summary and Further Development

Yield curve modeling essentially involves two processes. First, we attempt to fit a curve to observed yields over terms to maturity, as we discussed in Chapters 3 and 4. Second, as we discussed in this chapter, we focused on modeling how this yield curve shifts over time. Describing how yield curves shift over time is an obviously complex matter. In this chapter, we sought to provide a superficial introduction to a small fraction of the huge number of yield curve models and model types. Each of the models that we discussed above (Vasicek, Merton, and CIR) is of the single factor type in that each allows for a single instantaneous rate to be randomly generated and then builds the entire yield curve off of that rate. This is analogous to the central bank setting an overnight rate (e.g., the Fed setting the federal funds rate) and the market building a yield curve off of that rate. Then, the entire yield curve shifts as this instantaneous (fed funds) rate evolves over time.

All of the models that we discussed above are continuous models, not allowing for discrete jumps. The Vasicek and CIR models are both based on mean-reverting instantaneous rates and are examples of *exponential affine models*. This means that these two models produce bond prices that are exponentially related to spot rates, which, in turn, are linearly related to the instantaneous rate. These models take the following form: $B_{0,T} = e^{\Lambda - \kappa r(0)}$, or $\ln(B) = \Lambda - \kappa r_0 = -r_{0,T}T$. This means that, first, spot rates $r_{0,T}$ are simple linear functions of the instantaneous spot rate r_0. In effect, as we discussed earlier, Λ is the log of the bond price when $r_{0,T} = 0$ (the vertical intercept term) and κ depicts the rate of change in the bond price as the instantaneous spot rate increases. That is, κ is a slope term for the single factor r_0. In effect, κ tells us the impact on bond prices of a change in the overnight federal funds rate. These models are attractive because they tend to be easy to work with.

An attractive feature of exponential affine models is that most (e.g., Vasicek) require only a few inputs in addition to maturities such as λ, the pullback factor; σ, the instantaneous standard deviation; and μ, the long-term mean rate. The most difficult of these to calibrate is usually the pullback factor λ, typically calibrated with historical data and a statistical analysis such as least squares regression or maximum likelihood estimators. While such computational techniques are outside the scope of this book, calculations are often fairly straightforward despite other problems that often arise with interpretations and applications.

The Vasicek, CIR, and Merton models are also examples of equilibrium models that seek to price securities based on supply and demand conditions. Unfortunately, these models often do a poor job of explaining the current pricing structure of bonds; that is, they do not explain the current term structure well over the entire range of maturities. On the other hand, they might do a better job of predicting the evolution of prices and term structures. An alternative to the equilibrium models are the no-arbitrage models (e.g., Ho and Lee (1986); Hull and White (1990)), which examine currently existing pricing relationships and estimate prices so as to prevent arbitrage. Such no-arbitrage models often predict future time structures poorly. While the ideal models are those that both explain the current term structure and predict future term structures, we often have to choose between the two desirable qualities.

There are many other types of models, some very different from those introduced in this chapter. Multi-factor models (e.g., Longstaff and Schwartz (1992) and Hull and White (1994a,b)) allow for multiple sources of uncertainty. These models can allow for twisting of the yield curve; that is, where different segments of the yield curve shift in different directions. Some models are largely generalizations of the models discussed here, such as Chan et al. (1992) and Hull and White (1990), which allow for time-varying parameters.

8.5 WHERE DO WE GO FROM HERE?

This book is an introduction to the field of quantitative finance, a very broad field indeed, covering areas ranging from portfolio analysis and derivatives pricing to financial econometrics and computational methods. While our book does at least touch on each of these areas, our orientation has been less a survey approach and more a focus on risk-neutral pricing. Hence, martingales have played a key role here. We have discussed martingale processes of various types and several alternatives to them. A large number of important securities, instruments, and portfolios have properties that can be modeled with these processes, along with many other instruments that we have not even discussed in this book. Thus, we have barely skimmed the surface here. For example, with the tools presented in this book, we can price plain vanilla options whose underlying instruments prices follow binomial and Wiener processes, and we have discussed bond pricing under varying term structure models. We have also introduced, but not fully developed, discussions on topics including certain iterative techniques and non-normal distributions. As much as we can accomplish with the tools introduced here, there is much more work to do to be able to analyze the huge variety of instruments that trade in today's markets, not to mention those that do not yet have markets.

Fortunately, there are some outstanding books and materials to pursue upon completing this one. For example, Baxter and Rennie (1996) and Hirsa and Neftci (2014) are excellent general financial mathematics texts, whose coverage somewhat overlaps the coverage of this text, but provide for somewhat more rigor, depth, and breadth of coverage. Shreve (2004, 2005) provides an outstanding two volume introduction to stochastic calculus with a focus on pricing derivatives. The second volume introduces continuous time mathematics and can be read without first reading Volume I. His more advanced text with Karatzas (Karatzas and Shreve (1998)) is significantly more rigorous.

Derivatives texts such as the classic by Hull (2011), Wilmott, Howison, and Dewynne (1995), and Wilmott (2006) provide outlets for opportunities to apply the models covered here to derivatives and fixed income instruments as do risk management texts such as Christoffersen (2003). Numerical methods and other computational tools can be studied in Brandimarte (2002) and Hirsa (2012). Empirical procedures for testing these models can be reviewed in Campbell et al. (1996). Extensions and other advances in options analysis are covered in Lewis (2000). The best-selling text on portfolio analysis remains Elton et al. (2009). We do hope that this book will facilitate the reading of these and other excellent quantitative finance books.

8.6 EXERCISES

8.1. Suppose that short-term (instantaneous) interest rates follow an Ornstein–Uhlenbeck (Vasicek) process. The long-term mean interest rate μ equals 0.06, the short-term (instantaneous) rate r_0 currently equals 0.02, the pullback factor λ equals 0.4, and the instantaneous standard deviation σ of short rates equals 0.01.

   a. Write a stochastic differential equation with numerical values that expresses this Ornstein–Uhlenbeck process.

   b. Calculate expected instantaneous rates for 1, 2, 3, and 30 years.

   c. Now, assume that the instantaneous standard deviation σ of short rates equals 0 (not 0.01) and that other values remain unchanged. Recall from Chapter 3 that in the discreet time case, the long-term spot rate $r_{0,T}$ is merely a geometric mean of the shortest-term spot rate and the series of single-period forward rates extending to time T:

$$(1+r_{0,T})^T = \prod_{t=1}^{T}(1 + r_{t-1,t}); \quad r_{0,T} = \sqrt[T]{\prod_{t=1}^{T}(1 + r_{t-1,t})} - 1$$

   Based on your formulas and calculations from part b above, derive a formula for calculating a spot rate $r_{0,T}$ in an Ornstein–Uhlenbeck continuous time space environment.

   d. Calculate spot rates $r_{0,T}$ for 1, 2, 3, 10, 20, and 30 years based on problem details in part c.

   e. Based on your solutions to parts c and d, draw the yield curve for this economy.

8.2. Suppose that short-term (instantaneous) interest rates follow an Ornstein–Uhlenbeck process. The long-term mean interest rate μ equals 0.04, the short-term (instantaneous) rate r_0 currently equals 0.01, the pullback factor λ equals 0.2, and the instantaneous standard deviation σ of short rates equals 0.02: $dr_t = \lambda(\mu - r_t)\,dt + \sigma\,dZ_t = 0.2(0.04 - r_t)\,dt + 0.02\,dZ_t$.
  a. Calculate the expected instantaneous forward rates for 1 month, 1 year, and 2 years.
  b. Calculate 1-month, 1-year, and 2-year variances of instantaneous rates.

8.3. Suppose that short-term interest rates follow a Ornstein–Uhlenbeck process. The long-term mean interest rate μ equals 0.06, the short-term (instantaneous) rate r_0 currently equals 0.02, the pullback factor λ equals 0.4, and the instantaneous standard deviation σ of short rates equals 0.01:

$$dr_t = \lambda(\mu - r_t)\,dt + \sigma\,dZ_t = 0.4(0.06 - r_t)\,dt + 0.01\,dZ_t$$

  a. What is the value of a 1-year zero-coupon bond B_0 with a face value $F = 1000$, assuming the risk premium of the bond is zero?
  b. What is the 1-year ($T = 1$) spot rate $r_{0,1}$?
  c. What is the expected instantaneous 1-year forward rate?
  d. Calculate $\frac{\partial B}{\partial t}$ at time $t = 0$.
  e. What would be the value of a 2-year, 5% coupon bond in this economy?

8.4. Prove that the yield curve produced by the Vasicek model starts at the current instantaneous rate r_0. Hint: Use the following definition of the instantaneous spot rate at time 0 that is justified by the discussion at the beginning of Section 8.3.2:

$$\lim_{T \to 0} r_{0,T} = r_\infty - (r_\infty - r_0)\lim_{T \to 0}\frac{\kappa}{T} + \frac{\sigma_r^2}{4\lambda}\lim_{T \to 0}\frac{\kappa^2}{T}$$

8.5. Consider the time 0 instantaneous return on a bond with maturity date T and payoff F. In Section 4.2.3, we calculated the value of the bond to be $B_0 = B_T e^{-r_0 T} = F e^{-r_0 T}$. We calculated the bond's price sensitivity to interest rate changes to be $\frac{\partial B}{\partial r} = -TFe^{-rT} = -TB_0$ (duration times bond price) and this sensitivity to interest rate changes to be $\frac{\partial^2 B}{\partial r^2} = T^2 Fe^{-rT} = T^2 B_0$ (convexity).
  a. How would we calculate $\frac{\partial B}{\partial r}$ in this chapter in a Vasicek environment assuming the time is currently $t = 0$?
  b. How would we calculate $\frac{\partial^2 B}{\partial r^2}$ in this chapter in a Vasicek environment assuming the time is currently $t = 0$?
  c. How would we calculate $\frac{\partial B}{\partial r}$ and $\frac{\partial^2 B}{\partial r^2}$ in this chapter in a Merton (1973) environment assuming the time is currently $t = 0$?
  d. Show how you can obtain your solution for part c for the case of zero drift by adapting the Vasicek model to produce the Merton (1973) model.

8.6. Suppose that short-term interest rates follow a CIR process. The long-term mean interest rate μ equals 0.06, the short-term (instantaneous) rate r_0 currently equals 0.02, the pullback factor λ equals 0.4, and the instantaneous standard deviation σ of short rates equals 0.01:

$$dr_t = \lambda(\mu - r_t)\,dt + \sigma\sqrt{r_t}\,dZ_t = 0.4(0.06 - r_t)\,dt + 0.01\sqrt{r_t}\,dZ_t$$

Assume that the risk premium is equal to zero.

a. What is the value of a 1-year zero-coupon bond B_0 with a face value $F = 1000$?

b. What is the 1-year $(T = 1)$ spot rate $r_{0,1}$?

c. What is the 30-year spot rate?

8.7. Find the price of a bond at time t with face value F that matures at time T assuming that the short-term interest rate follows the Merton model $dr_t = \mu \, dt + \sigma_r \, dZ_t$. Assume that the bond has a constant risk premium equal to Θ.

8.8. Derive the expected value and variance at time 0 of the short-term rate r_t assuming that it follows the Cox–Ingersoll–Ross model $dr_t = \lambda(\mu - r_t) \, dt + \sigma_r \sqrt{r_t} \, dZ_t$.

NOTES

1. Following the derivation of Vasicek (1977), note that this underlying short-rate function is quite general. For example, the coefficient a can be $\lambda(\mu - r_t)$, so that the instantaneous short interest rate follows an underlying Ornstein–Uhlenbeck process.

2. This proportional relationship between excess return and risk is consistent with the Capital Asset Pricing Model and many other equilibrium financial relationships.

3. Here, we assume that the process $dr_t = \lambda(\mu - r_t) \, dt + \sigma_r \, dZ_t$ is risk neutral, occurring in probability space $\mathbb{Q}$. Alternatively, under probability measure $\mathbb{P}$, $r_\infty = \mu - \Theta \sigma_r / \lambda - \frac{\sigma_r^2}{2\lambda^2}$. Under probability measure $\mathbb{Q}$, Θ equals zero. Also, note that under measure $\mathbb{Q}$, if $\lambda = 0$, the underlying stochastic process is arithmetic Brownian motion.

4. See end-of-chapter Exercise 8.7 for this derivation.

References

Baxter, M., Rennie, A., 1996. Financial Calculus: An Introduction to Derivative Pricing. Cambridge University Press, Cambridge, UK.

Brandimarte, P., 2002. Numerical Methods in Finance. A MATLAB-Based Introduction. Wiley Series in Probability and Statistics. Wiley, New York.

Campbell, J.Y., Lo, A.W., Craig MacKinlay, A., 1996. The Econometrics of Financial Markets. Princeton University Press, Princeton, New Jersey.

Chan, K.S., Karolyi, A., Longstaff, F.A., Sanders, A.B., 1992. An empirical comparison of alternative models of the short-term interest rate. J. Finance. 47 (3), 1209–1227.

Christoffersen, P.F., 2003. Elements of Financial Risk Management. Academic Press, London.

Cox, J., Ingersoll, J., Ross, S., 1985a. An intertemporal general equilibrium model of asset prices. Econometrica. 53, 363–384.

Cox, J., Ingersoll, J., Ross, S., 1985b. A theory of the term structure of interest rates. Econometrica. 53, 385–408.

Elton, E.J., Gruber, M.J., Brown, S.J., Goetzmann, W.N., 2009. Modern Portfolio Theory and Investment Analysis. eighth ed. Wiley, New York, New York.

Hirsa, A., 2012. Computational Methods in Finance. Chapman & Hall/CRC.

Hirsa, A., Neftci, S., 2014. An Introduction to the Mathematics of Financial Derivatives. *third ed.* Academic Press, San Diego, California.

Ho, T.S.Y., Lee, S.B., 1986. Term structure movements and pricing interest rate contingent claims. J. Finance. 41 (4), 1011–1029.

Hull, J.C., 2011. Options, Futures, and Other Derivatives. eighth ed. Prentice Hall, Upper Saddle River, NJ.

Hull, J., White, A., 1990. Pricing interest rate derivative securities. Rev. Financ. Stud. 3 (4), 573–592.

Hull, J., White, A., 1994a. Numerical procedures for implementing term structure Models I. J. Derivatives. Fall, 7–16.

Hull, J., White, A., 1994b. Numerical procedures for implementing term structure Models II. J. Derivatives. Winter, 37–48.

Karatzas, I., Shreve, S.E., 1998. Brownian Motion and Stochastic Calculus. Springer Science, New York.

Lewis, A.L., 2000. Option Valuation under Stochastic Volatility: With Mathematica Code. Financial Press, Newport Beach, California.

Longstaff, F.A., Schwartz, E.S., 1992. Interest rate volatility and the term structure: a two-factor general equilibrium Model. J. Finance. 47 (4), 1259–1282.

Merton, R.C., 1973. Theory of rational option pricing. Bell J. Econ. Manage. Sci. 4, 141–183.

Rendleman Jr., R.J., Bartter, B.J., 1980. The pricing of options on debt securities. J. Financ. Quant. Anal. 15 (1), 11–24.

Shreve, S.E., 2004. Stochastic Calculus for Finance, Volume II: Continuous-Time Models. Springer, New York.

Shreve, S.E., 2005. Stochastic Calculus for Finance, Volume I: The Binomial Asset Pricing Model. Springer, New York.

Uhlenbeck, G.E., Ornstein, L.S., 1930. On the theory of Brownian motion. Phys. Rev. 36, 823–841.

Vasicek, O., 1977. An equilibrium characterization of the term structure. J. Financ. Econ. 5, 177–188.

Wilmott, P., 2006. Paul Wilmott on Quantitative Finance. second ed. Wiley, New York (selected chapters).

Wilmott, P., Howison, S., Dewynne, J., 1995. The Mathematics of Financial Derivatives: A Student Introduction. Cambridge University Press.

Appendix A: The z-table

The z-Table

z	0.00	0.01	0.02	0.03	0.04	0.05	0.06	0.07	0.08	0.09
0.0	0.0000	0.0040	0.0080	0.0120	0.0159	0.0199	0.0239	0.0279	0.0319	0.0358
0.1	0.0398	0.0438	0.0478	0.0517	0.0557	0.0596	0.0636	0.0675	0.0714	0.0753
0.2	0.0793	0.0832	0.0871	0.0909	0.0948	0.0987	0.1026	0.1064	0.1103	0.1141
0.3	0.1179	0.1217	0.1255	0.1293	0.1331	0.1368	0.1406	0.1443	0.1480	0.1517
0.4	0.1554	0.1591	0.1628	0.1664	0.1700	0.1736	0.1772	0.1808	0.1844	0.1879
0.5	0.1915	0.1950	0.1985	0.2019	0.2054	0.2088	0.2123	0.2157	0.2190	0.2224
0.6	0.2257	0.2291	0.2324	0.2356	0.2389	0.2421	0.2454	0.2486	0.2517	0.2549
0.7	0.2580	0.2611	0.2642	0.2673	0.2703	0.2734	0.2764	0.2793	0.2823	0.2852
0.8	0.2881	0.2910	0.2939	0.2967	0.2995	0.3023	0.3051	0.3078	0.3106	0.3133
0.9	0.3159	0.3186	0.3212	0.3238	0.3264	0.3289	0.3315	0.3340	0.3365	0.3389
1.0	0.3413	0.3437	0.3461	0.3485	0.3508	0.3531	0.3554	0.3577	0.3599	0.3621
1.1	0.3643	0.3665	0.3686	0.3708	0.3729	0.3749	0.3770	0.3790	0.3810	0.3830
1.2	0.3849	0.3869	0.3888	0.3906	0.3925	0.3943	0.3962	0.3980	0.3997	0.4015
1.3	0.4032	0.4049	0.4066	0.4082	0.4099	0.4115	0.4131	0.4147	0.4162	0.4177
1.4	0.4192	0.4207	0.4222	0.4236	0.4251	0.4265	0.4279	0.4292	0.4306	0.4319
1.5	0.4332	0.4345	0.4357	0.4370	0.4382	0.4394	0.4406	0.4418	0.4429	0.4441
1.6	0.4452	0.4463	0.4474	0.4484	0.4495	0.4505	0.4515	0.4525	0.4535	0.4545
1.7	0.4554	0.4564	0.4573	0.4582	0.4591	0.4599	0.4608	0.4616	0.4625	0.4633
1.8	0.4641	0.4649	0.4656	0.4664	0.4671	0.4678	0.4686	0.4693	0.4699	0.4706
1.9	0.4713	0.4719	0.4726	0.4732	0.4738	0.4744	0.4750	0.4756	0.4761	0.4767
2.0	0.4772	0.4778	0.4783	0.4788	0.4793	0.4798	0.4803	0.4808	0.4812	0.4817
2.1	0.4821	0.4826	0.4830	0.4834	0.4838	0.4842	0.4846	0.4850	0.4854	0.4857
2.2	0.4861	0.4864	0.4868	0.4871	0.4875	0.4878	0.4881	0.4884	0.4887	0.4890
2.3	0.4893	0.4896	0.4898	0.4901	0.4904	0.4906	0.4909	0.4911	0.4913	0.4916
2.4	0.4918	0.4920	0.4922	0.4925	0.4927	0.4929	0.4931	0.4932	0.4934	0.4936
2.5	0.4938	0.4940	0.4941	0.4943	0.4945	0.4946	0.4948	0.4949	0.4951	0.4952
2.6	0.4953	0.4955	0.4956	0.4957	0.4959	0.4960	0.4961	0.4962	0.4963	0.4964
2.7	0.4965	0.4966	0.4967	0.4968	0.4969	0.4970	0.4971	0.4972	0.4973	0.4974
2.8	0.4974	0.4975	0.4976	0.4977	0.4977	0.4978	0.4979	0.4979	0.4980	0.4981
2.9	0.4981	0.4982	0.4982	0.4983	0.4984	0.4984	0.4985	0.4985	0.4986	0.4986
3.0	0.4986	0.4987	0.4987	0.4988	0.4988	0.4989	0.4989	0.4989	0.4990	0.4990

Appendix B: Exercise Solutions

CHAPTER 1

1.1. The sum is as follows:

$$
\begin{bmatrix} 2 & 4 & 9 \\ 6 & 4 & 25 \\ 0 & 2 & 11 \end{bmatrix} + \begin{bmatrix} 3 & 0 & 6 \\ 2 & 1 & 3 \\ 7 & 0 & 4 \end{bmatrix} = \begin{bmatrix} 5 & 4 & 15 \\ 8 & 5 & 28 \\ 7 & 2 & 15 \end{bmatrix}
$$

$$\quad\quad\quad A \quad\quad\quad\quad B \quad\quad\quad\quad C$$

1.3. a.
$$
\begin{bmatrix} 1 & 8 & 9 \\ 6 & 4 & 25 \\ 3 & 2 & 35 \end{bmatrix} \quad \begin{bmatrix} 1 & 6 & 3 \\ 8 & 4 & 2 \\ 9 & 25 & 35 \end{bmatrix}
$$

$$\quad\quad A \quad\quad\quad\quad A^{\mathrm{T}}$$

b. The transpose of a column vector is a row vector:

$$
\begin{bmatrix} 9 \\ 6 \\ 3 \\ 7 \end{bmatrix} \quad [9 \; 6 \; 3 \; 7]
$$

$$\quad\quad y \quad\quad\quad\quad y^{\mathrm{T}}$$

Similarly, the transpose of a row vector is a column vector.

c. Note that the transpose V^{T} of a symmetric matrix V is V:

$$
V = \begin{bmatrix} 0.09 & 0.01 & 0.04 \\ 0.01 & 0.16 & 0.10 \\ 0.04 & 0.10 & 0.64 \end{bmatrix} V^{\mathrm{T}} = \begin{bmatrix} 0.09 & 0.01 & 0.04 \\ 0.01 & 0.16 & 0.10 \\ 0.04 & 0.10 & 0.64 \end{bmatrix} = V
$$

1.5. a. $2A = \begin{bmatrix} 2(-2) & 2(0) \\ 2(3) & 2(4) \end{bmatrix} = \begin{bmatrix} -4 & 0 \\ 6 & 8 \end{bmatrix}$

b. $A^{\mathrm{T}} = \begin{bmatrix} -2 & 3 \\ 0 & 4 \end{bmatrix}$

c. $A + B = \begin{bmatrix} -2+7 & 0+3 \\ 3+5 & 4-1 \end{bmatrix} = \begin{bmatrix} 5 & 3 \\ 8 & 3 \end{bmatrix}$

d.
$$\mathbf{AB} = \begin{bmatrix} -2(7) + 0(5) & -2(3) + 0(-1) \\ 3(7) + 4(5) & 3(3) + 4(-1) \end{bmatrix} = \begin{bmatrix} -14 & -6 \\ 41 & 5 \end{bmatrix}$$

e.
$$\mathbf{BA} = \begin{bmatrix} 7(-2) + 3(3) & 7(0) + 3(4) \\ 5(-2) - 1(3) & 5(0) - 1(4) \end{bmatrix} = \begin{bmatrix} -5 & 12 \\ -13 & -4 \end{bmatrix}$$

1.7. $\mathbf{AXB} = \mathbf{AB} - \mathbf{B}$. $\mathbf{AXBB}^{-1} = (\mathbf{AB} - \mathbf{B})\mathbf{B}^{-1}$, $\mathbf{AXI} = \mathbf{ABB}^{-1} - \mathbf{BB}^{-1}$, $\mathbf{AX} = \mathbf{AI} - \mathbf{I}$, $\mathbf{AX} = \mathbf{A} - \mathbf{I}$, $\mathbf{A}^{-1}\mathbf{AX} = \mathbf{A}^{-1}(\mathbf{A} - \mathbf{I})$, $\mathbf{IX} = \mathbf{A}^{-1}\mathbf{A} - \mathbf{A}^{-1}\mathbf{I}$, $\mathbf{X} = \mathbf{I} - \mathbf{A}^{-1}$.

1.9. We can write the system of equations as the following equivalent matrix equation:

$$\begin{bmatrix} 0.02 & 0.04 \\ 0.06 & 0.08 \end{bmatrix} \qquad \begin{bmatrix} x_1 \\ x_2 \end{bmatrix} \quad = \quad \begin{bmatrix} 0.03 \\ 0.01 \end{bmatrix}$$
$$\mathbf{C} \qquad\qquad\quad \cdot \quad \mathbf{x} \quad = \quad \mathbf{s}$$

In problem 1.6.e, we already found the inverse of the matrix $\mathbf{C}$:

$$\mathbf{C}^{-1} = \begin{bmatrix} -100 & 50 \\ 75 & -25 \end{bmatrix}$$

Solving for the vector $\mathbf{x}$ gives:

$$\begin{bmatrix} x_1 \\ x_2 \end{bmatrix} = \begin{bmatrix} -100 & 50 \\ 75 & -25 \end{bmatrix} \cdot \begin{bmatrix} 0.03 \\ 0.01 \end{bmatrix} = \begin{bmatrix} -2.5 \\ 2 \end{bmatrix}$$
$$\mathbf{x} \quad = \qquad \mathbf{C}^{-1} \qquad \cdot \quad \mathbf{s}$$

Thus: $x_1 = -2.5$ and $x_2 = 2$.

1.11. a. The first step to find the minimum (or maximum) value of y for the function $y = 5x^2 - 3x + 2$ is to set the first derivative of y with respect to x equal to zero and then solve for x:

$$\frac{dy}{dx} = 10x - 3 = 0$$

$$10x = 3$$

$$x = \frac{3}{10}$$

The x-value $x = 3/10$ is called a critical value. This x-value is where the graph may have a maximum or minimum value for its y-value. To determine the nature of this critical value, one can either use the first or the second derivative test. The first derivative test makes use of the facts that we know that when $dy/dx > 0$, y increases as x increases; and when $dy/dx < 0$, y decreases as x increases. In this problem, observe that the derivative dy/dx is positive when $x > 0.3$, negative when $x < 0.3$, and zero when $x = 0.3$. Thus, in this case, y is decreasing as x increases when $x < 0.3$, and y increases as x increases when $x > 0.3$. This means that we must have a minimum for y at $x = 0.3$. The minimum value of y is $y(0.3) = 5(0.3)^2 - 3(0.3) + 2 = 1.55$. So the lowest point on this graph is $(0.3, 1.55)$. Another method to determine maxima and minima is the second derivative test.

This test relies on the following properties of the second derivative. If the second derivative is greater than zero, we have a minimum value for y (the function is concave up). When the second derivative is less than zero, we have a maximum (the function is concave down). If the second derivative is zero or changes in sign, then we have insufficient information to draw any conclusion. The second derivative in the above example is given by: $d^2y/dx^2 = 10$, also written $f''(x) = 10$. Since the second derivative 10 is greater than zero, we have found a minimum value for y. Although it does not apply here, in many cases, more than one "local" minimum or maximum value will exist.

    b. When $y = -7x^2 + 4x + 5$, the first derivative is: $dy/dx = -14x + 4$. Setting the first derivative equal to zero, we find our maximum as follows:

$$-14x + 4 = 0$$

$$-14x = -4$$

$$x = \frac{2}{7}$$

We use the second derivative test to ensure that this is a maximum. The second derivative is: $d^2y/dx^2 = -14$. Since -14 is less than zero, we have a maximum at $x = 2/7$.

1.13. The derivatives are as follows:

    a. $dy/dx = 12(4x + 2)^2$

    b. $dy/dx = 3x/\sqrt{(3x^2 + 8)}$

    c. $dy/dx = 6x(12x^2 + 10x) + 6(4x^3 + 5x^2 + 3) = 96x^3 + 90x^2 + 18 = (26.25x - 41.575)(1.5x - 4)^2 (2.5x - 3.5)^3$

    d. $dy/dx = 4.5(1.5x - 4)^2(2.5x - 3.5)^4 + 10(1.5x - 4)^3(2.5x - 3.5)^3$

    e. $dy/dx = -50/x^3$

    f. $dy/dx = [(60x - 84) - (60x - 160)]/(10x - 14)^2 = 76/(10x - 14)^2$

1.15. Solve as follows:

$$\frac{dy}{dt} = \frac{dy}{dx}\frac{dx}{dt} = 3x^2(2t) = 3(t^2 + 1)^2(2t) = 6t(t^2 + 1)^2$$

1.17. If s denotes the length of the sides of the floor and A its area, then $A = s^2$. Since our measurement of the length of the side was $s = 20$ ft, we estimate that the area of the floor is $A = 20^2 = 400$ ft². The differential $dA = 2sds$. In this case $s = 20$ and $ds = \pm 0.1$, and so $dA = 2(20)(\pm 0.1) = \pm 4$ ft². This tells us that the error in approximating the area of the floor by 400 ft² can be off by ± 4 ft² due to the error in measuring its length. Furthermore the error in the area using the differential is itself an approximation of the possible error in the area. The exact possible error can be found without using calculus by the following calculations:

$$A(20.1) - A(20) = (20.1)^2 - (20)^2 = 4.01$$

and:

$$A(19.9) - A(20) = (19.9)^2 - (20)^2 = -3.99$$

Thus, the possible error in the area is in the range from -3.99 ft² to 4.01 ft². Observe that the differential provided a reasonable estimate of this range of possible errors.

1.19. First, set up the Lagrange function:

$$L = 50x^2 - 10x + \lambda(100 - 0.1x)$$

Next, find the first-order conditions:

$$dL/dx = 100x - 0.1\lambda = 10$$

$$dL/d\lambda = 100 - 0.1x = 0, \text{ or:}$$

$$\begin{bmatrix} 100 & 0.1 \\ -0.1 & 0 \end{bmatrix} \begin{bmatrix} x \\ \lambda \end{bmatrix} = \begin{bmatrix} 10 \\ -100 \end{bmatrix}$$

$$\mathbf{C} \qquad \mathbf{x} \quad = \quad \mathbf{s}$$

Regardless, we find that $x = 1000$ and $\lambda = 999{,}900$.

1.21. a. $\displaystyle \int (10x - x^2)\,dx = 5x^2 - \frac{1}{3}x^3 + C$

b. $\displaystyle \int_0^1 (10x - x^2)\,dx = F(1) - F(0) = \left(5x^2 - \frac{1}{3}x^3\right)\Big|_0^1 = 4\frac{2}{3}$

c. $\displaystyle \int_0^1 10x - x^2\,dx \approx \sum_{i=1}^5 f(x_i)\Delta x = \sum_{i=1}^5 \frac{1}{5}(10x_i - x_i^2) = 0.392 + 0.768 + 1.128 + 1.472 + 1.8 = 5.56$

d. $\displaystyle \int_0^1 10x - x^2\,dx \approx \sum_{i=1}^{10} f(x_i)\Delta x = \sum_{i=1}^{10} \frac{1}{10}(10x_i - x_i^2)$

$$= 0.099 + 0.196 + 0.291 + 0.384 + 0.475 + 0.564 + 0.651 + 0.736 + 0.819 + 0.9 = 5.115$$

1.23. a. We will use antiderivatives as follows:

$$\int_0^1 (8x - 9x^2)\,dx = F(1) - F(0) = (4x^2 - 3x^3)\Big|_0^1 = (4 \cdot 1^2 - 3 \cdot 1^3) - (4 \cdot 0^2 - 3 \cdot 0^3) = 1$$

b. 20 rectangles are plotted as follows:

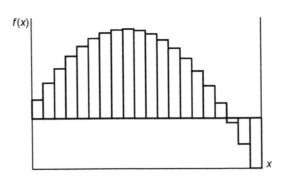

1.25. $\displaystyle \int_1^3 x^2\,dx = \frac{1}{3}x^3\Big|_1^3 = \frac{1}{3}3^3 - \frac{1}{3}1^3 = \frac{26}{3}$

CHAPTER 2

2.1. a. The sample space is $\Omega = \{1,2,3,4,5,6\}$.

b. There are $2^6 = 64$ possible subsets of Ω.

c. By rule 3 of the requirement to be a probability, we have $P(\{2,4,6\}) = P(\{2\}) + P(\{4\}) + P(\{6\}) = 1/6 + 1/6 + 1/6 = 1/2$. Thus, the probability of obtaining an even number is $1/2$.

2.3. a. Each outcome has a one-third or 0.33333 probability of being realized since the probabilities are equal and must sum to one.

b. $E[\text{SALES}] = (800,000 \cdot 0.33333) + (500,000 \cdot 0.33333) + (400,000 \cdot 0.33333)$

$E[\text{SALES}] = 566,667$

c. $\text{Var}[\text{SALES}] = [(800,000 - 566,667)^2 \times 0.33333 + (500,000 - 566,667)^2$

$\times 0.33333 + (400,000 - 566,667)^2 \times 0.33333] = 28,890,000,000 = \sigma^2_{\text{SALES}}$

d. Expected return of Project A $= (0.3 \times 0.33333) + (0.15 \times 0.33333) + (0.01 \times 0.33333) = 0.15333$

e. Variance of A's returns $= [(0.3 - 0.15333)^2 \times 0.33333 + (0.15 - 0.15333)^2 \times 0.33333 + (0.01 - 0.15333)^2 \times 0.33333] = 0.01402 = \sigma^2_A$

f. Expected return of Project B $= (0.2 \times 0.33333) + (0.13 \times 0.33333) + (0.09 \times 0.33333) = 0.14$.
Variance of B's returns $= [(0.2 - 0.14)^2 \times 0.33333 + (0.13 - 0.14)^2 \times 0.33333 + (0.09 - 0.14)^2 \times 0.33333] = 0.002067 = \sigma^2_B$

g. Standard deviations are square roots of variances.

$$\sigma_{\text{SALES}} = 170,000$$

$$\sigma_A = 0.1184$$

$$\sigma_B = 0.4546$$

h. $\text{Cov}[\text{SALES}, A] = \sum_{i=1}^{n}(\text{SALES}_i - E[\text{SALES}]) \times (R_{Ai} - E[R_A]) \times P_i$

$$\text{Cov}[\text{SALES}, A] = (800,000 - 566,670)$$
$$\times (0.3 - 0.15333) \times 0.33333 + (500,000 - 566,670)$$
$$\times (0.15 - 0.15333) \times 0.33333 + (400,000 - 566,670)$$
$$\times (0.01 - 0.15333) \times 0.33333 = 19,300 = \sigma_{\text{SALES},A}$$

i. $\rho_{s,A} = \frac{\sigma_{\text{SALES},A}}{\sigma_{\text{SALES}} \times \sigma_A} = \frac{19,300}{170,000 \times 0.1184} = 0.959$

j. First, find the covariance between sales and returns on B:

$$\text{Cov}[\text{SALES}, B] = (800,000 - 566,670)$$
$$\times (0.20 - 0.14) \times 0.33333 + (500,000 - 566,670)$$
$$\times (0.13 - 0.14) \times 0.33333 + (400,000 - 566,670)$$
$$\times (0.09 - 0.14) \times 0.33333$$
$$= 7667 = \sigma_{\text{SALES},B}$$

$$\rho_{\text{SALES},B} = \frac{\sigma_{\text{SALES},B}}{\sigma_{\text{SALES}} \times \sigma_B} = \frac{7667}{170,000 \times 0.04546} = 0.992$$

k. Coefficient of determination is simply coefficient of correlation squared: $0.992^2 = 0.985$

2.5. **a.** Since probabilities must sum to one, the probability must equal 0.15.

    b. First, note that there is a $0.25 = 0.10 + 0.05 + 0.10$ probability that the return will be 0.05 and 0.20 and 0.55 probabilities that the return will be 0.15. Thus, the expected return is $0.05 \times 0.25 + 0.10 \times 0.20 + 0.15 \times 0.55 = 0.115$. The variance is $0.25 \times (0.05 - 0.115)^2 + 0.20 \times (0.10 - 0.115)^2 + 0.55 \times (0.15 - 0.115)^2 = 0.001775$, which implies a standard deviation equal to 0.04213.

2.7. Simply reduce the standard deviations in the z scores in problem (2.6) to 0.05.

    a. 0.9544

    b. 0.4772

    c. 0.0228

    d. 0.0013

2.9. **a.** Var $= 0.0025$; std.dev. $= 0.05$

    b. -0.00125

2.11. **a.**

Date	Company X return	Company Y return	Company Z return
1/09	–	–	–
1/10	0	0	0.00207
1/11	0.00249	0.00625	−0.00413
1/12	0	0.00621	−0.00207
1/13	0.00249	0.00617	−0.00208
1/14	−0.002480	0	0.00208
1/15	0.03980	0.04907	0.04158
1/16	0.00239	−0.00585	−0.02994
1/17	−0.00239	0.00588	0
1/18	0.00239	0.00585	0.00206
1/19	0.00239	−0.00581	0
1/20	−0.00238	0.00585	0

    b. See part c.

    c.

Stock	Average return	Standard deviation
X	0.004064	0.011479
Y	0.006693	0.014150
Z	0.000869	0.015537

2.13. **a.** From this density function, we can obtain a distribution function by integrating as follows:

$$P(x) = \int_{-\infty}^{x} p(t)\, dt = \int_{0}^{x} 0.5t\, dt = 0.25t^2 |_0^x = 0.25x^2$$

for $0 \leq x \leq 2$. If $x < 0$, then $P(x) = 0$ and if $x > 2$, then $P(x) = 1$.

b. $P(0) = 0$, $P(2) = 1$ and $0 \leq P(x)$ for all x.

c. The random return on the stock is $r = f(X) = 0.8X - 0.2$. This means that $X = (r + 0.2)/0.8$. When $r = 0.1$, $X = 0.375$ and when $r = 0.4$, $X = 0.75$. Thus, the probability that stock's return r will be between 0.1 and 0.4 is the same as the probability that X is between 0.375 and 0.75.

$$P(0.1 < r < 0.4) = P(0.375 < X < 0.75) = \int_{0.375}^{0.75} 0.5x \, dx = 0.25x^2 \Big|_{0.375}^{0.75} = 0.14063 - 0.03516 = 0.1055$$

Thus, there is a 10.55% probability that x_i will be between 0.375 and 0.75 and that the stock return will range from 0.1 to 0.4.

d. By Eq. (2.9), the expected return for this stock is:

$$E[r] = \int_{-\infty}^{\infty} f(x)p(x) \, dx = \int_{0}^{2} (0.8x - 0.2)p(x) \, dx = \int_{0}^{2} (0.8x - 0.2) \cdot 0.5x \, dx$$

$$= \left(\frac{2}{15}x^3 - \frac{1}{20}x^2 \right) \Big|_{0}^{2} = \left(\frac{16}{15} - \frac{1}{5} \right) - 0 = \frac{13}{15}$$

e. By Eq. (2.10) and using the expectation result above, the variance for this stock is:

$$\sigma^2 = E[r^2] - (E[r])^2 = \int_{-\infty}^{\infty} (f(x))^2 p(x) \, dx - \left(\tfrac{13}{15} \right)^2$$

$$= \int_{0}^{2} (0.8x - 0.2)^2 \cdot 0.5x \, dx - \left(\tfrac{13}{15} \right)^2$$

$$= \int_{0}^{2} (0.32x^3 - 0.16x^2 + 0.02x) dx - \left(\tfrac{13}{15} \right)^2$$

$$= \left(\frac{0.32}{4}x^4 - \frac{0.16}{3}x^3 + \frac{0.02}{2}x^2 \right) \Big|_{0}^{2} - 0.75111 = 0.89333 - 0.75111 = 0.1422$$

f. Using part c, we calculate the conditional expected value as follows:

$$E[r | 0.1 < r < 0.4] = \frac{\int_{0.375}^{0.75} f(x)p(x) \, dx}{\int_{0.375}^{0.75} p(x) \, dx} = \frac{\int_{0.375}^{0.75} (0.8x - 0.2)(0.5x) \, dx}{0.1055}$$

$$\int_{0.375}^{0.75} (0.8x - 0.2)(0.5x) \, dx = \left(\frac{2}{15}x^3 - \frac{1}{20}x^2 \right) \Big|_{0.375}^{0.75} = 0.028125 - 0 = 0.028125$$

$$E[r | 0.1 < r < 0.4] = \frac{0.028125}{0.1055} = 0.267$$

2.15. $P[0.5 \leq X \leq 1] = \int_{.5}^{1} 3 e^{-3X} \, dx = -e^{-3X} \Big|_{.5}^{1} = -e^{-3} + e^{-1.5} = 0.173$

2.17.

$$\text{Var}[aX + bY] = E[\{aX + bY - E[aX + bY]\}^2] = E[\{aX - aE[X] + bY - bE[Y]\}^2]$$

$$= E[\{a(X - E[X]) + b(Y - E[Y])\}^2]$$

$$= E[a^2(X - E[X])^2 + 2ab(X - E[X])(Y - E[Y]) + b^2(Y - E[Y])^2]$$

$$= a^2 E[(X - E[X])^2] + 2abE[(X - E[X])(Y - E[Y])] + b^2 E[(Y - E[Y])^2]$$

$$= a^2 \text{Var}[X] + 2abE[XY - XE[Y] - YE[X] + E[X]E[Y]] + b^2 \text{Var}[Y]$$

$$= a^2 \text{Var}[X] + 2ab\{E[XY] - E[X]E[Y] - E[Y]E[X] + E[X]E[Y]\} + b^2 \text{Var}[Y]$$

By independence $E[XY] = E[X]E[Y]$, which means that the middle term above equals zero, and $\text{Var}[aX + bY] = a^2\text{Var}[X] + b^2\text{Var}[Y]$.

$$P(45 < S_1 < 60) = P\left(0.9 < \frac{S_1}{50} < 1.2\right) = P(\ln(0.9) < \ln\left(\frac{S_1}{50}\right) < \ln(1.20))$$

$$= P(-0.10536 < 0.1 + 0.5Z < 0.18232) = P\left(\frac{-0.10536 - 0.1}{0.5} < Z < \frac{0.18232 - 0.10}{0.5}\right)$$

2.19. a.

$$= P(-0.41072 < Z < 0.16464) = N(0.16464) - N(-0.41072)$$

$$= 0.56537 - 0.34064 = 0.2247$$

b. We find the expected return over one period for our security using Eq. (2.26) as follows:

$$E\left[\frac{S_1}{50}\right] - 1 = E[e^X] - 1 = e^{\mu + \frac{1}{2}\sigma^2} - 1 = e^{0.1 + \frac{1}{2} \times 0.5^2} - 1 = 0.0126$$

c. The return cannot fall below -100% because $S_1/50$ is lognormally distributed and so $S_1/50 > 0$. The log of the price relative (it is normally distributed) could potentially fall as low as $-\infty$. Since the standard deviation is substantially larger than the mean, a very low price relative is quite possible. Since returns are "capped" from below (they are lognormally distributed), their expected value is higher.

2.21. To determine the probability that a microchip is of type 1 given that it is defective, we use the general version of Bayes theorem:

$$P[M_1|D] = \frac{P[D|M_1]P[M_1]}{\sum_{j=1}^3 P[D|M_j]P[M_j]} = \frac{0.001(0.15)}{0.001(0.15) + 0.003(0.3) + 0.004(0.55)} = 0.0462$$

Similarly, we calculate that $P[M_2|D] = 0.2769$ and $P[M_3|D] = 0.6769$.

2.23. a. $E[R_p] = 0.06, \sigma_p^2 = 0.0081, \sigma_p = 0.09$

     b. $E[R_p] = 0.095, \sigma_p^2 = 0.014231, \sigma_p = 0.1193$

     c. $E[R_p] = 0.165, \sigma_p^2 = 0.055181, \sigma_p = 0.2349$

     d. $E[R_p] = 0.20, \sigma_p^2 = 0.09, \sigma_p = 0.3$

2.25. a. $E[R_p] = 0.075, \sigma_p = 0.16$

     b. $E[R_p] = 0.075, \sigma_p = 0.14$

     c. $E[R_p] = 0.075, \sigma_p = 0.1166$

     d. $E[R_p] = 0.075, \sigma_p = 0.08718$

     e. $E[R_p] = 0.075, \sigma_p = 0.04$

2.27. Let X equal the number of 7% magnitude price jumps per month. In this case $\lambda = 1.25$.

     a.
$$P(X \geq 2) = 1 - P(X \leq 1) = 1 - P(X = 0) - P(X = 1)$$
$$= 1 - e^{-1.25}\frac{1.25^0}{0!} - e^{-1.25}\frac{1.25^1}{1!} = 1 - 0.6446 = 0.3554$$

     b.
$$P(X \leq 2) = P(X = 0) + P(X = 1) + P(X = 2)$$
$$= e^{-1.25}\frac{1.25^0}{0!} + e^{-1.25}\frac{1.25^1}{1!} + e^{-1.25}\frac{1.25^2}{2!} = 0.8685$$

2.29. a. We wish to find the distribution function $P(X \leq 0, Y \leq 0)$. Note that:

$$P(X \leq 0, Y \leq 0) = P\left(\frac{X - 5\%}{10\%} \leq \frac{0 - 5\%}{10\%}, \frac{Y - 10\%}{20\%} \leq \frac{0 - 10\%}{20\%}\right) = P(Z_1 \leq -0.5, Z_2 \leq -0.5)$$

where $Z_1 = (X - 5\%)/10\%$ and $Z_2 = (Y - 10\%)/20\%$ are the standardized normal random variables for X and Y. From Section 2.8.2 on the bivariate normal distribution, we know that this probability is equal to the joint distribution function $M(-0.5, -0.5; 0.8)$. We would find from an appropriate calculator that this cumulative density is approximately 0.219, implying that the probability is 0.219 that the random variables X and Y are both less than 0.

     b. We need to calculate the joint distribution function $M(-0.5, -0.5; 0)$. We would find from an appropriate calculator that this cumulative density is approximately 0.095, implying that the probability is 0.095 that the random variables X and Y are both less than 0.

     c. It is interesting to note that it can be shown that if we have a bivariate normal distribution with a correlation coefficient of 0 then the random variables X and Y are independent. Thus, in this case, the probability that both X and Y are less than 0 equals the product of the probabilities that each random variable is less than 0, which is $(0.309)(0.309) = 0.095$.

2.31. First note that $E[cZ^2] = c \times \text{Var}(Z) = c$. Thus:

$$\text{Var}(Z^2 c) = \frac{1}{\sqrt{2\pi}}\int_{-\infty}^{\infty}(z^2 c - c)^2\, e^{-z^2/2}\, dz = c^2\frac{1}{\sqrt{2\pi}}\int_{-\infty}^{\infty}(z^4 - 2z^2 + 1)e^{-z^2/2}\, dz$$

We already know that:

$$\frac{1}{\sqrt{2\pi}}\int_{-\infty}^{\infty} e^{-z^2/2}\, dz = 1 \quad \text{and} \quad \text{Var}(Z) = \frac{1}{\sqrt{2\pi}}\int_{-\infty}^{\infty} z^2\, e^{-z^2/2}\, dz = 1$$

So there is left to evaluate:

$$\frac{1}{\sqrt{2\pi}}\int_{-\infty}^{\infty} z^4\, e^{-z^2/2}\, dz = \frac{1}{\sqrt{2\pi}}\int_{-\infty}^{\infty} z^3\, e^{-z^2/2}z\, dz$$

Choose $u = z^3$ and $dv = e^{-z^2/2}z\, dz$ and integrate by parts. Since $du = 3z^2 dz$ and $v = -e^{-\frac{z^2}{2}}$, then:

$$\frac{1}{\sqrt{2\pi}}\int_{-\infty}^{\infty} z^3\, e^{-z^2/2}z\, dz = -\frac{1}{\sqrt{2\pi}}z^3\, e^{-\frac{z^2}{2}}\Big|_{-\infty}^{\infty} + \frac{3}{\sqrt{2\pi}}\int_{-\infty}^{\infty} z^2\, e^{-\frac{z^2}{2}}\, dz = 3\,\text{Var}(Z) = 3$$

This gives $\text{Var}(Z^2 c) = (3 - 2 + 1)c^2 = 2c^2$.

2.33. To derive the variance of portfolio p as a function of security variances, covariances, and weights, we use distributive property from algebra and the linearity properties of the expectation operator:

$$E[(R_p - E[R_p])^2] = E\left(\sum_{i=1}^{m} w_i R_i - E\left[\sum_{i=1}^{m} w_i R_i\right]\right)^2$$

$$= E\left[\left(\sum_{i=1}^{m} w_i R_i - E\left[\sum_{i=1}^{m} w_i R_i\right]\right)\left(\sum_{j=1}^{m} w_j R_j - E\left[\sum_{j=1}^{m} w_j R_j\right]\right)\right]$$

$$= E\left[\sum_{i=1}^{m}\sum_{j=1}^{m} w_i w_j (R_i - E[R_i])(R_j - E[R_j])\right]$$

$$= \sum_{i=1}^{m}\sum_{j=1}^{m} w_i w_j E[(R_i - E[R_i])(R_j - E[R_j])]$$

$$= \sum_{i=1}^{m}\sum_{j=1}^{m} w_i w_j \sigma_{i,j} = \mathbf{w}^T \mathbf{V} \mathbf{w}$$

2.35. a. Since this is an equally weighted portfolio, the weight of each security i would be $w_i = 1/n$. Let σ denote the standard deviation shared by all securities. Portfolio variance is determined as follows:

$$\sigma_p^2 = \sum_{\substack{i=1 \\ i \neq j}}^{n}\sum_{j=1}^{n} w_i w_j \sigma_{ij} + \sum_{i=1}^{n}\sum_{j=1}^{n} w_i w_j \sigma_{ij} = 2\left(\sum_{\substack{i=1 \\ i<j}}^{n}\sum_{j=1}^{n} w_i w_j \sigma_{ij}\right) + \sum_{i=1}^{n} w_i^2 \sigma_i^2$$

$$\sigma_p^2 = 2\left(\sum_{\substack{i=1 \\ i<j}}^{n}\sum_{j=1}^{n}(1/n)^2 \cdot 0 \cdot \sigma \cdot \sigma)\right) + \sum_{i=1}^{n}(1/n)^2 \sigma^2 = 0 + (1/n)^2 \cdot n \cdot \sigma^2 = \sigma^2/n$$

b. Since this is now a perfectly diversified portfolio, the weight of each security i, $w_i = 1/n$ approaches zero as $n \to \infty$. As $n \to \infty$, $\sigma_p^2 \to 0$.

CHAPTER 3

3.1. a. $PV = 10{,}000/1.1^{20} = 10{,}000/6.7275 = 1486.44$

  b. $PV = 10{,}000/1.1^{10} = 10{,}000/2.5937425 = 3855.43$

  c. $PV = 10{,}000/1.1^{1} = 10{,}000/1.1 = 9090.91$

  d. $PV = 10{,}000/1.1^{0.5} = 10{,}000/1.1^{05} = 10{,}000/1.0488088 = 9534.63$

  e. $PV = 10{,}000/1.1^{0.2} = 10{,}000/1.0192449 = 9811.18$; note: 73 days is 0.2 of 1 year

3.3. $CF_n = CF_1(1 + g)^{n-1}$

  a. $CF_2 = 10{,}000\,(1 + 0.1)^{2-1} = 10{,}000\,(1 + 0.1) = 10{,}000 \times 1.1 = 11{,}000$

  b. $CF_3 = 10{,}000\,(1 + 0.1)^{3-1} = 10{,}000 \times 1.21 = 12{,}100$

  c. $CF_5 = 10{,}000\,(1 + 0.1)^{5-1} = 10{,}000 \times 1.4641 = 14{,}641$

  d. $CF_{10} = 10{,}000\,(1 + 0.1)^{10-1} = 10{,}000 \times 2.3579477 = 23{,}579.48$

3.5. $PV_{gp} = CF_1/(r - g) = 100/(0.12 - 0.05) = 1428.57$

3.7. a. First, the monthly discount rate is $0.1 \div 12 = 0.008333333$

$$PV = \frac{1000}{0.00833333}\left[1 - \frac{1}{(1+0.00833333)^{360}}\right] = \$113{,}951$$

  b. Yes, since the $PV = \$113{,}951$ exceeds the \$100,000 price

  c. $100{,}000 = \frac{1000}{r/12}\left[1 - \frac{1}{(1+r/12)^{360}}\right]$

  Solve for r; by process of substitution, we find that $r = 0.11627$. One can also find r by using Newton's method for approximating roots, covered in any standard calculus text. We also cover this method in Section 7.4.2, or one can use the method of bisection covered in Section 7.4.1.

3.9. a. Bond yields or spot rates will be determined simultaneously to avoid associating contradictory rates for the annual coupons on each of the three bills:

$$
\begin{bmatrix} 50 & 1050 & 0 \\ 60 & 60 & 1060 \\ 90 & 90 & 1090 \end{bmatrix}
\begin{bmatrix} d_1 \\ d_2 \\ d_3 \end{bmatrix}
\begin{bmatrix} 947.376 \\ 904.438 \\ 981 \end{bmatrix}
$$

$$\mathbf{CF} \qquad\qquad \times \quad \mathbf{d} \quad = \quad \mathbf{b_0}$$

First, invert Matrix **CF**, then use this inverse to pre-multiply Vector $\mathbf{b_0}$ to obtain Vector **d**:

$$
\begin{bmatrix} -0.001 & -0.03815 & 0.0371 \\ 0.001 & 0.00181667 & -0.00177 \\ 0 & 0.003 & -0.002 \end{bmatrix}
\begin{bmatrix} 947.376 \\ 904.438 \\ 981 \end{bmatrix}
=
\begin{bmatrix} d_1 \\ d_2 \\ d_3 \end{bmatrix}
=
\begin{bmatrix} 0.94338 \\ 0.85734 \\ 0.75132 \end{bmatrix}
$$

$$\mathbf{CF}^{-1} \qquad\qquad\qquad \mathbf{b_0} \qquad = \qquad \mathbf{d}$$

Thus, spot rates are calculated as follows:

$$\frac{1}{d_1} - 1 = 0.06 = r_{0,1}$$

$$\frac{1}{\frac{1}{\frac{1}{2}}{d_2}} - 1 = 0.08 = r_{0,2}$$

$$\frac{1}{\frac{1}{\frac{1}{3}}{d_3}} - 1 = 0.10 = r_{0,3}$$

b. Prices for Bonds 1, 2, and 3 above will lead to the following forward rates:

$$r_{1,2} = \frac{(1+0.08)^2}{(1+0.06)} - 1 = 0.10$$

$$r_{2,3} = \frac{(1+0.10)^3}{(1+0.06)(1+0.10)} - 1 = 0.14$$

$$r_{1,3} = \sqrt{\frac{(1+0.10)^3}{(1+0.06)}} - 1 = 0.12$$

c. Any other bond with cash flows paid at the ends of some combination of years 1, 2, and 3 must have a value that is consistent with these three spot rates. Thus, the 3-year 2% coupon bond is valued as follows:

$$PV_4 = 20d_1 + 20d_2 + 1020d_3 = \frac{20}{(1+0.06)^1} + \frac{20}{(1+0.08)^2} + \frac{1020}{(1+0.10)^3} = 802.36$$

3.11. Replicating portfolio combinations are found by solving for γ as follows:

$$\begin{bmatrix} 50 & 80 & 110 \\ 50 & 80 & 1110 \\ 1050 & 1080 & 0 \end{bmatrix} \begin{bmatrix} \gamma_A \\ \gamma_B \\ \gamma_C \end{bmatrix} = \begin{bmatrix} 30 \\ 30 \\ 1030 \end{bmatrix}$$

$$\mathbf{CF^T} \qquad \times \quad \gamma \quad = \quad \mathbf{cf_D}$$

Solving for γ, we obtain $\gamma_A = 1.6667$, $\gamma_B = -0.6667$, and $\gamma_C = 0$.

3.13. First, find the values of pure securities 1, 2, and 3 as follows:

$$\begin{bmatrix} 5 & 7 & 9 \\ 2 & 4 & 8 \\ 9 & 1 & 3 \end{bmatrix} \begin{bmatrix} \psi_1 \\ \psi_2 \\ \psi_3 \end{bmatrix} = \begin{bmatrix} 5 \\ 3 \\ 5 \end{bmatrix}$$

$$\begin{bmatrix} 0.0227272 & -0.0681818 & 0.11363636 \\ 0.375 & -0.375 & -0.125 \\ -0.1931818 & 0.3295454 & 0.03409090 \end{bmatrix} \begin{bmatrix} 5 \\ 3 \\ 5 \end{bmatrix} = \begin{bmatrix} \psi_1 \\ \psi_2 \\ \psi_3 \end{bmatrix} = \begin{bmatrix} 0.47727 \\ 0.125 \\ 0.19318 \end{bmatrix}$$

$$\qquad \mathbf{CF}^{-1} \qquad\qquad \mathbf{v} \;\; = \;\; \boldsymbol{\psi} \;\; = \;\; \boldsymbol{\psi}$$

Thus, we find that $\psi_1 = 0.47727$, $\psi_2 = 0.125$, and $\psi_3 = 0.19318$. The value of security D equals $1 \times 0.47727 + 1 \times 0.125 + 1 \times 0.19318 = 0.79545$.

3.15. Under probability measure $\mathbb{Q}$, the expected present value of the stock is:

$$S_0 = quS_0/(1+r) + (1-q)dS_0/(1+r); \text{multiply both sides by } (1+r)/S_0$$
$$1 + r = qu + (1-q)d = qu + d - qd = q(u-d) + d$$
$$q = (1 + r - d)/(u - d)$$

3.17. a. Since the riskless return rate is 0.125, the current value of a security guaranteed to pay $1 in 1 year would be $1/1.125 = 0.88889$. The security payoff vectors are as follows:

$$h = \begin{bmatrix} 10 \\ 16 \\ 25 \end{bmatrix}, b = \begin{bmatrix} 1 \\ 1 \\ 1 \end{bmatrix}, c_{15} = \begin{bmatrix} 0 \\ 1 \\ 10 \end{bmatrix}, c_9 = \begin{bmatrix} 1 \\ 7 \\ 16 \end{bmatrix}$$

Portfolio holdings are determined as follows:

$$\begin{bmatrix} 10 & 1 & 0 \\ 16 & 1 & 1 \\ 25 & 1 & 10 \end{bmatrix} \cdot \begin{bmatrix} \gamma_h \\ \gamma_b \\ \gamma_{c15} \end{bmatrix} = \begin{bmatrix} 1 \\ 7 \\ 16 \end{bmatrix}$$

The following includes the inverse of the securities payoff matrix:

$$\begin{bmatrix} \gamma_h \\ \gamma_b \\ \gamma_{c15} \end{bmatrix} = \begin{bmatrix} -0.2 & 0.22222 & -0.02222 \\ 3 & -2.2222 & 0.22222 \\ 0.2 & -0.33333 & 0.133333 \end{bmatrix} \cdot \begin{bmatrix} 1 \\ 7 \\ 16 \end{bmatrix} = \begin{bmatrix} 1 \\ -9 \\ 0 \end{bmatrix}$$

We find that a portfolio replicating a call with an exercise price of 9 can be constructed with the following numbers of shares, T-bills, and calls with an exercise price of 15: $\gamma_h = 1$ and $\gamma_b = -9$ and $\gamma_{c15} = 0$. Thus, the payoff structure of a single call with an exercise price of 9 can be replicated with a portfolio comprising 1 share of stock, short-selling 9 T-bills, and selling 0 calls with an exercise price of 15. This portfolio requires a net investment of $1 \times 14 - 9 \times 0.88889 + 0 \times 15 = \$6$. Since the call has the same payoff structure as this portfolio, its current value must be $6.

b. To find the pure security prices, we need to solve the matrix equation:

$$\begin{bmatrix} 10 & 16 & 25 \\ 1 & 1 & 1 \\ 0 & 1 & 10 \end{bmatrix} \begin{bmatrix} \psi_1 \\ \psi_2 \\ \psi_3 \end{bmatrix} = \begin{bmatrix} 14 \\ 0.88889 \\ 3 \end{bmatrix}$$

Pure security prices are $\psi_1 = 0.46667$, $\psi_2 = 0.1358$, and $\psi_3 = 0.28642$. Thus, synthetic probabilities are $0.46667/0.88889 = 0.525$, $0.1358/0.88889 = 0.1528$, and $0.28642/0.88889 = 0.3222$.

3.19. **a.** We will have a set of two payoff vectors in a two-outcome economy. The set is linearly independent. Hence, this set forms the basis for the two-outcome space. Since we have market prices for these two securities, we can price all other securities in this economy. First, we solve for the value of the $22-exercise price call as follows:

$$\begin{bmatrix} \gamma_s \\ \gamma_{c18} \end{bmatrix} = \begin{bmatrix} 15 & 0 \\ 25 & 7 \end{bmatrix}^{-1} \begin{bmatrix} 0 \\ 3 \end{bmatrix}$$

$$\begin{bmatrix} \gamma_s \\ \gamma_{c18} \end{bmatrix} = \begin{bmatrix} 0.066667 & 0 \\ -0.238095 & 0.142857 \end{bmatrix} \begin{bmatrix} 0 \\ 3 \end{bmatrix} = \begin{bmatrix} 0 \\ 0.42857 \end{bmatrix}$$

Thus, the call with an exercise price equal to $22 can be replicated with 0.42857 calls with an exercise price equal to $18. The value of this call equals $0.42857 \times 7 = \$3$.

b. The riskless return rate is determined as follows:

$$\begin{bmatrix} 0.066667 & 0 \\ -0.238095 & 0.142857 \end{bmatrix} \begin{bmatrix} 1 \\ 1 \end{bmatrix} = \begin{bmatrix} 0.066667 \\ -0.095238 \end{bmatrix}$$

Since the riskless asset is replicated with 0.066667 shares of stock and short positions in 0.095238 calls, the value of the riskless asset is $0.066667 \times 20 - 0.095238 \times 7 = 0.66667$, implying a riskless return rate equal to $1/0.66667 - 1 = 0.50$.

c. Solve for the value of the put as follows:

$$\begin{bmatrix} 0.066667 & 0 \\ -0.238095 & 0.142857 \end{bmatrix} \begin{bmatrix} 25 \\ 15 \end{bmatrix} = \begin{bmatrix} 1.66667 \\ -3.8095 \end{bmatrix}$$

implying that its value is $1.66667 \times 20 - 3.8095 \times 7 = \$6.67$. Note that this put value is lower than either of the two potential cash flows that it may generate. This is due to the particularly high riskless return rate.

CHAPTER 4

4.1. In this problem, $F = 1000$, $T = 12$, and $r(t) = 0.007 - 0.00003t^2$. Using Eq. (4.3), the price of the bond is:

$$B_0 = 1000e^{\left(-\int_0^{12}(0.007-0.00003t^2)\,dt\right)} = 1000e^{-(0.007t-0.00001t^3|_0^{12})} = 1000e^{-0.06672} = \$935.46$$

4.3. First, separate by multiplying both sides of the differential equation by dt and divide both sides by y. Then integrate to obtain:

$$\int \frac{dy}{y} = \int t\,dt$$

$$\ln y = \frac{1}{2}t^2 + C$$

since $y(0) = 100 > 0$:

$$y = e^{\frac{1}{2}t^2 + C} = K e^{\frac{1}{2}t^2}$$

The initial condition implies:

$$100 = K e^0 = K$$

Thus:

$$y = 100\, e^{\frac{1}{2}t^2}$$

Alternatively, it is sometimes convenient to work out the solution by using a definite integral rather than an indefinite integral. So, we could also find the solution by the following steps:

$$\int_{100}^{y_T} \frac{dy}{y} = \int_0^T t\, dt$$

Observe that to integrate in the variable t from 0 to T means that we are integrating in the variable y from $y(0) = 100$ to y_T.

$$\ln y_T - \ln 100 = \frac{1}{2} T^2$$

$$\ln\left(\frac{y_T}{100}\right) = \frac{1}{2} T^2$$

$$\frac{y_T}{100} = e^{\frac{1}{2}T^2}$$

$$y_T = 100 e^{\frac{1}{2}T^2}$$

4.5. Let $V = V_t$ denote the value of the investment at time t. The problem states that:

$$\frac{dV}{dt} = k\sqrt{V} \text{ and } V(0) = V_0$$

Multiplying both sides of the first equation by dt and dividing both sides by $\sqrt{V}$, we have:

$$\frac{dV}{\sqrt{V}} = k\, dt$$

Next, we integrate both sides of the equation from 0 to T:

$$\int_{V_0}^{V_T} \frac{dV}{\sqrt{V}} = \int_0^T k\, dt$$

Evaluating the integrals, we see that:

$$2\sqrt{V_T} - 2\sqrt{V_0} = kT$$

Algebraically solving for V_T gives the solution to be:

$$V_T = \left(\frac{kT}{2} + \sqrt{V_0}\right)^2$$

4.7. The solution to this differential equation gives the bond's price at time t will be obtained by the following:

$$\int \frac{dB_t}{B_t} = \int r_0\, dt$$

These integrals are solved as follows:

$$\ln B_t = r_0 t + C$$

We write the anti-logs of the results of both sides as:

$$e^{\ln B_t} = e^{r_0 t + C}$$
$$B_t = Ke^{r_0 t}$$

where $K = e^C$. This equation represents a general solution to our differential equation. Since $B_T = F$, evaluating the solution at $t = T$ gives: $F = Ke^{r_0 T}$.
Solving for K we have:

$$K = Fe^{-r_0 T}$$

Substituting the value for K into the general solution for B_t we obtain the desired solution:

$$B_t = Fe^{-r_0 T}e^{r_0 t} = Fe^{-r_0(T-t)}$$

4.9. First, divide both sides of the differential by $M - S_t$ to obtain:

$$\frac{dS_t}{M - S_t} = \mu\, dt$$

Since, by the chain and log rules, the integral of $dS_t/(M - S_t)$ equals $- \ln(M - S_t)$, we will use the expression $\ln(M - S_t)$ to obtain the solution for S_t. The differential of the function $f(S_t) = \ln(M - S_t)$ is:

$$d[\ln(M - S_t)] = \frac{-1}{M - S_t}dS_t = -\mu\, dt$$

Changing the variable from t to s and integrating from 0 to t results in:

$$\ln(M - S_t) - \ln(M - S_0) = -\mu t$$

or:

$$\ln\left(\frac{M - S_t}{M - S_0}\right) = -\mu t$$

Exponentiating we have:

$$\frac{M - S_t}{M - S_0} = e^{-\mu t}$$

Solving for S_t gives:

$$S_t = M - (M - S_0)e^{-\mu t}$$

4.11. Since $r_0 = 6.5 > \mu = 6$, then $|\mu - r_t| = -(\mu - r_t) = r_t - \mu$. So the solution takes the form:

$$r_t = \mu + e^K e^{\lambda t}$$

We evaluate the solution at $t = 0$ and use our initial condition in order to determine e^K:

$$r_0 = \mu + e^K e^{\lambda \times 0}$$
$$6.5 = 6 + e^K$$

Thus, $e^K = 0.5$. Next, we solve for $\lambda = -0.074381184$ by using the exchange rate drift over the past 3 months:

$$r_3 = 6.4 = 6 + 0.5 \times e^{3\lambda}$$

$$\frac{6.4 - 6}{0.5} = e^{3\lambda} = 0.8$$

Finally, we find that it takes $t = 21.638$ days from 3 days ago for the exchange rate to drop to 6.1:

$$r_t = 6.1 = 6 + 0.5 \times e^{-0.07438t}$$

$$t = \frac{1}{-0.07438} \ln \frac{0.1}{0.5} = 21.638$$

4.13. a. The equation is separable. We will divide both sides by $r(\mu - r)$ to separate and integrate:

$$\int \frac{dr}{r(\mu - r)} = \int \lambda \, dt$$

b. Using partial fractions, this is written:

$$\int \left(\frac{1}{\mu r} + \frac{1}{\mu(\mu - r)}\right) dr = \int \lambda \, dt$$

It is easy to start with the equation immediately above, find a common denominator for the terms being added, and verify that it equals the one above it in part a. For now, $0 < r < \mu$, leading to a general solution for the logistic equation:

$$\frac{1}{\mu}\ln|r| - \frac{1}{\mu}\ln|\mu - r| = \lambda t + K$$

$$\ln r - \ln(\mu - r) = \lambda \mu t + \mu K$$

$$e^{\ln r - \ln(\mu - r)} = e^{\lambda \mu t} \, e^{\mu K}$$

$$\frac{r}{(\mu - r)} = e^{\mu K} \, e^{\lambda \mu t}$$

$$r = (\mu - r)e^{\mu K} \, e^{\lambda \mu t}$$

$$r(1 + e^{\mu K} \, e^{\lambda \mu t}) = \mu e^{\mu K} e^{\lambda \mu t}$$

$$r = \frac{\mu e^{\mu K} \, e^{\lambda \mu t}}{(1 + e^{\mu K} \, e^{\lambda \mu t})}$$

c. Our solution to the logistic equation is:

$$r_t = \frac{6 e^{\mu K} \, e^{0.007 \times 6t}}{(1 + e^{\mu K} \, e^{0.007 \times 1000t})}$$

The initial condition is:

$$r_0 = 5 = \frac{6 e^{\mu K} \, e^{0.007 \times 6 \times 0}}{(1 + e^{\mu K} \, e^{0.007 \times 6 \times 0})} = \frac{6 e^{\mu K}}{(1 + e^{\mu K})}$$

which implies that $e^{\mu K} = 5/(6 - 5) = 5$, which means that:

$$r_t = \frac{30 e^{0.007 \times 6t}}{1 + 5 e^{0.007 \times 6t}}$$

d. Simply substitute 10 for t in part c: $r_{10} = \dfrac{30 e^{0.007 \times 610}}{1 + 5 e^{0.007 \times 6 \times 10}} = 5.30312$

4.15. The dividend stream is evaluated as follows:

$$PV[0, 2] = 10,000 \int_0^2 e^{(0.03-0.05)t} dt = \$10,000 \left[-\frac{e^{(0.03-0.05)t}}{0.05-0.03} \right] \Big|_0^2 = \$19,605.28$$

4.17. a. Let g denote the annual rate of increase of the investor's withdrawals. The change in the value of the account over the time interval $[t, t + dt]$ is:

$$dFV_t = rFV_t \, dt - PMTe^{gt} \, dt$$

Dividing by dt gives the differential equation:

$$\frac{dFV_t}{dt} = rFV_t - PMTe^{gt}$$

b. Note that the differential equation is not separable. We find the solution using an integrating factor. First, rewrite the differential equation in the form:

$$\frac{dFV_t}{dt} - rFV_t = -PMTe^{gt}$$

Similar to problem 4.16, the integrating factor is:

$$\rho(t) = e^{\int(-r)dt} = e^{-rt}$$

Multiply both sides of the differential equation by the integrating factor to get:

$$e^{-rt}\frac{dFV_t}{dt} - re^{-rt}FV_t = -PMT\,e^{(g-r)t}$$

By the product rule for differentiation, we can rewrite the equation above in the form:

$$\frac{d}{dt}(e^{-rt}FV_t) = -PMTe^{(g-r)t}$$

Integrating both sides of the equation from 0 to T, we have:

$$\int_0^T \frac{d}{dt}(e^{-rt}FV_t)\,dt = -\int_0^T PMTe^{(g-r)t}\,dt$$

$$e^{-rt}FV_t\big|_0^T = \frac{PMT}{r-g}e^{(g-r)t}\big|_0^T$$

$$e^{-rT}FV_T - FV_0 = \frac{PMT}{r-g}(e^{(g-r)T} - 1)$$

Solving for FV_T gives:

$$FV_T = FV_0e^{rT} + \frac{PMT}{r-g}(e^{gT} - e^{rT})$$

c. In this case, $FV_0 = 1{,}000{,}000$, $PMT = 100{,}000$, $r = 0.04$, and $g = 0.02$. Thus, the amount of money in the account after 8 years will be:

$$FV_8 = 1{,}000{,}000e^{0.04\times8} + \frac{100{,}000}{0.04-0.02}(e^{0.02\times8} - e^{0.04\times8}) = \$359{,}043.30$$

d.

$$1{,}000{,}000e^{0.04T} + \frac{100{,}000}{0.04-0.02}(e^{0.02T} - e^{0.04T}) = 0$$

Solving for T, we obtain:

$$T = \frac{\ln 1.25}{0.02} = 11.16 \text{ years}$$

4.19. a. Using the continuous time framework introduced in the chapter, the current market value, duration and convexity for this bond are computed as follows:

$$B_0 = 108.58 + 98.24 + 88.89 + 80.43 + 679.31 = 1055.48$$

$$\text{Dur} = 108.58 + 196.49 + 266.69 + 321.75 + 3396.57 = 4.064$$

$$\text{Conv} = \frac{1}{2}(108.58 + 392.99 + 800.08 + 1287.01 + 16982.85) = 9.2715$$

b. First- and second-order approximations are as follows:

$$B_{t+dt} \approx 1055.479 \times [1 - 0.03(4.064)] = 926.78$$

$$B_{t+dt} \approx 987.2569 \times [1 - 0.03(4.064) + (0.03)^2(9.2715)] = 935.58$$

c. The actual bond value under the new interest rate is 935.18.

4.21. a. The probability that the put will be exercised is computed as follows:

$$P[S_T < 16] = \int_{10}^{16} \frac{1}{10} dS_T = \frac{1}{10} S_T \Big|_{10}^{16} = 1.6 - 1 = 0.6$$

b. The expected value of the stock, contingent on its price exceeding 16 is:

$$E[S_T | S_T > 16] = \frac{\int_{16}^{20} \frac{S_T}{10} dS_T}{\int_{16}^{20} \frac{1}{10} dS_T} = \frac{\frac{1}{20} S_T^2 \Big|_{16}^{20}}{0.4} = \frac{20 - 12.8}{0.4} = 18$$

c. The expected value of the put, contingent on it being exercised is:

$$E[p_T | S_T < X] = \frac{\int_0^X (X - S_T) q[S_T] dS_T}{\int_0^X q[S_T] dS_T} = \frac{\int_{10}^{16} \frac{16 - S_T}{10} dS_t}{\int_{10}^{60} \frac{1}{10} dS_T} = \frac{1.6 S_T - \frac{1}{20} S_T^2 \Big|_{10}^{16}}{0.6}$$

$$= \frac{(25.6 - 12.8) - (16 - 5)}{0.6} = \frac{1.8}{0.6} = 3$$

d. The expected value of the put at time T is:

$$E[p_T] = \int_{10}^{16} (16 - S_T) \frac{1}{10} dS_T = 1.6 S_T - \frac{1}{20} S_T^2 \Big|_{10}^{16} = (25.6 - 12.8) - (16 - 5) = 1.8$$

e. One important note in this problem is that the problem statement provides no reason to believe that the distribution of stock payoffs in the second period is any different from the distribution in the first period. Whether this might be a reasonable assumption might be a matter open for debate. We assume here that the parameters of the uniform distribution in the second period are identical to those of the first period, with the outcome of the first period not affecting the distribution in the second. However, waiting for two periods for the terminal cash flow will reduce the present value of the payoff. Thus, the present value of this put would simply be its discounted value:

$$c_0 = \int_X^\infty (S_T - X)q[S_T]\,dS_T e^{-rT} = 1.8e^{-1.1 \times 2} = 1.4737$$

CHAPTER 5

5.1. a. All of the possible distinct price two-period change outcomes for the stock to time 2 are $(1,1)$, $(1,-1)$, $(-1,1)$, and $(-1,-1)$.
  b. The sample space for this process is $\Omega = \{(1,1), (1,-1), (-1,1), (-1,-1)\}$.
  c. The time 0 events are Ω and $\emptyset$. $P(X_0 = 0) = P(\Omega) = 1$ and $P(X_0 \neq 0) = P(\emptyset) = 0$.
  d. p.
  e. $P(\{(1,1),(1,-1)\}) = p$.
  f. Yes.
  g. No. $P(X_1 = 1 \text{ or } X_1 = -1) = P(\Omega) = 1$.
  h. $P(X_2 = 0 \text{ or } X_2 = 2) = P(\{(1,1),(1,-1),(-1,1)\}) = 2p(1-p) + p^2 = p(2-p)$.
  i. Since the σ-algebra $\mathscr{F}_1$ is defined such that $\mathscr{F}_1 \subset \mathscr{F}_2$, this allows us to compute probabilities that involve random variables at different times such as:

$$P(X_1 = 1 \text{ and } X_2 = 0) = P(\{(1,1),(1,-1)\} \cap \{(1,-1),(-1,1)\}) = P(\{(1,-1)\}) = p(1-p).$$

5.3. a. Yes: For every fixed number of cards t that have been dealt, the sum S_t of the values of the t dealt cards is a random variable. Furthermore, this random variable is indexed by time (technically, by the number of cards that have been dealt). The common probability space is the probability space that arises from the set of outcomes of all possible ways to deal 52 cards from the deck.
  b. No: It is enough to show the following one case: $P(S_3 = 9 \mid S_0 = 0, S_1 = 2, S_2 = 6) \neq P(S_3 = 9 \mid S_2 = 6)$ to justify that the Markov property is violated. Note that $Z_t = S_t - S_{t-1}$ denotes the value of the tth card that is dealt. Thus, $P(S_3 = 9 \mid S_0 = 0, S_1 = 2, S_2 = 6) = P(Z_3 = 3 \mid Z_1 = 2, Z_2 = 4)$ and $P(S_3 = 9 \mid S_2 = 6) = P(Z_3 = 3 \mid Z_1 + Z_2 = 6)$. We calculate that:

$$P(Z_3 = 3 \mid Z_1 = 2, Z_2 = 4) = \frac{P(Z_1 = 2, Z_2 = 4, Z_3 = 4)}{P(Z_1 = 2, Z_2 = 4)} = \frac{\dfrac{4 \times 4 \times 4}{52 \times 51 \times 50}}{\dfrac{4 \times 4}{52 \times 51}} = \frac{2}{25}$$

In order that $Z_1 + Z_2 = 6$, either $Z_1 = 2$ and $Z_2 = 4$, or $Z_1 = 4$ and $Z_2 = 2$, or $Z_1 = 3$ and $Z_2 = 3$. Thus:

$$P(Z_3 = 3 | Z_1 + Z_2 = 6) = \frac{P(Z_1 + Z_2 = 6, Z_3 = 3)}{P(Z_1 + Z_2 = 6)}$$

$$= \frac{P(Z_1 = 2, Z_2 = 4, Z_3 = 3) + P(Z_1 = 4, Z_2 = 2, Z_3 = 3) + P(Z_1 = 3, Z_2 = 3, Z_3 = 3)}{P(Z_1 = 2, Z_2 = 4) + P(Z_1 = 4, Z_2 = 2) + P(Z_1 = 3, Z_2 = 3)}$$

$$= \frac{\dfrac{4 \times 4 \times 4}{52 \times 51 \times 50} + \dfrac{4 \times 4 \times 4}{52 \times 51 \times 50} + \dfrac{4 \times 3 \times 2}{52 \times 51 \times 50}}{\dfrac{4 \times 4}{52 \times 51} + \dfrac{4 \times 4}{52 \times 51} + \dfrac{4 \times 3}{52 \times 51}} = \frac{19}{275} \neq \frac{2}{25}$$

   c. Yes: Since $E[S_t | S_{t-1}] = S_{t-1} + E[S_t - S_{t-1} | S_{t-1}]$ and $S_t - S_{t-1} > 0$, then $E[S_t | S_{t-1}] > S_{t-1}$.

   d. Yes: Since the cards are dealt with replacement, then each card is dealt from the originally randomly shuffled deck. This implies that the choice of each card dealt is independent of the choice of any other card that is dealt. Thus, the random variables $\{Z_t\}$ are independent of one another. As we proved in Section 5.1.2.2, the resulting process is Markov.

   e. Yes: By the same reason as in part c.

   f. Yes: By the same reason as in part d.

   g. Yes: First, we write $E[S_t | S_{t-1}] = S_{t-1} + E[S_t - S_{t-1} | S_{t-1}]$. Note that since $Z_t = S_t - S_{t-1}$, $S_0 = 0$, and $S_{t-1} = Z_1 + \ldots + Z_{t-1}$, then the random variable $S_t - S_{t-1}$ is independent of the random variable S_{t-1}. This implies that $E[S_t - S_{t-1} | S_{t-1}] = E[S_t - S_{t-1}] = E[Z_t]$. As the tth card is being dealt, the probability is $1/52$ of it being any given card in the deck. There are $4 \times 5 = 20$ cards that have a point value of 1. There are $4 \times 3 = 12$ cards that have a point value of 0. There are $4 \times 5 = 20$ cards that have a point value of -1. Thus:

$$E[Z_t] = 1 \times \frac{20}{52} + 0 \times \frac{12}{52} + (-1) \times \frac{20}{52} = 0$$

This means that $E[S_t | S_{t-1}] = S_{t-1} + E[S_t - S_{t-1} | S_{t-1}] = S_{t-1}$. We conclude that S_t is a martingale.

5.5. a. $2^4 = 16$

   b.

+	+	+	+,	−	+	+	+,	−	+	+	−,	−	+	−	−
+	+	+	−,	+	+	−	−,	−	+	−	+,	−	−	+	−
+	+	−	+,	+	−	+	+,	−	−	+	+,	−	−	−	+
+	−	+	+,	+	−	−	+,	+	−	−	−,	−	−	−	−

   c. Since each transaction is equally likely to result in a price increase or decrease, then each ordering has equal probability ($p = \frac{1}{2}$). The probability that any particular ordering will be realized is $P[\text{ordering}] = p^n = 1/2^4 = 0.0625$.

d. The probability of three price increases followed by a single decrease $(+ + + -)$ equals 0.0864:

$$P[+, +, +, -] = P[\text{ordering}] = p^{y*}(1-p)^{n-y*} = p^3(1-p)^{4-3} = 0.0864$$

e. The probability that exactly $y* = 3$ price increases will result from $n = 4$ transactions where $p = 0.6$ is calculated from the following:

$$P(y*) = \binom{n}{y*} p^{y*}(1-p)^{n-y*}$$

$$P(3) = \binom{4}{3} 0.6^3(1-0.6)^{4-3} = \frac{4 \cdot 3 \cdot 2 \cdot 1}{3 \cdot 2 \cdot 1 \cdot 1} \cdot 0.216 \cdot 0.4 = 4 \cdot 0.0864 = 0.3456$$

f. $P[y* > 3] = P[y* = 4] = 0.6^4 = 0.1296$.

5.7. a. Under physical probability measure $\mathbb{P}$, the expected value of $S_{t+\Delta t}$ at time $t + \Delta t$ given S_t is

$$E_{\mathbb{P}}[S_{t+t}|S_t] = p\left(S_t + a\sqrt{\Delta t}\right) + (1-p)\left(S_t - a\sqrt{\Delta t}\right) = S_t + (2p-1)\left(a\sqrt{\Delta t}\right)$$

b. This Markov process is also a martingale in the case where $p = 1/2$.

c. Since the variances of $S_{t+\Delta t}$ and $S_{t+\Delta t} - S_t$ are equal if S_t is given, the variance of $S_{t+\Delta t}$ at time $t + \Delta t$ given S_t is:

$$\mathrm{Var}_{\mathbb{P}}[S_{t+t} - S_t|S_t] = E_{\mathbb{P}}[(S_{t+\Delta t} - S_t)^2] - (E_{\mathbb{P}}[S_{t+\Delta t} - S_t])^2$$

$$= p\left(a\sqrt{\Delta t}\right)^2 + (1-p)\left(-a\sqrt{\Delta t}\right)^2 - \left[p\left(a\sqrt{\Delta t}\right) + (1-p)\left(-a\sqrt{\Delta t}\right)\right]^2$$

$$= p\left(a\sqrt{\Delta t}\right)^2 + (1-p)\left(-a\sqrt{\Delta t}\right)^2 - p^2(a^2\Delta t) + (1-p)^2(a^2\Delta t) + 2p(1-p)a^2\Delta t$$

$$= 2pa^2\Delta t - 4p^2a^2\Delta t - p^2(a^2\Delta t) + a^2\Delta t + p^2a^2\Delta t - 2p^2a^2\Delta t + 2pa^2\Delta t - 2p^2a^2\Delta t$$

$$= 2pa^2\Delta t - 4p^2a^2\Delta t + 2pa^2\Delta t = 4pa^2\Delta t - 4p^2a^2\Delta t$$

$$= 4p(1-p)a^2\Delta t$$

d. If $p = 0.5$, the expected value and variance of $S_{t+\Delta t}$ given S_t are S_t and $a^2\Delta t$, respectively.

e. The standard deviation of $S_{t+\Delta t}$ given S_t equals $a\sqrt{\Delta t}$.

5.9. The expected value of the stock is computed as follows:

$$E[S_4] = S_0[pu + (1-p)d]^t = [0.75 \times 1.2 + 0.25 \times 0.8]^4$$

$$= \$146.41$$

5.11. a. As we showed at the end of Section 5.2.1.1, we have the conditional expectation:

$$E_{\mathbb{P}}[S_t|S_s] = [pu + (1-p)d]^{t-s} S_s$$

for $s < t$. If $pu + (1-p)d = 1$, then $[pu + (1-p)d]^{t-s} = 1$ and $E_{\mathbb{P}}[S_t|S_s] = 1 \times S_s = S_s$.

b. From Section 5.2.1, we know that $E_Q[S_t|S_s] = [qu+(1-q)d]^{t-s}S_s$ for $s < t$. From time s to time t, we need to discount the expected value by the amount $(1+r)^{t-s}$. With respect to the bond, the stock expected value is:

$$\frac{E_Q[S_t|S_s]}{[1+r]^{t-s}} = \frac{[qu+(1-q)d]^{t-s}S_s}{[1+r]^{t-s}} = \left[\frac{qu+(1-q)d}{1+r}\right]^{t-s}S_s = 1 \times S_s = S_s$$

if $(qu + (1-q)d)/(1+r) = 1$ or equivalently $qu + (1-q)d = 1+r$. Thus, S_t is a martingale with respect to the bond in this case.

c. $qu + (1-q)d = 1+r$. Rewrite as $qu + d - qd = 1+r$. Thus, $q(u-d) + d = 1+r$ and $q = (1+r-d)/(u-d)$.

5.13. a. Risk-neutral probabilities are computed as follows:

$$\begin{bmatrix} 30 & 70 \\ 100 & 100 \end{bmatrix}\begin{bmatrix} \psi_{0,1;d} \\ \psi_{0,1;u} \end{bmatrix} = \begin{bmatrix} 50 \\ 90 \end{bmatrix}; \quad \begin{array}{l} \psi_{0,1;d} = 0.325 \\ \psi_{0,1;u} = 0.575 \end{array}$$

$$q_{0,1;d} = \psi_{0,1;d}/(\psi_{0,1;d} + \psi_{0,1;u}) = 0.36111$$

$$q_{0,1;u} = \psi_{0,1;u}/(\psi_{0,1;d} + \psi_{0,1;u}) = 0.63889$$

b. Yes: There are no arbitrage strategies in this payoff space.

c. Yes: The probability measure is unique because the number of priced securities equals the number of potential outcomes, and their set of payoff vectors is complete.

d. $c_T = Max[0, S_T - X]$; $c_T = \$0$ or $\$15$

e. $\$100/\$90 - 1 = 0.1111$

f. The hedge ratio is computed as follows:

$$\alpha = \frac{c_u - c_d}{S_0(d-u)} \quad \alpha = \frac{15-0}{50(0.6-1.4)} = -0.375$$

g.

$$c_0 = \frac{c_u q + c_d(1-q)}{1+r} = \frac{15.63889 + 0(1-0.63889)}{1+0.11111} = 8.625$$

h.

$$p_0 = \frac{p_u q + p_d(1-q)}{1+r} = \frac{0.63889 + 25(1-0.63889)}{1+0.11111} = 8.125$$

5.15. First, we will estimate q_u from u, d (which equals $1/u = 0.69252$) and r:

$$d = \frac{1}{1.444} = 0.69252; q_{u,2} = \frac{1+0.08 \times 0.75/2 - 0.69252}{1.444 - 0.69252} = 0.449087$$

$$q_{u,3} = \frac{1+0.08 \times 0.75/3 - 0.69252}{1.444 - 0.69252} = 0.435779; q_{u,8} = \frac{1+0.08 \times 0.75/8 - 0.69252}{1.444 - 0.69252} = 0.4191$$

For the two-time period framework, call valuation calculations proceed as follows:

$$a = \text{Int}\left[\text{Max}\left[\frac{\ln\left(\frac{80}{50 \cdot 0.69252^2}\right)}{\ln\left(\frac{1.444}{0.69252}\right)}, 0\right] + 1\right] = 2$$

$$c_0 = \frac{0.449087^2 \times 0.550913^{2-2} \times [1.444^2 \times 0.69252^{2-2} \times 50 - 80]}{(1 + 0.08 \cdot 0.75/2)^2} = 4.61$$

For the three-time period framework, call valuation calculations proceed as follows:

$$a = \text{Int}\left[\text{Max}\left[\frac{\ln\left(\frac{80}{50 \cdot 0.69252^3}\right)}{\ln\left(\frac{1.444}{0.69252}\right)}, 0\right] + 1\right] = 3$$

$$c_0 = \frac{0.435779^3 \times 0.564221^{3-3} \times [1.444^3 \times 0.69252^{3-3} \times 50 - 80]}{(1 + 0.08 \cdot 0.75/3)^3} = 5.50$$

For the eight-time period framework, call valuation calculations proceed as follows:

$$a = \text{Int}\left[\text{Max}\left[\frac{\ln\left(\frac{80}{50 \cdot 0.69252^8}\right)}{\ln\left(\frac{1.444}{0.69252}\right)}, 0\right] + 1\right] = 5$$

$$c_0 = \frac{0.4191^8 \times 0.5809^{8-8} \times [1.444^8 \times 0.6925^{8-8} \times 50 - 80]}{(1 + 0.08 \cdot 0.75/8)^8}$$

$$+ 8 \times \frac{0.4191^7 \times 0.5809^{8-7} \times [1.444^7 \times 0.6925^{8-7} \times 50 - 80]}{(1 + 0.08 \cdot 0.75/8)^8}$$

$$+ 28 \times \frac{0.4191^6 \times 0.5809^{8-6} \times [1.444^6 \times 0.6925^{8-6} \times 50 - 80]}{(1 + 0.08 \cdot 0.75/8)^8}$$

$$+ 56 \times \frac{0.4191^5 \times 0.5809^{8-5} \times [1.444^5 \times 0.6925^{8-5} \times 50 - 80]}{(1 + 0.08 \cdot 0.75/8)^8} = 14.39$$

Put values are found with put–call parity. The following are call and put values for the two-, three-, and eight-period frameworks:

n	c_0	p_0
2	4.61	30.02
3	5.50	30.89
8	14.39	39.75

The calls are easily valued in the two- and three-step models because the maximum number of upjumps (two and three) are required for exercise. In the eight-period framework, five to eight upjumps were required for exercise.

5.17. Simply regard time s as time 0, so that S_0 becomes the value S_s. Thus, time t becomes time $t - s$ since it takes time $t - s$ to go from time s to time t. The value S_t remains S_t, since to go from the value S_s in $t - s$ steps of time will give us S_t. So keeping S_t on the left side of the equation, replacing S_0 with S_s and t with $t - s$ on the right side of the given equation gives the desired result.

5.19.

$$P\left\{T^*_{Low} \le T^*_{High}\right\} = \frac{S^*_{High} - S_0}{S^*_{High} - S^*_{Low}} = \frac{5 - 0}{5 + 2} = 0.7143;$$

changing the variance will not change the probabilities.

5.21. We are going to make use of Eq. (5.28) with $S_0 = 40$, $X = 30$, $\sigma = 0.5$, $r = 0.05$, and $T = 1$. We first need to find:

$$d_2 = \frac{\ln\left(\frac{40}{30}\right) + (0.05 - \frac{1}{2}0.5^2) \times 1}{0.5 \times \sqrt{1}} = 0.4254$$

Using Eq. (5.28) we obtain:

$$P(S_T < 30) = 1 - P(S_T \ge 30) = 1 - N(0.4254) = 1 - 0.6647 = 0.3353$$

5.23. Minor simplification and expanding the square yields:

$$\hat{\mu}T + \sigma\sqrt{T}z - \frac{z^2}{2} = \left(\hat{\mu} + \frac{\sigma^2}{2}\right)T - \frac{(z - \sigma\sqrt{T})^2}{2}$$

$$\hat{\mu}T + \sigma\sqrt{T}z - \frac{z^2}{2} = \hat{\mu}T + \frac{\sigma^2}{2}T - \frac{z^2}{2} + 2\frac{\sigma\sqrt{T}z}{2} - \frac{\sigma^2}{2}T$$

$$\sigma\sqrt{T}z = \frac{\sigma^2}{2}T + 2\frac{\sigma\sqrt{T}z}{2} - \frac{\sigma^2}{2}T$$

$$\sigma\sqrt{T}z = +\sigma\sqrt{T}z$$

5.25. a. The probability of option exercise:

$$P[S_T > 50] = N(d_2) = N\left[\frac{\ln\left(\frac{S_0}{X}\right) + \left(r - \frac{\sigma^2}{2}\right)T}{\sigma\sqrt{T}}\right] = N\left[\frac{\ln\left(\frac{45}{50}\right) + \left(0.05 - \frac{0.16}{2}\right)}{0.4}\right] = N(-0.3384) = 0.36753$$

b.

$$E\left[S_T | S_T > X\right] = \frac{S_0\, e^{rT} N(d_1)}{N(d_2)} = \frac{45 \times e^{0.05 \times 1} \times N\left[\frac{\ln\left(\frac{45}{50}\right) + \left(0.05 + \frac{0.16}{2}\right)}{0.4}\right]}{0.36753}$$

$$= \frac{45 \times e^{0.05} \times 0.52456}{0.36753} = 67.52$$

c.

$$E[c_{GAP,T}] = (E[S_T|S_T > X] - M)N(d_2) = (67.52 - 40) \times 0.36753 = 10.114$$

d. $c_{GAP,0} = \frac{E[c_{GAP,T}]}{e^{rT}} = \frac{10.114}{e^{0.05 \times 1}} = 9.62$

CHAPTER 6

6.1. By the product rule, $dX_t = t^2 dZ_t + Z_t d(t^2) = t^2 dZ_t + 2tZ_t\, dt$.

6.3. a. $\int_0^t 5\, dX_s = 5X_s|_0^t = 5s^2|_0^t = 5(t)^2 - 5(0)^2 = 5t^2$

b. $\int_0^t 5\, dZ_s = 5Z_s|_0^t = 5Z_t - 5Z_0 = 5Z_t$

The solution for part a is a real-valued function of time. The solution for part b is the stochastic process five times standard Brownian motion.

6.5. a. First, note that the integral $\int_s^T f_t\, dZ_t$ can be approximated by the sum $\sum_{k=0}^{n-1} f_{t_k}\Delta Z_{t_k}$ where $s = t_0 < t_1 < t_2 < \cdots < t_n = T$ and $\Delta Z_{t_k} = Z_{t_{k+1}} - Z_{t_k}$. This means that the expectation in our result can be approximated by:

$$E\left[\sum_{k=0}^{n-1} f_{t_k}\Delta Z_{t_k}\Big|\mathscr{F}_s\right] = \sum_{k=0}^{n-1} E[f_{t_k}\Delta Z_{t_k}|\mathscr{F}_s]$$

By the tower property from Section 5.5, we have:

$$E[f_{t_k}\Delta Z_{t_k}|\mathscr{F}_s] = E[E[f_{t_k}\Delta Z_{t_k}|\mathscr{F}_{t_k}]|\mathscr{F}_s]$$

However, at time t_k, the stochastic process f_{t_k} is determined. Thus, we have:

$$E[E[f_{t_k}\Delta Z_{t_k}|\mathscr{F}_{t_k}]|\mathscr{F}_s] = E[f_{t_k}E[\Delta Z_{t_k}|\mathscr{F}_{t_k}]|\mathscr{F}_s] = E[f_{t_k}(0)|\mathscr{F}_s] = 0$$

Using the results from the last two equations, and replacing the right hand of the first equation gives:

$$\sum_{k=0}^{n-1} E[f_{t_k}\Delta Z_{t_k}|\mathscr{F}_s] = 0$$

Referring to the first equation, which must equal zero, we see that our result in the problem statement must hold.

b. Integrating $dX_t = f_t\, dZ_t$ from s to T, $s < T$, we obtain:

$$X_T - X_s = \int_s^T f_t\, dZ_t$$

and solving for X_T gives:

$$X_T = X_s + \int_s^T f_t\, dZ_t$$

Next, we calculate the required expectation:

$$E[X_T|\mathcal{F}_s] = E[X_s|\mathcal{F}_s] + E\left[\int_s^T f_t \, dZ_t|\mathcal{F}_s\right] = X_s + E\left[\int_s^T f_t \, dZ_t|\mathcal{F}_s\right] = X_s$$

where the last equality follows from the result in part a of this problem.

6.7. Consider any subset $A \subseteq \Omega$. The set A must also have at most a finite number of elements: $A = \{\omega_1, \omega_2, \ldots, \omega_n\}$. Observe that:

$$Q(A) = \sum_{i=1}^n q(\omega_i) = \sum_{i=1}^n \frac{q(\omega_i)}{p(\omega_i)} p(\omega_i)$$

which proves the result.

6.9. Values for Radon–Nikodym derivatives cannot be computed because $\mathbb{P}$ is not equivalent to Q. Notice that $p_3 > 0$ while $q_3 = 0$, hence, $\mathbb{P}$ is not absolutely continuous with respect to Q.

6.11. a. Risk-neutral probabilities are computed as follows:

$$\begin{bmatrix} 10 & 16 \\ 100 & 100 \end{bmatrix}\begin{bmatrix} \psi_{0,1,1} \\ \psi_{0,1,2} \end{bmatrix} = \begin{bmatrix} 12 \\ 88.888889 \end{bmatrix}; \quad \begin{aligned} \psi_{0,1,1} &= 0.37037 \\ \psi_{0,1,2} &= 0.51852 \end{aligned}$$

$$q_{0,1,1}/(\psi_{0,1,1} + \psi_{0,1,2}) = 0.4166667$$

$$q_{0,1,2}/(\psi_{0,1,1} + \psi_{0,1,2}) = 0.5833333$$

b. $\xi(d) = \dfrac{0.41667}{0.25} = 1.66667; \quad \xi(u) = \dfrac{0.58333}{0.75} = 0.77778$

6.13. a. If there were exactly k increases and $n - k$ decreases in the price from time 0 to n then:

$$\xi_n = \left(\frac{q}{p}\right)^k \left(\frac{1-q}{1-p}\right)^{n-k}$$

We know that ξ_n gives the correct multiplicative factor to the change the measure from $\mathbb{P}$ to Q since:

$$\xi_n p^k (1-p)^{n-k} = \left(\tfrac{q}{p}\right)^k \left(\tfrac{1-q}{1-p}\right)^{n-k} p^k (1-p)^{n-k} = q^k (1-q)^{n-k}$$

b. Recall that there are $\dbinom{n}{k}$ possible permutations with exactly k upjumps and $n - k$ downjumps in the price from time 0 to n. So, the probability of obtaining exactly k upswings and $n - k$ downswings in the price from time 0 to n equals:

$$\binom{n}{k} p^k (1-p)^{n-k}$$

with respect to probability measure $\mathbb{P}$, and equals:

$$\binom{n}{k} q^k (1-q)^{n-k}$$

with respect to probability measure $\mathbb{Q}$. Obviously, this implies:

$$\frac{d\mathbb{Q}}{d\mathbb{P}} \binom{n}{k} p^k (1-p)^{n-k} = \binom{n}{k} q^k (1-q)^{n-k}$$

6.15. We will follow and generalize the change of measure procedure introduced at the start of Section 6.2.5. Note that the exponent of our converted expression can be expanded so that:

$$-\frac{1}{2}\left(\frac{x-(\mu-\lambda)}{\sigma}\right)^2 = -\frac{1}{2}\left(\frac{(x-\mu)+\lambda}{\sigma}\right)^2 = -\frac{1}{2}\left[\left(\frac{x-\mu}{\sigma}\right)^2 + \frac{2(x-\mu)\lambda}{\sigma^2} + \frac{\lambda^2}{\sigma^2}\right]$$

$$= -\frac{1}{2}\left(\frac{x-\mu}{\sigma}\right)^2 - \frac{\lambda x}{\sigma^2} + \frac{\lambda}{\sigma^2} - \frac{\lambda^2}{2\sigma^2}$$

Thus:

$$f_\mathbb{Q}(x) = \frac{1}{\sigma\sqrt{2\pi}} e^{-\frac{1}{2}\left(\frac{x-(\mu-\lambda)}{\sigma}\right)^2} = \frac{1}{\sigma\sqrt{2\pi}} e^{-\frac{1}{2}\left(\frac{x-\mu}{\sigma}\right)^2} e^{\left(-\frac{\lambda x}{\sigma^2} + \frac{\lambda}{\sigma^2} - \frac{\lambda^2}{2\sigma^2}\right)} = f_\mathbb{P}(x)\xi(x)$$

6.17. a. First note that $E[Z^2] = \text{Var}[Z] = 1$. Thus:

$$\text{Var}(Z^2) = \frac{1}{\sqrt{2\pi}} \int_{-\infty}^{\infty} (z^2-1)^2 e^{-z^2/2}\, dz = \frac{1}{\sqrt{2\pi}} \int_{-\infty}^{\infty} (z^4 - 2z^2 + 1) e^{-z^2/2}\, dz$$

We already know that:

$$\frac{1}{\sqrt{2\pi}} \int_{-\infty}^{\infty} e^{-z^2/2}\, dz = 1 \text{ and } \text{Var}(Z) = \frac{1}{\sqrt{2\pi}} \int_{-\infty}^{\infty} z^2\, e^{-z^2/2}\, dz = 1$$

So there is left to evaluate:

$$\frac{1}{\sqrt{2\pi}} \int_{-\infty}^{\infty} z^4\, e^{-z^2/2}\, dz = \frac{1}{\sqrt{2\pi}} \int_{-\infty}^{\infty} z^3\, e^{-z^2/2} z\, dz$$

Choose $u = z^3$ and $dv = e^{-z^2/2} z\, dz$ and integrate by parts. Since $du = 3z^2\, dz$ and $v = -e^{-\frac{z^2}{2}}$, then:

$$\frac{1}{\sqrt{2\pi}} \int_{-\infty}^{\infty} z^3\, e^{-z^2/2} z\, dz = -\frac{1}{\sqrt{2\pi}} z^3\, e^{-\frac{z^2}{2}} \Big|_{-\infty}^{\infty} + \frac{3}{\sqrt{2\pi}} \int_{-\infty}^{\infty} z^2\, e^{-\frac{z^2}{2}}\, dz = 3\, \text{Var}(Z) = 3$$

This gives:

$$\text{Var}(Z^2) = 3 - 2 + 1 = 2$$

b. First note that $E[Z^2 c] = c$. It is now straightforward to calculate $\text{Var}[Z^2 c]$:

$$\text{Var}[Z^2 c] = E[(Z^2 c - c)^2] = c^2 E[(Z^2 - 1)^2] = c^2\, \text{Var}[Z^2] = 2c^2$$

6.19. The mean and variance of arithmetic returns are computed as follows:

$$E[r] = E\left[\frac{S_T}{S_o} - 1\right] = e^{\alpha T + \frac{1}{2}\sigma^2 T} - 1 = e^{0.05 + 0.045} - 1 = 0.09966$$

$$\text{Var}[r] = (e^{\sigma^2 T} - 1)e^{2\alpha T + \sigma^2 T} = (e^{0.09} - 1)e^{2 \times 0.05 + 0.09} = 0.11388$$

6.21. a.

$$S_T = S_0 e^{\left[\left(0.001 - \frac{0.02^2}{2}\right)T + 0.02 Z_T\right]} = S_0 e^{(0.0008T + 0.02 Z_T)}$$

b.

$$E\left[\ln\frac{S_{52}}{S_0}\right] = \left[\left(0.001 - \frac{0.02^2}{2}\right)0.52\right] = 0.0416$$

$$\text{Var}\left[\ln\frac{S_{52}}{S_0}\right] = \sigma^2 T = 0.02^2 \cdot 52 = 0.0208$$

6.23. First, divide both sides of the differential by $M - S_t$ to obtain:

$$\frac{dS_t}{M - S_t} = \mu \, dt + \sigma \, dZ_t$$

Since the integral of $dS_t/(M - S_t)$ for real-valued functions S_t equals $-\ln(M - S_t)$, we will use the expression $\ln(M - S_t)$ to obtain the correct solution for the stochastic process S_t. Recall that Itô's formula is:

$$dy = \left[a\frac{\partial y}{\partial x} + \frac{\partial y}{\partial t} + \frac{1}{2}b^2\frac{\partial^2 y}{\partial x^2}\right]dt + b\frac{\partial y}{\partial x}dZ_t$$

Define $y(x,t) = \ln(M - x)$ so that $Y_t = y(S_t,t)$. The partial derivatives of y evaluated at $(x,t) = (S_t,t)$ are as follows:

$$\frac{\partial y}{\partial x} = \frac{-1}{M - S_t}; \frac{\partial^2 y}{\partial x^2} = \frac{-1}{(M - S_t)^2}; \frac{\partial y}{\partial t} = 0$$

We now use Itô's lemma, where $a = \mu(M - S_t)$ and $b = b = (M - S_t)$:

$$d(\ln(M - S_t)) = \mu(M - S_t)\left(\frac{-1}{M - S_t}\right)dt + 0 + \frac{1}{2}\sigma^2(M - S_t)^2\left(\frac{-1}{(M-S_t)^2}\right)dt$$

$$+\sigma(M - S_t)\left(\frac{-1}{M - S_t}\right)dZ_t$$

$$= -\mu \, dt - \frac{1}{2}\sigma^2 \, dt - \sigma dZ_t = -\left(\mu + \frac{\sigma^2}{2}\right)dt - \sigma dZ_t$$

Changing the variable from t to s and integrating from 0 to t results in:

$$\ln(M - S_t) - \ln(M - S_0) = -\left(\mu + \frac{\sigma^2}{2}\right)t - \sigma Z_t$$

or:

$$\ln\left(\frac{M - S_t}{M - S_0}\right) = -\left(\mu + \frac{\sigma^2}{2}\right)t - \sigma Z_t$$

Exponentiating we have:

$$\frac{M - S_t}{M - S_0} = e^{-\left(\mu + \frac{\sigma^2}{2}\right)t - \sigma Z_t}$$

Solving for S_t gives:

$$S_t = M - (M - S_0)e^{-\left(\mu + \frac{\sigma^2}{2}\right)t - \sigma Z_t}$$

6.25. Employ the three-step technique for solving stochastic differential equations as follows:
1. Attempt the ordinary calculus solution:
$\int_0^T (\mu \, dt + Z_t \, dZ_t) = \mu T + \frac{1}{2}Z_T^2$.
2. Find the differential of $\mu T + \frac{1}{2}Z_T^2$ using Itô's lemma. First, define the function
$F(x,t) = \mu t + \frac{1}{2}x^2$, so that $F(Z_t, t) = \mu t + \frac{1}{2}Z_t^2$. Invoking Itô's lemma with $a = 0$ and $b = 1$, we have:

$$dF(Z_t, t) = \left(\frac{\partial F}{\partial t} + + \frac{1}{2}\frac{\partial^2 F}{\partial x^2}\right)dt + \frac{\partial F}{\partial x}dZ_t = \left(\mu + \frac{1}{2}\right)dt + Z_t \, dZ_t = dS_t + \frac{1}{2}dt$$

3. Integrating both sides of this equation yields:

$$\int_0^T dF(Z_t, t) = \int_0^T dS_t + \frac{1}{2}\int_0^T dt$$

$$\mu T + \frac{1}{2}Z_T^2 = S_T - S_0 + \frac{1}{2}T$$

Solving for S_T results in $S_T = S_0 + (\mu - \frac{1}{2})T + \frac{1}{2}Z_T^2$.

CHAPTER 7

7.1. By the general product rule for stochastic processes in Section 6.1.1, the change in the value of the portfolio equals:

$$dV_t = d(\gamma_{s,t} S_t) + d(\gamma_{b,t} B_t) = \gamma_{s,t}\, dS_t + S_t\, d\gamma_{s,t} + dS_t\, d\gamma_{s,t} + \gamma_{b,t}\, dB_t + B_t\, d\gamma_{b,t} + dB_t\, d\gamma_{b,t}$$
$$= (S_t + dS_t)d\gamma_{s,t} + (B_t + dB_t)\, d\gamma_{b,t} + \gamma_{s,t}\, dS_t + \gamma_{b,t}\, dB_t$$

   The infinitesimal transactions are buying or shorting $d\gamma_{s,t}$ shares of the stock at a price of $S_t + dS_t$ per share and $d\gamma_{b,t}$ units of the bond at price of $B_t + dB_t$ per unit. Thus, the change in value resulting from these transactions is $(S_t + dS_t)d\gamma_{s,t} + (B_t + dB_t)d\gamma_{b,t}$. From the equation above, we see that $dV_t = \gamma_{s,t}\, dS_t + \gamma_{b,t}\, dB_t$ if and only if $(S_t + dS_t)d\gamma_{s,t} + (B_t + dB_t)\, d\gamma_{b,t} = 0$ as we set out to prove.

7.3. The options are valued with the Black–Scholes model in a step-by-step format in the following table:

	OPTION 1	OPTION 2	OPTION 3	OPTION 4
$d(1)$	0.957739	−0.163836	0.061699	0.131638
$d(2)$	0.657739	−0.463836	−0.438301	−0.292626
$N[d(1)]$	0.830903	0.434930	0.524599	0.552365
$N[d(2)]$	0.744647	0.321383	0.330584	0.384904
Call	7.395	2.455	4.841	4.623
Put	0.939	5.416	7.803	5.665

7.5. Put–call parity states the first relation generally, and the second in a Black–Scholes environment:

$$p_0 = c_0 + Xe^{-rT} - S_0$$

$$p_0 = S_0 N(d_1) - \frac{X}{e^{rT}} N(d_2) + Xe^{-rT} - S_0$$

   With some algebra, and given the symmetry of the normal distribution about its mean, we rewrite as follows:

$$p_0 = S_0(N(d_1) - 1) - \frac{X}{e^{rT}}(N(d_2) - 1) = Xe^{-rT}N(-d_2) - S_0 N(-d_1)$$

7.7. Differentiating the call function with respect to σ and using the chain rule we have:

$$\frac{\partial c}{\partial \sigma} = S_0 \frac{\partial N(d_1)}{\partial \sigma} - Xe^{-rT}\frac{\partial N(d_2)}{\partial \sigma} = S_0 \frac{dN(d_1)}{d(d_1)}\frac{\partial d_1}{\partial \sigma} - Xe^{-rT}\frac{dN(d_2)}{d(d_2)}\frac{\partial d_2}{\partial \sigma}$$

Since $d_1 = d_2 + \sigma\sqrt{T}$, then:

$$\frac{\partial d_1}{\partial \sigma} = \frac{\partial d_2}{\partial \sigma} + \sqrt{T}$$

This gives:

$$\frac{\partial c}{\partial \sigma} = S_0 \frac{dN(d_1)}{d(d_1)}\left(\frac{\partial d_2}{\partial \sigma} + \sqrt{T}\right) - Xe^{-rT}\frac{dN(d_2)}{d(d_2)}\frac{\partial d_2}{\partial \sigma}$$

$$= S_0\sqrt{T}\frac{dN(d_1)}{d(d_1)} + \frac{\partial d_2}{\partial \sigma}\left(S_0\frac{dN(d_1)}{d(d_1)} - Xe^{-rT}\frac{dN(d_2)}{d(d_2)}\right) = \frac{S_0\sqrt{T}}{\sqrt{2\pi}}e^{-\frac{1}{2}d_1^2}$$

and we are done.

7.9. Zero. This is because the futures contract can be replicated with a long position in a call and a short position in a put with the same exercise terms. The gammas of the call and the put are the same. A long position in a futures contract is replicated with a single long position in a call and a single short position in a put. Thus, the gamma of the long call position offsets the gamma of the short put position. We can also solve this problem by computation. By the call–put parity we have the futures contract has the value $V_0 = c_0 - p_0 = S_0 - Xe^{-rT}$. Thus, the gamma of this portfolio equals:

$$\frac{\partial^2 V_0}{\partial S_0^2} = 0$$

7.11. a. The Black–Scholes model is written as follows:

$$c_0 = S_0 N(d_1) - Xe^{-rT}N(d_2)$$

We will use a combination of the product and chain rules to differentiate c_0 with respect to T to obtain the option theta:

$$\theta = \frac{\partial c_0}{\partial T} = S_0\frac{d(N(d_1))}{d(d_1)}\frac{\partial d_1}{\partial T} - Xe^{-rT}\frac{d(N(d_2))}{d(d_2)}\frac{\partial d_2}{\partial T} - Xe^{-rT}N(d_2)(-r)$$

We will rewrite this derivative to exploit a trick described in the Chapter 7 additional reading on the companion website that will enable us to group and eliminate some terms afterwards. Since $d_1 = d_2 + \sigma\sqrt{T} = d_2 + \sigma T^{\frac{1}{2}}$, then $\frac{\partial d_1}{\partial T} = \frac{\partial d_2}{\partial T} + \frac{1}{2}\sigma T^{-\frac{1}{2}} = \frac{\partial d_2}{\partial T} + \frac{\sigma}{2\sqrt{T}}$. Putting this result into the equation above and simplifying slightly gives:

$$\frac{\partial c_0}{\partial T} = S_0\frac{d(N(d_1))}{d(d_1)}\left[\frac{\partial d_2}{\partial T} + \frac{\sigma}{2\sqrt{T}}\right] - Xe^{-rT}\frac{d(N(d_2))}{d(d_2)}\frac{\partial d_2}{\partial T} + rXe^{-rT}N(d_2)$$

$$\frac{\partial c_0}{\partial T} = \left[S_0\frac{d(N(d_1))}{d(d_1)} - Xe^{-rT}\frac{d(N(d_2))}{d(d_2)}\right]\frac{\partial d_2}{\partial T} + \frac{S_0\sigma}{2\sqrt{T}}\frac{d(N(d_1))}{d(d_1)} + rXe^{-rT}N(d_2)$$

Using equation (3) in the Chapter 7 additional reading on the companion website shows that the expression in the square brackets above equals zero, and so:

$$\theta = \frac{\partial c_0}{\partial T} = \frac{S_0\sigma}{2\sqrt{2\pi T}}e^{-\frac{1}{2}d_1^2} + rXe^{-rT}N(d_2)$$

b. As we derived in the Chapter 7 additional reading on the companion website, the solution for the call at time t is given by equation (2)

$$c = SN(d_1(S, T-t)) - Xe^{-r(T-t)}N(d_2(S, T-t))$$

where the functions:

$$d_1(S, T-t) = \frac{\ln\left(\frac{S}{X}\right) + (r + \frac{1}{2}\sigma^2)(T-t)}{\sigma\sqrt{T-t}} \text{ and } d_2(S, T-t) = d_1(S, T-t) - \sigma\sqrt{T-t}$$

Notice that the price c of the call at time t is the same as the price of the call at time 0, $c_0(S_0, T)$, if one replaces S_0 with S and T with $T-t$ in the expression for c_0. Essentially, it is as if time t becomes time 0, so that S is now the initial stock price and the time to reach the exercise date is $T-t$. This can be expressed as: $c = c_0(S, T-t)$. Thus, it must be the case by the chain rule that:

$$\frac{\partial c}{\partial t} = \frac{\partial c_0(S, T-t)}{\partial(T-t)} \frac{d(T-t)}{dt} = -\frac{\partial c_0(S, T-t)}{\partial(T-t)}$$

$$= -\theta(S, T-t)$$

Using part a, we obtain the required derivative:

$$\frac{\partial c}{\partial t} = \frac{-S\sigma}{2\sqrt{2\pi(T-t)}}e^{-\frac{1}{2}d_1^2} - rXe^{-r(T-t)}N(d_2)$$

where d_1 and d_2 are evaluated at S and $T-t$.

c. In order to show that the Black–Scholes pricing equation we obtained for the call is a solution of the Black–Scholes differential equation, we first substitute $c - SN(d_1)$ for $Xe^{-r(T-t)}N(d_2)$ into the right side of the equation for $\frac{\partial c}{\partial t}$ above and rearrange terms as follows:

$$\frac{\partial c}{\partial t} = r(c - SN(d_1)) - \frac{1}{2}S^2\sigma^2\frac{e^{-\frac{1}{2}d_1^2}}{\sqrt{2\pi}} \cdot \frac{1}{S\sigma\sqrt{T-t}}$$

As derived in the Chapter 7 additional reading on the companion website, we have the following equalities:

$$\frac{\partial c}{\partial S} = N(d_1)$$

$$\frac{\partial^2 c}{\partial S^2} = \frac{e^{\left(\frac{-d_1^2}{2}\right)}}{\sqrt{2\pi}} \cdot \frac{1}{S_0\sigma\sqrt{T}}$$

which applied at time 0. If we consider these values at time t, then:

$$N(d_1) = \frac{\partial c}{\partial S}$$

and:

$$\frac{\partial^2 c}{\partial S^2} = \frac{1}{\sqrt{2\pi}} e^{-\frac{1}{2}d_1^2} \cdot \frac{1}{S\sigma\sqrt{T-t}}$$

where d_1 is evaluated at S and $T - t$. Substituting these equalities into the theta equation (actually its negative) above produces the Black–Scholes differential equation, which verifies that the Black–Scholes model is a solution to the differential equation:

$$\frac{\partial c}{\partial t} = rc - rS\frac{\partial c}{\partial S} - \frac{1}{2}S^2\sigma^2\frac{\partial^2 c}{\partial S^2}$$

The essential difference between the two is the sign change, arising from the fact that a decrease in T means that the option expires sooner and an increase in t means that the option expires sooner; $\Theta = -\partial c/\partial t$. More importantly, this result verifies that the Black–Scholes model is a solution to the Black–Scholes differential equation (7.23) in Section 7.3.2.

7.13. First, we calculate the critical value S_{T1}^* for underlying option exercise at time T_1. A search process reveals that this equals 66.578906:

$$c_{u,T1} = S_{T1}^* \times N\left(\frac{\left[\ln\left(\frac{S_{T1}^*}{75}\right) + \left(0.05 + \frac{1}{2}\times .16\right)\times(1-0.5)\right]}{0.4\sqrt{0.5}-0.25}\right)$$

$$-\frac{75}{e^{0.05\times(1-0.5)}}N\left(\frac{\left[\ln\left(\frac{S_{T1}^*}{75}\right) + \left(0.05 - \frac{1}{2}\times 0.0016\right)\times(1-0.5)\right]}{0.4\times\sqrt{1-0.5}}\right) = X_1 = 5; S_{T1}^*$$

$$= 66.578906$$

Note that the correlation coefficient between our two random variables is $\rho = \sqrt{0.5/1} = 0.70710$. We use the reasoning from Section 5.4 and the bivariate normal distribution to calculate the compound call value as follows:

$$c_0 = 70 \times 0.4834 - 75 \times e^{-(0.05\times 1)} \times 0.3382 - 5e^{-(0.05\times 0.5)} \times 0.05494 = 7.03314$$

$$d_1 = \frac{\left[\ln\left(\frac{70}{66.578906}\right) + \left(0.05 + \frac{1}{2}\times 0.16\right)\times 0.5\right]}{0.4\times\sqrt{0.5}} = 0.4070$$

$$d_2 = 0.4070 - 0.4\times\sqrt{0.5} = 0.1241$$

$$y_1 = \frac{\left[\ln\left(\frac{70}{75}\right) + \left(0.05 + \frac{1}{2}\times 0.16\right)\times 1\right]}{0.4\times\sqrt{1}} = 0.1525$$

$$y_2 = 0.1525 - 0.4\times\sqrt{1} = -0.2475$$

Thus, we find that the value of this compound call equals 7.03314. The bivariate probabilities were calculated using a spreadsheet-based multinomial cumulative distribution calculator.

7.15. a. First, we work through the European known-dividend call model:

$$d_1 = \frac{\ln\left(\frac{30 - 2\,e^{0.03 \times 0.25}}{35}\right) + \left(0.03 + \frac{1}{2} \times 0.16\right) \times .5}{0.4 \times \sqrt{0.5}} = -0.5926; N(d_1) = 0.2767$$

$$d_2 = -0.5926 - 0.4 \times \sqrt{0.5} = -0.0875; N(d_2) = 0.1907$$

$$c_0 = (30 - 2\,e^{0.03 \times 0.25}) \times 0.2767 - \frac{35}{e^{0.03 \times .5}} \times 0.1907 = 1.18$$

Since the call is currently selling for $2, it is not a good investment; it should be sold.

b. Based on put–call parity and the known European dividend model, the put is worth $p_0 = 1.18 + 35(0.985) - 30 + 2e^{-0.03 \times 0.25} = 7.64$.

c. If the call were presumed to be exercised on its ex-dividend date, it would be worth 0.869:

$$d_1 = \frac{\ln\left(\frac{30}{35}\right) + \left(0.03 + \frac{1}{2} \times 0.16\right) \times 0.25}{0.4 \times \sqrt{0.25}} = -0.6333; N(d_1) = 0.2633$$

$$d_2 = -0.6333 - 0.4 \times \sqrt{0.25} = -0.8333; N(d_2) = 0.2024$$

$$c_0 = 30 \times 0.2633 - \frac{35}{e^{0.03 \times 0.25}} \times 0.2024 = 0.869$$

Thus, the value of the call is Max[1.18, 0.869] = 1.18.

d. Under the Geske–Roll–Whaley model, we first calculate S_{tD}^*, the critical stock value on the ex-dividend date required for early exercise of the call:

$$c_{tD}^* = S_{tD}^* + D - X$$

$$4.38954 = 37.38954 + 2 - 35$$

The value of the American call is found to be 1.23, calculated as follows:

$$d_1 = \frac{\left[\ln\left(\frac{30 - 2 \cdot e^{-0.03 \times 0.25}}{35}\right) + \left(0.03 + \frac{1}{2} \cdot 0.16\right) \cdot 0.5\right]}{0.4 \cdot \sqrt{0.5}} = -0.5926$$

$$d_2 = -0.5926 - 0.4 \cdot \sqrt{0.5} = -0.8754$$

$$y_1 = \frac{\left[\ln\left(\frac{30 - 2 \cdot e^{-0.03 \times 0.25}}{37.38954}\right) + \left(0.03 + \frac{1}{2} \cdot 0.16\right) \cdot 0.25\right]}{0.4 \cdot \sqrt{0.25}} = -1.3058$$

$$y_2 = -1.3058 - 0.4 \cdot \sqrt{0.25} = -1.5058$$

$$c_o = \left(30 - 2 \cdot e^{-0.03 \times 0.25}\right) \cdot 0.09581 + \left(30 - 2 \cdot e^{-0.03 \times 0.25}\right) \cdot 0.1994$$

$$- 35 \cdot e^{-0.03 \times 0.5} \cdot 0.1415 - (35 - 2) \cdot e^{-0.03 \times 0.25} \cdot 0.06606 = 1.23$$

7.17. To answer this question, we first calculate d_1:

$$d_1 = \frac{\ln\left(\frac{0.032}{0.03}\right) + \left(0.03 - 0.1 + \frac{0.01^2}{2}\right) \cdot 0.5}{0.01\sqrt{0.5}} = 4.180913$$

Next we calculate d_2:

$$d_2 = d_1 - s\sqrt{T} = 4.180913 - 0.007071 = 4.173842$$

Next, find cumulative normal density functions (z-values) for d_1 and d_2:

$$N(d_1) = N(4.180913) = 0.99998548$$
$$N(d_2) = N(4.173842) = 0.99998502$$

Finally, we value the call as follows:

$$c_0 = 0.032 \times e^{-0.1 \times 0.5} \times 0.99998548 - 0.03 \times e^{-0.03 \times 0.5} \times 0.99998502 = \$0.00088598$$

We can evaluate the European put using put–call parity as follows:

$$p_0 = c_0 + Xe^{-r(d)T} - s_0 e^{-r(f)T} = 0.00088598 + 0.03e^{-0.03 \times 0.5} - 0.032e^{-0.1 \times 0.5} = 0.000000001$$

The put is valued at slightly more than 0. The estimates for the cumulative normal distributions will be great enough to make it appear as though the put has negative value. We will accept that we estimated with error and note that the put value always exceeds zero.

7.19. a. The time t_c payoff function for the chooser option equals $\text{Max}\left[c_{t_c}, p_{t_c}\right]$.
   b. By put–call parity, $p_{t_c} = c_{t_c} + X\,e^{-r(T-t_c)} - S_{t_c}$
   c. $\text{Max}\left[c_{t_c}, p_{t_c}\right] = \text{Max}[c_{t_c}, c_{t_c} + Xe^{-r(T-t_c)} - S_{t_c}] = c_{t_c} + \text{Max}\left[0, Xe^{-r(T-t_c)} - S_{t_c}\right]$. Pay special attention to the far right side of this equality because we will use it to value the chooser option. The far right side of the equality implies that the time t_c payoff function for a chooser option is the equivalent of a portfolio of a call on the underlying, with exercise price X and expiration date T plus a put on the underlying with exercise price equal to $Xe^{-r(T-t_c)}$ and expiration date t_c.
   d. The owner of the chooser option selects the call on the choice date if $c_{t_c} > p_{t_c}$, which will hold if $c_{t_c} > c_{t_c} + Xe^{-r(T-t_c)} - S_{t_c}$, or if $Xe^{-r(T-t_c)} < S_{t_c}$. Otherwise, he selects the put.
   e. Based largely on our solution to part c above (in particular, the far right term), we value the chooser option as follows:

$$V_{\text{chooser}} = \left[S_0 N(d_1) - \frac{X}{e^{rT}}N(d_2)\right] + \left[Xe^{-r(T-t_c)}N(-d_4) - S_0 N(-d_3)\right]$$

$$d_1 = \frac{\ln\left(\frac{S_0}{X}\right) + \left(r + \frac{1}{2}\sigma^2\right)T}{\sigma\sqrt{T}}$$

$$d_2 = d_1 - \sigma\sqrt{T}$$

$$d_3 = \frac{\ln\left(\frac{S_0}{Xe^{-r(T-t_c)}}\right) + \left(r + \frac{1}{2}\sigma^2\right)t_c}{\sigma\sqrt{t_c}} = \frac{\ln\left(\frac{S_0}{Xe^{r(-T+t_c)}}\right) + rt_c + \frac{1}{2}\sigma^2 t_c}{\sigma\sqrt{t_c}}$$

$$= \frac{\ln\left(\frac{S_0}{X}\right) - \ln\left(e^{r(-T+t_c)}\right) + rt_c + \frac{1}{2}\sigma^2 t_c}{\sigma\sqrt{t_c}} = \frac{\ln\left(\frac{S_0}{X}\right) + rT + \frac{1}{2}\sigma^2 t_c}{\sigma\sqrt{t_c}} \quad d_4 = d_3 - \sigma\sqrt{t_c}$$

f. Value the chooser option as follows:

$$V_{\text{chooser}} = 50 \times N(d_1) - \frac{50}{e^{0.05 \times 1}} N(d_2) + 50e^{-0.05 \times (1-0.5)} \times N(-d_4) - 50 \times N(-d_3)$$

$$d_1 = \frac{\ln\left(\frac{50}{50}\right) + \left(0.05 + \frac{1}{2} \times 0.16\right) \times 1}{0.4 \times \sqrt{1}} = 0.325$$

$$d_2 = 0.325 - 0.4\sqrt{1} = -0.075$$

$$d_3 = \frac{\ln\left(\frac{50}{50}\right) + 0.05 \times 1 + \frac{1}{2} \times 0.16 \times 0.5}{0.4 \times \sqrt{0.5}} = 0.3182$$

$$d_4 = 0.3182 - 0.4 \times \sqrt{0.5} = 0.0354$$

$$V_{\text{chooser}} = 50 \times 0.6274 - \frac{50}{e^{0.05 \times 1}} \times 0.4701 + 50e^{-0.05 \times 0.5} \times 0.4859 - 50 \times 0.3752 = 9.011 + 4.352 = 13.363$$

CHAPTER 8

8.1. a. $dr_t = \lambda(\mu - r_t)\,dt + 0.01\,dZ_t = 0.4(0.06 - r_t)\,dt + 0.01\,dZ_t$.

b. Our calculations are done with the following equation, such as the case in the equation that follows it:

$$E[r_t] = \mu + e^{-\lambda t}(r_0 - \mu)$$

$$E[r_1] = 0.06 + e^{-0.41}(0.02 - 0.06) = 0.033187$$

The following table reflects appropriate calculations:

T	$E[r_t]$	t	$E[r_t]$
1	0.033187	3	0.047952
2	0.042027	30	0.060000

c. We need a formula to obtain a mean rate from time 0 to time T. In the discreet setting, we obtain the spot rate from $(1 + r_{0,T})^T = \prod_{t=1}^{T}(1 + r_{t-1,t})$. Analogously, in the continuous case, $e^{Tr_{0,T}} = e^{\int_0^T r_t\,dt} = e^{\int_0^T \mu + e^{-\lambda t}(r_0 - \mu)dt}$. Thus, we see that the T-period spot rate is a mean of forward rates, and compute spot rates as follows:

$$r_{0,T} = \int_0^T (\mu + e^{-\lambda t}(r_0 - \mu))\frac{1}{T}dt$$

$$r_{0,T} = \frac{\mu t}{T} - \frac{1}{T}e^{-\lambda t}(r_0 - \mu)\Big|_0^T$$

$$r_{0,T} = \mu - \frac{1}{T}e^{-\lambda T}(r_0 - \mu) + \frac{1}{\lambda T}(r_0 - \mu)$$

$$r_{0,T} = \mu - \frac{1}{\lambda T}(e^{-\lambda T} - 1)(r_0 - \mu)$$

d. We will use the formula from part c to solve these problems. The following table lists spot rates $r_{0,T}$:

T	$E[r_{0,T}]$	T	$E[r_{0,T}]$
1	0.02703	10	0.05018
2	0.03247	20	0.05500
3	0.03671	30	0.05667

e. The following is the yield curve for our example:

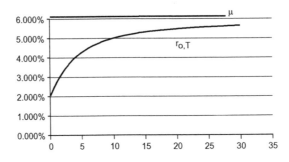

8.3. a. The following are our calculations:

$$\kappa = \frac{1 - e^{-\lambda T}}{\lambda} = \frac{1 - e^{-0.4 \times 1}}{0.4} = 0.8242 \text{ and } r_\infty = \mu - \frac{\sigma_r \theta}{\lambda} - \frac{\sigma_r^2}{2\lambda^2} = 0.06 - 0 - \frac{0.01^2}{2 \times 0.4^2} = 0.0596875$$

We use Eq. (8.20) to calculate the price of the bond:

$$B_0 = Fe^{\left([r_\infty - r_0] - r_\infty T - \frac{[\sigma_r \kappa]^2}{4\lambda}\right)} = 1000e^{\left(0.8242[0.0596875 - 0.02] - 0.0596875 \times 1 - \frac{[0.01 \times 0.8242]^2}{4 \times 0.4}\right)} = 973.342$$

b. $r_{0,1} = -\ln\left(\frac{B_0}{F}\right)/T = -\frac{\ln\left(\frac{973.342}{1000}\right)}{1} = 0.02702.$

c. $E[r_t] = \mu + e^{-\lambda t}(r_0 - \mu) = 0.06 + e^{-0.4}(0.02 - 0.06) = 0.033187.$

d. From equation (8) in the additional reading for Chapter 8 on the companion website, we see that the bond price at time t is exactly the same as the bond price at time 0 if, in the time 0 bond formula, we replace T with $T - t$ and r_0 with r_t. At time $t = 0$, it follows that $\frac{\partial B}{\partial t}$ equals the negative of $\frac{\partial B_0}{\partial T}$. Thus:

$$\frac{\partial B}{\partial t}(t = 0) = -\frac{\partial B_0}{\partial T} = -Fe^{\left(\kappa[r_\infty - r_0] - r_\infty T - \frac{[\sigma_r \kappa]^2}{4\lambda}\right)}\left[(r_\infty - r_0)e^{-\lambda T} - r_\infty - \frac{2\sigma_r \kappa}{4\lambda}\sigma_r e^{-\lambda T}\right]$$

$$= -B_0\left[(r_\infty - r_0)e^{-\lambda T} - r_\infty - \frac{\sigma_r^2 \kappa}{2\lambda}e^{-\lambda T}\right]$$

In this case:

$$\frac{\partial B}{\partial t}(t = 0) = -973.342\left[(0.0596875 - 0.02)e^{-0.4} - 0.0596875 - \frac{0.01^2 \times 0.8242}{2 \times 0.4}e^{-0.4}\right] = 32.269$$

e. From part a, we see that the value of a $c = 100$ payment in the first year is 97.3342. Now, we perform calculations for the 2-year, 1100 payment:

$$\kappa = \frac{1 - e^{-\lambda T}}{\lambda} = \frac{1 - e^{-0.4 \times 2}}{0.4} = 1.3766776 \text{ and } r_\infty = \mu - \frac{\sigma_r \theta}{\lambda} - \frac{\sigma_r^2}{2\lambda^2} = 0.06 - 0 - \frac{0.01^2}{2 \times 0.4^2} = 0.0596875$$

Once again, we use Eq. (8.20) to calculate the price associated with the 1100 payment:

$$Fe^{\left(\kappa[r_\infty - r_0] - r_\infty T - \frac{[\sigma_r \kappa]^2}{4\lambda}\right)} = 1100e^{\left(1.3766776\,[0.0596875 - 0.02] - 0.0596875 \times 2 - \frac{[0.01 \times 1.3766776]^2}{4 \times 0.4}\right)} = 1030.92207$$

$$B_0 = 1030.92207 + 97.3342 = 1128.26$$

8.5. a. First, note that the current value of a bond is calculated in a Vasicek framework as follows:

$$B_0 = Fe^{\left(\kappa[r_\infty - r_0] - r_\infty T - \frac{[\sigma_r \kappa]^2}{4\lambda}\right)}$$

This means that $\frac{\partial B}{\partial r} = -\kappa B_0 = \frac{e^{-\lambda T} - 1}{\lambda} B_0$.

b. $\frac{\partial^2 B}{\partial r^2} = \kappa^2 B_0$

c.

$$\frac{\partial B}{\partial r} = \frac{\partial Fe^{-Tr_0 - \left[\frac{T^2[\mu - \Theta \sigma_r]}{2}\right] + \left[\frac{T^3 \sigma_r^2}{6}\right]}}{\partial r} = -TB_0$$

$$\frac{\partial^2 B}{\partial r^2} = T^2 B_0$$

d. Note that the only between the underlying stochastic process in the zero drift Merton (1973) model and the Vasicek model is that $\lambda = 0$ in Merton. Note that as $\lambda \to 0$, the underlying stochastic process stochastic process in the Vasicek model will approach arithmetic Brownian motion with zero drift (the Merton (1973) model with zero drift), and $\kappa \to T$. In fact, using L'Hopital's rule, we calculate that $\lim_{\lambda \to 0} \frac{1 - e^{-\lambda T}}{\lambda} = \lim_{\lambda \to 0} \frac{T e^{-\lambda T}}{1} = T$. This means that $\lambda \to 0$, we have $\frac{\partial B}{\partial r} = -\kappa B_0 \to -TB_0$ just as in Chapter 4. Similarly, $\frac{\partial^2 B}{\partial r^2} = \kappa^2 B_0 \to T^2 B_0$, just as in Chapter 4.

8.7. Since $dr_t = \mu\,dt + \sigma_r\,dZ_t$, $r_t = r_0 + \mu t + \sigma_r Z_t$. If we shift the time variable, then $r_s = r_t + \mu(s - t) + \sigma_r \tilde{Z}_{s-t}$ where $\tilde{Z}_{s-t}$ is Brownian motion. By equation (1) in the additional reading for Chapter 8 on the companion website, we know that the bond price equals the following, which we develop further:

$$B(t, T) = E_\mathbb{P}\left[Fe\left(-\int_t^T r_s\,ds - \frac{1}{2}\int_t^T \Theta^2\,ds - \int_t^T \Theta\,dZ_s\right) | \mathcal{F}_t\right]$$

$$= E_\mathbb{P}\left[Fe\left(-\int_t^T (r_t + \mu(s - t) + \sigma_r \tilde{Z}_{s-t})\,ds - \frac{1}{2}\int_t^T \Theta^2\,ds - \int_t^T \Theta\,dZ_s\right) | \mathcal{F}_t\right]$$

$$= Fe\left(-\left(r_t + \frac{1}{2}\Theta^2\right)(T - t) - \frac{\mu}{2}(T - t)^2\right) E_\mathbb{P}\left[e\left(-\sigma_r \int_t^T \tilde{Z}_{s-t}\,ds - \int_t^T \Theta\,dZ_s\right) | \mathcal{F}_t\right]$$

Next, we note that:

$$E_\mathbb{P}\left[-\sigma_r \int_t^T \tilde{Z}_{s-t}\,ds - \int_t^T \Theta\,dZ_s | \mathcal{F}_t\right] = -\sigma_r \int_t^T E_\mathbb{P}[\tilde{Z}_{s-t} | \mathcal{F}_t]\,ds - E_\mathbb{P}\left[\int_t^T \Theta\,dZ_s | \mathcal{F}_t\right] = 0$$

This result follows from the fact that the expectation operation and integration operation commute in this case, which we will not prove. Furthermore, we also used Eq. (6.1) in Section 6.1.5 and the fact that standard Brownian motion has the property $E_{\mathbb{P}}[\tilde{Z}_{s-t}|\mathscr{F}_t] = 0$ for $s \geq t$. Using the fact that the expectation of the process above equals zero means that the variance of the same process will be:

$$\text{Var}\left[-\sigma_r\int_t^T \tilde{Z}_{s-t}\,ds - \int_t^T \Theta\,dZ_s \middle| \mathscr{F}_t\right] = E_{\mathbb{P}}\left[\left(-\sigma_r\int_t^T \tilde{Z}_{s-t}\,ds - \int_t^T \Theta\,dZ_s\right)^2 \middle| \mathscr{F}_t\right]$$

$$= E_{\mathbb{P}}\left[\left(\sigma_r\int_t^T\int_0^{s-t} d\tilde{Z}_u\,ds + \int_t^T \Theta\,dZ_s\right)^2 \middle| \mathscr{F}_t\right]$$

Interchanging the order of integration in the double integral gives:

$$\text{Var}\left[-\sigma_r\int_t^T \tilde{Z}_{s-t}\,ds - \int_t^T \Theta\,dZ_s \middle| \mathscr{F}_t\right] = E_{\mathbb{P}}\left[\left(\sigma_r\int_0^{T-t}\int_{u+t}^T ds\,d\tilde{Z}_u + \int_t^T \Theta\,dZ_s\right)^2 \middle| \mathscr{F}_t\right]$$

$$= E_{\mathbb{P}}\left[\left(\int_0^{T-t}(\sigma_r(T-t-u)+\Theta)d\tilde{Z}_u\right)^2 \middle| \mathscr{F}_t\right]$$

By the Itô Isometry:

$$\text{Var}\left[-\sigma_r\int_t^T \tilde{Z}_{s-t}\,ds - \int_t^T \Theta\,dZ_s \middle| \mathscr{F}_t\right] = \int_0^{T-t}(\sigma_r(T-t-u)+\Theta))^2\,du = \frac{1}{3\sigma_r}(\sigma_r(T-t)+\Theta)^3 - \frac{\Theta^3}{3\sigma_r}$$

Note that:

$$-\sigma_r\int_t^T \tilde{Z}_{s-t}\,ds - \int_t^T \Theta\,dZ_s = -\sigma_r\int_t^T \tilde{Z}_{s-t}\,ds - \Theta(Z_T - Z_t)$$

By the theorem in Section 6.1.5.3, this process has a normal distribution. By Eq. (2.26) in Section 2.6.4 and after some algebraic simplification, we have:

$$B(t,T) = Fe\left(-r_t(T-t) - \frac{(\mu - \Theta\sigma_r)}{2}(T-t)^2 + \frac{\sigma_r^2}{6}(T-t)^3\right)$$

Appendix C: Glossary of Symbols

LOWER CASE LETTERS

a	(1) drift term in an Itô process; (2) constant
b	(1) volatility term in an Itô process; (2) constant
$\mathbf{b_0}$	vector of bond prices at time 0
c	(1) call value; (2) constant; (3) coupon rate
c_t	price of call at time t
c_i	ith constant
$\mathbf{c}$	vector of call prices
$\mathbf{cf_i}$	payoff cash flow vector for security i
d	downjump
$\mathbf{d}$	vector of discount functions
$\mathbf{\hat{d}}$	vector of discount functions including time 0
df, dt, dy	differentials
dS_t, dX_t, dY_t	differentials of a stochastic process
d_t	discount function for time t payoff
d_1, d_2	(1) parameters in option pricing solutions; (2) discount functions
$\mathbf{e_i}$	payoff vector for pure security i
f	function
$f(t,T)$	forward rate
f_t	stochastic process
g	(1) function; (2) growth rate of an annuity
h	function
i, j, k	counters
ln	natural logarithm
m	(1) number of rows in a matrix; (2) number of times interest compounded
max	maximum
min	minimum
n	(1) number of columns in a matrix; (2) ending number in a summation
p	(1) physical probability; (2) put value; (3) portfolio
$\mathbf{p}$	vector of put prices
q	synthetic probability
r	(1) riskless return or interest rate; (2) return
r_B	instantaneous return on a bond
$r(d)$	domestic riskless rate
$r(f)$	foreign riskless rate
$r(t)$	interest rate as a deterministic function of time
r_t	instantaneous short-term interest rate
$r_{0,T}$	yield to maturity, effective rate over time period 0 to T, T-year spot rate
$r_{t,T}$	forward rate
r_∞	infinite long rate for a bond following the Vasicek model
s	time value that is usually less than time value t
t	time value
t_D	ex-dividend date
t_i	discrete time value i
u	upjump
w, w_i	weights (normally for a security)
$\mathbf{x}$	vector of security or exercise prices
y_1, y_2	parameters in compound option pricing solutions

UPPER CASE LETTERS

A	set
$\mathbf{A, B,} \ldots$	matrices
$\mathbf{A^T}$	transpose of a matrix
$B, B_0, B_t, B_{t,T}, \ldots$	bond prices
C	constant of integration
$\mathbf{CF}$	payoff cash flow matrix
$CF_{i,j}, CF_{i,t}$	entries (elements) of payoff cash flow matrix
$\widehat{\mathbf{CF}}$	cash flow matrix including time 0
D	(1) dividend amount; (2) constant
Conv	convexity
Dur	duration
E	expected value operator
F	(1) face value of a bond; (2) function; (3) cumulative distribution function
F_T	forward price
FV	future value
$\mathbf{I}$	identity matrix
K	constant
L	Lagrange function
M	bivariate (cumulative) normal distribution function
M_t	(1) martingale; (2) stochastic process
Max	maximum
N	normal (as in normal distribution)
P	(1) probability; (2) portfolio value
PMT	payment on an annuity
PV	present value
P_i	ith probability
R_p	portfolio return
S_t	stock price modeled as a stochastic process
S_t^*	critical value of stock price for pricing compound options
T	time, often at expiration, settlement, or maturity
$\mathbf{V}$	matrix of covariances
Var	variance
V_t	value of a (financial) derivative at time t
X	(1) random variable; (2) exercise price of an option
$\overline{X}$	sample mean
X_t	stochastic process
Z	standard normal random variable
$Z_t, \hat{Z}_t$	(standard) Brownian motion

GREEK LETTERS

α	(1) mean logarithmic return of security per unit time; (2) shares of stock
γ	vector representing a portfolio of already priced bonds forming a basis
$\hat{\gamma}$	vector representing any portfolio of bonds
γ_i	units of ith security in a portfolio
$\gamma_{s,t}, \gamma_{b,t}$	portfolio holding for a stock, respectively, a riskless bond
δ	dividend yield
Δ	change of (usually small change of)
ϵ	error

η_t	stochastic process
Θ	market price of risk
ι	vector of 1's
κ	parameter in bond price solution for Vasicek model
λ	parameter in Ornstein–Uhlenbeck process
Λ	parameter in bond price solution for Vasicek and CIR models
μ	mean of random variable, drift in an Itô process
ν_t	stochastic process
ξ	Radon–Nikodym derivative
Π	product
$\rho, \rho_{X,Y}$	correlation coefficient
ρ_t	stochastic process
ϱ	parameter in bond price solution for CIR model
σ	standard deviation, volatility in an Itô process
$\sigma_{X,Y}$	covariance
Σ	sum
τ	stopping time
τ_{S^*}	hitting time to first reach S^*
φ, ϕ	event
Φ	set of all possible events
$\psi_{i,j;k}$	pure security price
Ψ	vector of pure security prices
$\omega, \omega_u, \ldots$	state or outcome
Ω	sample space

SPECIAL SYMBOLS

$\lvert x \rvert$	absolute value of x
$<<$	absolutely continuous
$\approx$	approximately equal to
$\sim$	(1) has the probability distribution; (2) is equivalent to
$\binom{n}{k}$	binomial coefficient (n choose k)
$\subset$	contained in
$\in$	element of, member of
$\varnothing$	empty set
$\mathcal{F}$	filtration
$\mathcal{F}_i$	ith σ-algebra in the filtration
$\forall$	for all
$\Rightarrow$	implies
∞	infinity
$\cap$	intersection
$\mathbb{N}$	set of natural numbers
$\mathbb{P}$	physical probability measure
$\mathbb{Q}$	risk-neutral probability measure
$\mathbb{R}$	set of real numbers
$\times$	times (multiplication)
$\cup$	union

Glossary of Terms

Absolutely continuous $\mathbb{Q}$ is said to be *absolutely continuous* with respect to $\mathbb{P}$ if for each event ϕ in $\mathscr{F}$, $p(\phi) = 0 \Rightarrow q(\phi) = 0$.

Adapted process A stochastic process X_t is said to be adapted to a filtration $\{\mathscr{F}_t\}$ if its values only depend upon information up to time t (cannot see into the future).

American option Can be exercised any time prior to its expiry.

Annuity A series of equal payments made at regular intervals.

Arbitrage (1) The simultaneous purchase and sale of the same asset. (2) The near simultaneous purchase and sale of assets generating nearly identical cash flow structures.

Arbitrage-free Characterizes a market with no opportunities for a risk-free profit.

Arbitrage opportunity Arbitrage transaction producing a profit, by purchasing at a price that is less than the selling price.

Arbitrageur Trader who seeks to profit from arbitrage opportunities.

Arithmetic return Where time T and time 0 prices are, respectively, S_T and S_0, the arithmetic return is $r = S_T / S_0 - 1$.

Arrow–Debreu security See pure security.

Asset Anything of value that can be owned.

Barrier option An option contract with the same payoff characteristics as a "plain vanilla" call or put option except that it expires or is activated once the underlying asset value reaches a pre-specified hitting price. These are often referred to as either *knock out* or *knock in* options.

Basis point 0.0001 of a unit. Typically used in foreign exchange and fixed income markets quotes.

Basis risk Risk that markets might move too slowly to profit from an apparent arbitrage, or that markets might move opposite to the arbitrageur's expectations, at least in the short run.

Bernoulli trial A single random experiment with two possible outcomes (e.g., 0, 1) whose outcomes depend on a probability.

Beta Coefficient that measures the risk of a security relative to the risk of some factor (usually the market).

Binomial option pricing model Valuation model based on the assumption that the underlying stock follows a binomial return generating process.

Binomial process A process based upon a series of n statistically independent Bernoulli (0,1) trials, all with the same outcome probability.

Bivariate normal distribution Joint probability distribution in random variables X and Y with the property that any linear combination of the random variables X and Y has a normal distribution.

Black–Scholes option pricing model A continuous time–space option pricing model in a complete and arbitrage-free market that is based upon creating a perfectly hedged portfolio.

Bond Financial security that makes fixed payment(s) at specified interval(s).

Broker A security market participant who acts as an agent for investors, buying and selling on their behalf on a commission basis.

Brownian motion A Newtonian non-differentiable stochastic continuous time–space process whose increments are normally distributed and independent over time.

Call Security or contract granting its owner the right to purchase a given asset at a specified price on or before the expiration date of the contract.

Cameron–Martin–Girsanov theorem Shows how a change in the probability measure to an equivalent measure induces a shift in the drift of an adapted Itô process.

Cardinality The number of members of a set.

Cash flow Movement of cash in or out.

Central limit theorem States that the mean of a random sample of any i.i.d. random variables approaches a normal distribution as the sample size increases.

CIR process See Cox–Ingersoll–Ross process.

Claim Right of ownership or other share on asset.

Coefficient of correlation A measure of the strength and direction of the relationship between two sets of variables. Ranges between zero and one and may be regarded as a "standardized" covariance (dividing covariance by the product of the standard deviations of the two variable sets).

Common stock Security held and potentially traded by the residual claimants or "owners" of the firm.

Complete market   Market in which the set of priced security payoff vectors spans the set of all security payoff vectors in the market. Requires that the number of priced securities be as large as the number of potential outcomes.

Conditional expectation   The expected value of a random variable given a specified history.

Continuous   Defined over an interval of an infinite number of infinitesimal time periods or changes in state (a connected graph of points).

Continuously compounded   Interest that is compounded instantaneously, rather than at discrete times such as annually.

Convexity   (1) The slope of the slope of a function. (2) The sensitivity of the duration of a bond to changes in the market rate of interest.

Countably infinite   A set that can be put in a one-to-one correspondence with the set of natural numbers.

Counterparty   Person or institution with whom one trades.

Coupon   The interest rate on debt as a percentage of its face value.

Covariance   A statistical measure of the co-movement between two sets of variables.

Covered call writing   Writing or selling a call while owning the underlying asset (or appropriate combination of other assets) needed to generate a hedge.

Covered put writing   Writing or selling a put while shorting the underlying asset (or appropriate combination of other assets) needed to generate a hedge.

Cox–Ingersoll–Ross process   A mean-reverting stochastic process defined by the differential $dr_t = \lambda(\mu - r_t)\, dt + \sigma_r \sqrt{r_t}\, dZ_t$.

Cumulative distribution function   A function F of a random variable so that $F(x)$ equals the probability that the random variable is no larger than x.

Current yield   Measure of the annual interest payments made by a bond relative to its initial investment.

Currency option   See foreign exchange option.

Delta   Sensitivity of an option's price to changes in the price of the underlying security.

Density function   A function f of a continuous random variable so that (non-rigorous definition) $f(x)\, dx$ is the probability that the random variable lies in the interval $[x, x + dx]$ for infinitesimally small values of dx.

Derivative security   Security whose payoff function is derived from the value of some other security, rate, or index.

Differential   The infinitesimal change of a function resulting from the infinitesimal change(s) of the independent variable(s) upon which the function depends, ignoring any higher-order effects as long as they are negligible.

Diffusion process   Continuous Markov process.

Discount function   See present value.

Discount rate   A rate used to discount (usually reduce) future cash flows to express their values relative to current cash flows.

Discounted expected value   The expected value discounted by an appropriate rate.

Discrete process   A process whose variable can be assigned only a countable number of values.

Distribution   See probability distribution.

Distribution function   See cumulative distribution function.

Diversification   Holding multiple assets whose returns are not perfectly correlated.

Dividend leakage   Dividends paid or received at a continuous rate, typically by an index or other type of fund.

Dividend-protected   Call option provision that allows for the option holder to receive the underlying stock and any dividends paid during the life of the option in the event of exercise.

Doob's decomposition theorem   Every discrete sub- and supermartingale can be decomposed into a martingale and a predictable drift process.

Doob–Meyer decomposition theorem   Every continuous sub- and supermartingale can be decomposed into a martingale and a predictable drift process.

Dow Jones Industrial Average (DJIA)   A price-weighted average of 30 actively traded blue chip stocks.

Drift   The predictable change component of a stochastic process.

Duration   Measures the proportional sensitivity of a bond value or price to changes in the market rate of interest.

Dynamic hedge   Portfolio strategy designed to neutralize risk that is updated each period.

Equity   Security that represents ownership in a business or corporation. Normally called stock.

Equivalent martingale measure   In a complete market with physical probability measure $\mathbb{P}$, probability measure $\mathbb{Q}$ is said to be an equivalent martingale measure to $\mathbb{P}$ if both are equivalent probability measures, and every discounted security in the market is a martingale with respect to $\mathbb{Q}$.

Equivalent probability measure   A probability measure $\mathbb{Q}$ is said to be equivalent to measure $\mathbb{P}$ ($\mathbb{Q} \sim \mathbb{P}$) when it assigns probability $q = 0$ to each event that $\mathbb{P}$ assigns probability $p = 0$ and when it assigns to an event probability $0 < q \leq 1$ to each event that $\mathbb{P}$ assigns probability $0 < p \leq 1$.

ETF   See exchange-traded fund.

European option   Can be exercised only when it expires.

Exchange   A physical or electronic marketplace for trading securities.

Exchange-traded fund (ETF)   Closed end investment company whose shares are traded on an exchange.

Execution The process of selling or buying a security.

Exercise price See striking price.

Expectation See expected value.

Expected value Weighted average of a random variable, where weights are probabilities associated with outcomes.

Eurodollar Freely convertible dollar-denominated time deposit or debt instrument issued outside the United States and outside of the jurisdiction of the U.S. Federal Reserve.

European option Can be exercised only on its expiration date.

Exchange option See foreign exchange option.

Exercise date The future date that an option must be either be exercised or not exercised.

Exercise price Price at which an option can be exercised; that is, the price that the call owner has the right to pay for the underlying asset or put owner has the right to sell the underlying asset for. Also called striking price.

Exotic option Option with one or more exotic features such as non-constant exercise price, multiple underlying assets, etc.

Expiration date See exercise date.

Event Any subset in the sample space of a probability space that has a well-defined probability of occurring (technical definition: any member of the σ-algebra of the probability space).

Expectations theory See pure expectations theory.

Face value The principal or par value of debt.

Filtration A sequence of information sets indexed over time, so that each information set contains the required history for valuation at that time. More formally, a filtration is a sequence of σ-fields $\mathscr{F}_t$ such that $\mathscr{F}_0 \subset \mathscr{F}_1 \subset \cdots \subset \mathscr{F}_t \subset \cdots \subset \mathscr{F}$.

Foreign exchange Pertaining to trading of currencies.

Foreign exchange option Option on currency.

Forward contract Instrument that obliges its participants to either purchase (long) or sell (short) a given asset at a specified price on the future settlement date of that contract.

Forward rate Locked-in interest rate on loans originating after time 0 or in the future.

Fractal A process that self-replicates in its main characteristics at ever smaller scales in either time or measurement of its values.

Futures contract Instrument that obliges participants to either purchase or sell a given asset at a specified price on the future settlement date of that contract. Normally traded on an exchange, with standardized contractual terms and providing for margin and marking to the market.

Gamma Sensitivity of an option's delta to changes in the price of the underlying security.

Geometric Brownian motion A stochastic process whose logarithm is a sum of Brownian motion and drift.

Girsanov's theorem See Cameron–Martin–Girsanov theorem.

Greeks Sensitivities of options, other derivatives and portfolios thereof to Black–Scholes or other pricing model inputs and parameters.

Hedge fund A private fund that allows investors to pool their investment assets. To avoid Securties and Exchange Commission (SEC) registration and certain regulations, hedge funds usually only accept funds from small numbers (often less than 100) of accredited investors, typically high net worth individuals and institutions.

Hedge To take a position to reduce or neutralize risk.

Hedge ratio Defines the number of units of one security required to offset the position in one unit of another security in order to form or maintain a riskless portfolio.

Hedging probability See risk-neutral probability.

Hitting time The length of time required for a process with particular parameters to first realize or hit some given value.

Identically distributed Refers to two or more random variables that have the same probability distribution.

I.i.d. Abbreviation for independent and identically distributed (random variables).

Immunization Strategies concerned with matching the present values of asset portfolios with the present values of cash flows associated with future liabilities. More specifically, immunization strategies are primarily concerned with matching asset durations with liability durations.

Implementation risk Risk taken by an arbitrageur that one or more arbitrage transactions might fail to execute or be executed at prices that differ from what the arbitrageur anticipated.

Index A portfolio of stocks or other instruments intended to reflect performance of a particular market or sector.

Independent For each possible value of $X = x$ and $Y = y$, discrete random variables X and Y are said to be *independent* if $P[X = x, Y = y] = P[X = x]P[Y = y]$, or for continuous random variables if their joint density function $f(x,y)$ satisfies $f(x,y) = g(x)h(y)$, where $g(x)$ is the density function for X and $h(y)$ is the density function for Y.

Infinitesimal Value approaching zero.

Innovation The random component of a stochastic process.

Instantaneous forward rate The limit of the forward rate as the maturity date approaches the origination date.

Instantaneous rate Expressed as a percentage or proportion of principal, and typically annualized, the interest imposed on a loan maturing the immediately after it is originated.

Instrument Tradable asset, security, currency, or contract.

Integrand The function that is being integrated in either a definite or indefinite integral.

Investment company Institution that accepts funds from investors for the purpose of investing on their behalf.

Itô's lemma Gives an expression in terms of partial derivatives for the differential of an Itô process.

Itô process A stochastic process whose differential can be written as a sum of a drift and volatility term.

Joint probability distribution Probability distribution associated with two or more random variables studied jointly.

Kernel A time- and outcome-specific stochastic discount factor used to multiply a future potential cash flow associated with an asset to determine its present value.

Law of one price States that securities or portfolios of securities offering the same cash flow characteristics or baskets offering the same commodities must sell for the same price.

Lebesgue measure The Euclidean (ordinary) volume of a solid shape, Euclidean area of a surface, or Euclidean length of a line segment or curve.

Liquidity An asset's ability to be easily purchased or sold without causing significant change in the price of the asset.

Logarithmic return Where time T and time 0 prices are, respectively, S_T and S_0, the logarithmic return is $\ln(S_T/S_0)$.

Long Purchase or purchased.

Long-term rate (1) The interest rate on an instrument with a finite maturity or holding period. (2) The interest rate on an instrument with a maturity holding period exceeding 1 year.

Market The arena for buying and selling.

Market architecture The set of rules governing the trading process.

Market price of risk Increased return or value adjustment attributable to a unit of risk.

Markov process A stochastic process in which to make predictions of its future given its past history up to the present is equivalent to basing the predictions only on its present state. A Markov process has no memory.

Mark-to-market Accounting for the fair value of an asset or liability based on its current market price, or, if unavailable, based on either similar instruments or an appropriate valuation model.

Martingale A stochastic process whose expected value at any future time equals its current value.

Maturity The date at which payments cease on a debt security.

Measure In a probability setting, associates a probability with each event in the σ-algebra of the probability space.

Merton term structure model A stochastic process defined by the differential $dr_t = \mu \ dt + \sigma_r \ dZ_t$ and applied to interest rates.

Method of bisection An iterative method of bisecting intervals to find the roots of an equation.

Metric A non-negative computed value obtained from a data set (e.g., mean, variance).

Model risk Failure to fully understand the implications of a trading model.

Multiplicative movement The proportion by which an asset changes over a given time period.

Mutual fund (open-end investment company): An institution registered with the SEC under the Investment Act of 1940 that pools investors' funds into a single portfolio.

Natural number Positive integer.

Net present value (NPV) The discounted value of all cash flows net of the initial investment.

Newton–Raphson method An iterative method based on slopes to find the roots of an equation.

NPV See net present value.

Numeraire Unit in which values are expressed.

Option Security that grants its owner the right to buy (call) or sell (put) an asset (underlying asset) at a specific price (exercise or striking price) on or before the expiration date of the option contract.

Ornstein–Uhlenbeck process A mean reverting stochastic process defined by the differential $dr_t = \lambda(\mu - r_t) dt + \sigma_r \ dZ_t$.

Outcome An element in the sample space of a probability space.

Pairs trading An arbitrage strategy that involves taking offsetting positions of two different stocks (perhaps options or index contracts) with correlated returns, one long and one short, such that gains in one position are expected to more than offset losses in the other position.

Payoff A payment from a security, instrument, or institution.

PDE Abbreviation for partial differential equation.

Physical probability Probability measure assigned to events that reflect the likelihood or frequency of occurrence.

Plain vanilla option Option with no exotic features such as non-constant exercise price, multiple underlying assets, etc.

Poisson process A discrete random variable with probability distribution $P(X = k) = e^{-\lambda t}(\lambda t)^k/k!$.

Portfolio A collection of investment holdings.

Position An asset held or portfolio obligation incurred, normally as a result of a transaction.

Present value The value of a future cash flow or series of cash flows expressed in terms of money received now.

Previsible A stochastic process that is adapted and left-continuous (continuous from the left side).

Pricing kernel See kernel.

Probability distribution A rule or an equation that assigns a probability to every possible value or range of values of a random variable.

Probability measure A mapping from an event space to the real number space such that all events ϕ have probabilities $P(\phi) \in [0, 1]$ and the probability of the sample space equals 1.

Pullback factor A constant that reflects the speed of the mean-reverting adjustment for the short-term rate towards the long-term mean rate μ.

Pure discount note A debt security paying no interest; it only pays its face value or principal.

Pure expectations theory Term structure theory that states that long-term spot rates can be explained as a geometric mean of short-term spot and forward rates.

Pure security An imaginary security (useful for pricing real securities) that pays 1 unit if and only if a particular state of the market occurs.

Put Security or contract granting its owner the right to sell a given asset at a specified price on or before the expiration date of the contract.

Put–call parity An equation that expresses the pricing relationship between a put and a call.

Radon–Nikodym derivative The multiplicative factor that is used to convert from one probability measure to another probability measure.

Random walk A discrete stochastic process with independent increments.

Real option Application of option pricing methodology to non-financial assets.

Recombining Multiple paths lead to the same event or outcome.

Replicating portfolio A portfolio of securities that has the same payoff structure over time as the given security.

Residual claim The right to take possession of assets unclaimed by others.

Return Profit relative to initial investment amount.

Rho Sensitivity of an option's price to changes in the riskless return rate.

Risk Uncertainty.

Risk-free rate The return or interest rate on an asset with no risk.

Risk-neutral probability The probabilities in a complete market that guarantee arbitrage-free pricing and in which the discounted value of any security is a martingale.

Risk premium Return offered or demanded as compensation for accepting uncertainty.

Sample space Set of all possible outcomes.

SDE Abbreviation for stochastic differential equation.

Security Tradable claim on the assets of an institution or individual.

Self-financing portfolio strategy Portfolio strategy in which the net cash flow in or out of a portfolio is zero; purchases are financed with proceeds of sales.

Set A collection of well-defined objects called elements x such that any given x either (but not both) belongs to set A ($x \in A$) or does not belong to A ($x \notin A$).

Sharpe ratio Ratio of excess return (risk premium) to risk, usually as measured by standard deviation.

Short (1) To sell, to have sold, or to have short sold. (2) A position that obligates or enables the investor or contract participant to sell a given or underlying asset on the expiration or settlement date of a derivative contract.

Short sell To sell a security without actually owning it. Normally requires its purchase at a later date and may involve borrowing the security from another investor first, returning it when it is repurchased at a later date.

Short-term rate (1) The interest rate on an instrument with an infinitesimal maturity or holding period. (2) The interest rate on an instrument with a maturity holding period less than 1 year.

Sigma algebra A collection of sets that includes all possible complements and countable unions of the sets in the collection. Usually written as "σ-algebra."

Sigma field See sigma algebra.

Speculate To take a position of risk based on a forecast of the direction of a security price change.

Speculator Trader who speculates or takes positions of risk.

Spot rates Interest rate on a loan, bond, note, or other instrument originating at time 0 or now and producing the observed price of the instrument.

Spot transaction Occurs at the time of the agreement to make the exchange.

Spread The difference between the best offer and bid prices.

Standard deviation Square root of the variance.

State space The set of all possible values in a process.

Stochastic Random.

Stochastic differential equation An equation for the differential of a stochastic process.

Stochastic discount factor See kernel.

Stochastic process A sequence of random variables X_t indexed by time t. Roughly, a stochastic process is a random infinite list of values X_t sequenced over time.

Strike price See striking price.

Striking price Price at which an option can be exercised; that is, the price that the call owner has the right to pay for the underlying asset or put owner has the right to sell the underlying asset for. Also called exercise price.

Strip Single payment fixed income instrument issued through the US Treasury's *Separate Trading of Registered Interest and Principle Securities* (STRIPS) program.

Submartingale A stochastic process whose expected value at any future time is greater than or equal to its current value.

Supermartingale A stochastic process whose expected value at any future time is less than or equal to its current value.

Swap contract Instrument that provides for the exchange of cash flows associated with one asset, rate, or index for the cash flows associated with another asset, rate, or index.

Synthetic probability See risk-neutral probability.

Term structure of interest rates How interest rates on debt securities vary with respect to varying dates of maturity on the debt.

Theta Sensitivity of an option's price to changes in the option's time to expiry.

Time value The value of money at a given time measured in terms of money at some other point in time; normally pertains to the present value of a future cash flow.

Tower property The expectation result that $E[E[X_T|\mathscr{F}_t] |\mathscr{F}_s] = E[X_T|\mathscr{F}_s]$ for $T \geq t$ and $s \leq t$.

Trader Security market participant who trades, competing with each other to generate profits, seeking compatible counterparties in trade, and seeking with superior order placement and timing.

Transaction cost The administrative and/or broker cost associated with executing a transaction.

Treasury bill Short-term zero-coupon note issued by the Treasury of the United States federal government. Considered to be relatively free of risk.

Treasury bond Long-term (2–30 years) coupon-bearing debt instrument issued by the Treasury of the United States federal government. Considered to be relatively free of default risk.

Treasury note Intermediate-term (2–8 years) coupon-bearing debt instrument issued by the Treasury of the United States federal government. Considered to be relatively free of default risk.

Underlying asset Security or asset on which an option is written.

Uniform random variable A continuous random variable with a constant density function over a specified interval.

Variance A statistical measure of dispersion, risk, and uncertainty. It is the expected value of the squared deviation of a data point from the expected value of the data set. It is the square of standard deviation.

Vasicek model An interest rate model based on an Ornstein–Uhlenbeck process.

Vega Sensitivity of an option's price to changes in the underlying security's standard deviation of returns. Also known as kappa or zeta.

Volatility The unpredictable part of the stochastic process that is a measure of dispersion of the process from its predictable part.

Warrant An option on the treasury stock of a firm.

Wiener process See Brownian motion.

Yield The internal rate of return of a bond or group of bonds with a given level of risk and term to maturity. Sometimes refers to the market interest rate associated with a particular maturity date and/or risk level.

Yield curve Plotting of bond yields or market interest rates with respect to terms to maturity.

Yield to maturity The internal rate of return for a bond.

Zero-coupon bond A bond that makes no explicit interest payments.

Index